Applied Systems Theory

Rob Dekkers

Applied Systems Theory

2nd Edition

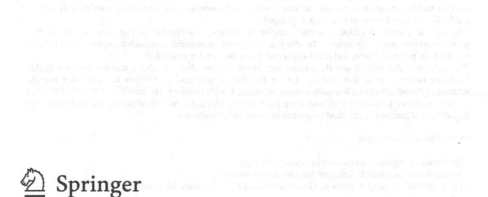

∑ Springer

Rob Dekkers
Adam Smith Business School
University of Glasgow
Glasgow
UK

ISBN 978-3-319-86185-2   ISBN 978-3-319-57526-1   (eBook)
DOI 10.1007/978-3-319-57526-1

Printed on acid-free paper

This Springer imprint is published by Springer Nature
The registered company is Springer International Publishing AG
The registered company address is: Gewerbestrasse 11, 6330 Cham, Switzerland

To Nil for her continued support and patience all the way through. Without her this book would not have been possible.

Synopsis

Systems theories aim at describing objects and entities, whether they are physical or human constructs of the mind to undertake studies. This synopsis introduces the most important concepts and models of Applied Systems Theory, a specific systems theory, to describe and to analyse objects and entities with the intent of understanding, modifying or predicting their structure and their behaviour; more detailed and expanded descriptions are found in the chapters of this book.

I   Systems, Entities as Part of a Whole

Machinery, houses, companies, computers, organisms and ecological networks as examples receive the label of systems when we want to isolate these objects of study from their environment. Whether it concerns organisational systems, such as companies, business processes and value chains, or technical systems, for example, ships and control systems, or ecological systems, we look at the object of study separated from its environment, perform an analysis and search for solutions to enhance its performance. This search is driven by unique problem definitions as a leading theme. Consequently, systems are essentially never the same; they depend on the problem and sometimes on the person performing the analysis. When we aim at improving the real-time response from an industrial robot or when the study focuses on designing the mechanical structure, each of these models of the robot as a system will differ according to its meaningful purpose denoted by the one who executes the study. Hence, the identification of a system is entirely dependent on the perspective (or problem definition) chosen.

Within a system, the elements do have mutual relationships between each other and with its environment (see Figure S.1). For instance, the quality system of a company might exist out of quality procedures, policies and guidelines; at the same time this system will link to the environment through relations with stakeholders, customers and suppliers. These structures describe the relationships elements do have within the system, which is called the internal structure, as well as with elements outside the system, which is named the external structure. For the case of the external structure, the external elements should be directly connected to elements within the system. For example, a manufacturing system might consist of pieces of equipment performing processes and it is coupled with the environment by the materials and parts delivered to it by suppliers and through the products it distributes to customers. Although a study attempts to isolate a system from its environment, the relationships with its environment, called the external structure, defines its purpose within the whole or universe.

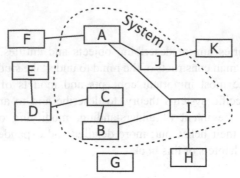

Figure S.1 *System with its elements and relationships. Each of the internal elements has at least one internal relationship to other elements within the boundary (A, B, C, I and J). The environment consists of those external elements that have direct relationships with internal elements (D, F, H and K). Some elements outside the system boundary have no or no direct relationship with internal elements (E and G) and should not be considered as part of the system's environment.*

Systems as Objects of Study

The aspects of a study will determine which relationships to explore and attribute values to the features of the aspect and the underlying parameters, see Figure S.2. An example is height; it is one of the parameters for dimensions as part of geometrical aspects, such as volume and position of objects of study. However, a specific aspect is always part of the larger set of system properties and relationships. Take for example a building; next to geometrical aspects, there are the aesthetic aspects (how the building looks) and the functional aspects (how it can be used). So, eventually we describe systems through specifying the particular relationships – aspects – they have with other elements, within the systems and their environments.

This distinction of aspects means also that when a search or an analysis extends to the properties of a system, there are two options for exploring it. Either an investigator concentrates on certain elements of the system (subsystem) or focuses on certain types of relationships within the system (aspectsystem), see Figure S.3. Look at the financial system of a company, being the filing and recording of financial transactions and mutations, and the

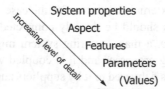

Figure S.2 *Interrelation between properties, aspects, features and parameters. The system properties may be broken down into aspects, which reflects the type of relationships investigated. When decomposing aspects, the investigator of a system looks at features and parameters (to which values can be attributed).*

generating of overviews about the current financial position. The overviews in the financial administration related to deliveries by suppliers represent a subsystem, whereas the cash flow is an aspectsystem. Both a subsystem and an aspectsystem should fulfil the condition that they are a system in their own right. Note that concentrating on certain relationships by means of an aspectsystem also implies discarding elements that have no mutual relationships of the specific type with other elements. A case in point would be a study that originally included the elements related to quality as part of a company's system, but when focusing on logistic relationships in a later stage, these elements may be better left out. The progress and results of the analysis and search determines which option for a closer look at a system, as a subsystem or an aspectsystem, will be taken for analysing and solving a problem.

Behaviour of Systems

Most studies of systems look at how properties of systems, elements and relationships change over time. Biologists wish to understand the emergence and the extinction of species, psychologists are seeking for the correct interventions in family units, managers want to improve performance of an

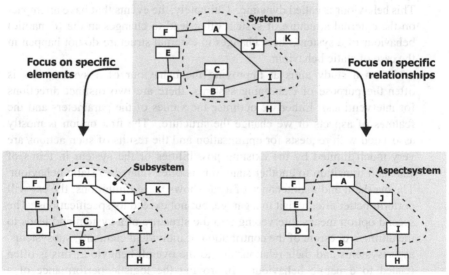

Figure S.3 *Two principles for investigating a system. When focusing on specific elements a subsystem may be distinguished; in this case, this subsystem is consisting of elements A, C and J. When focusing on a specific relationship only, the elements that have interrelationships of this type are looked at; in the figure these elements of the aspectsystem are A, B, I and J. Note that elements in the system that do not have this particular type of relationship with other elements are omitted. This also means that element D is no part of the environment of the aspectsystem.*

organisational system and engineers want to know under which conditions a technological system will still perform its tasks. In the case of static systems merely describing and denoting the elements and their relationships will fulfil the needs of a study. Differently, when a study concerns a dynamic system, the state of elements and relationships vary over time and the interests of the investigation will turn to the causes of variation and perhaps to how to deal with those variations. For instance, vehicles are exposed to different road and weather conditions, companies face the challenges of competitive forces and the biodiversity of species on earth evolves over time. Studies for these three examples might focus on the response of a vehicle to disturbance and maintaining direction of travel, the resilience of companies to changes in competitive forces and the sustainability of ecological systems.

The behaviour of a system as a whole, the changes in its relationships with the environment (the external structure), is triggered by events in the systems' boundaries. For the examples in the previous paragraph that might be: the road and weather conditions, the changing competition and the changes in the climate. Some of these changes lead to repetitive behaviour. Such is the case of the car; though the roads themselves might be very bumpy, the actual trajectory and position on the road are relatively predictable. In these cases, the behaviour is called static. Contrastingly, changes in the climate may cause extinction of species and adaptations by others. Thus, some systems will hardly ever return to a similar set of elements and relationships. This behaviour is called dynamic. Ultimately, the events that have an impact on the external structure of system often lead to changes in the (dynamic) behaviour of a system; these changes in external structure do not happen in the case of static behaviour.

When a study aims at improving the behaviour of a system, this is often the purpose of examining systems, there are two distinct directions for interventions. Either we optimise the values of the parameters and the features of aspects or we change the structure. The first option is mostly associated with requests for optimisation and the results of such actions are very much limited by the existing possibilities of the system in terms of state and transition to another state; this is often related to static behaviour. The overhaul and maintenance of a car shows these limitations; the car will perform better after a visit to a garage, but not exceed its specifications. The second option means intervening into the structure of a system and leads to the ultimate question of the contribution of individual elements, subsystems, aspectsystems and their relationships to the overall behaviour; this is often related to dynamic behaviour. Improving the logistic performance of a company might lead to a total new approach to the underlying concepts introducing new points for holding inventory and different methods for planning and scheduling; this leads to a new internal and possibly external structure. Both organisational systems and technical systems then face design requirements, as set by the relationships with their environment, to evaluate existing and alternative structures.

II   System Approaches

For analysing the structure and behaviour of a system, there are two points of departure for examining it in addition to the concepts of subsystems and aspectsystems:

- The system as a whole (without looking at the internal elements);
- The individual elements and the individual relationships (mostly the internal ones).

When starting with the elements a study will encounter the fundamental problem that we can not clearly distinguish the relationship the system has with its environment; this may lead to considering the system as being closed and ignoring the interaction with the environment. However, this relationship with the environment determines strongly the requirements imposed on the subsystems and the elements in case of open systems (which are the majority of cases). For example, the cargo load and speed as design parameters of a ship, derived from demand for this transport mode, will determine to a large extent the selection and dimensions of its propulsion system. The same would be the case for an organisation when determining how to distribute its products and services to consumers, i.e. the logistic system. So, starting by examining systems and then moving to subsystems, aspectsystems and elements creates the opportunity to define the relationships an element (or subsystem) has to its environment and the requirements imposed by those relationships.

Blackbox Approach

As a distinctive principle for examining systems, the blackbox approach focuses on the external relationships, starts from the system as a whole

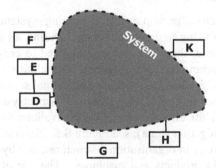

Figure S.4 *System as blackbox. The blackbox approach allows examining the external structure and the behaviour of a system. The elements and the internal structure are not considered for this type of analysis. The blackbox approach for analysing systems supports deductive reasoning by examining the behaviour of the system in response to external stimuli.*

and does not consider the internal elements and relationships. By solely concentrating on the external relationships of the system with its environment it studies the interactions between these relationships to understand the behaviour of the system. One could say that the blackbox approach is equivalent to looking at a system as being a single element (see Figure S.4). For example, an organisation is viewed as a blackbox when making enquiries about the relationships between orders and delivery dates without examining the internal processes for order processing. This might appear as a simple relationship revealing that orders for standard products have a lead-time of 24 hours, whereas the delivery of special products might mount up to 6 weeks; if it is found that generally these two product categories do not differ much, then this finding raises questions why special products take much longer. In general, it is not easy to obtain insight into the functioning of a blackbox since the possibilities of linking external relationships increase exponentially with the number of relationships. Nevertheless, this complexity of external relationships, the advantage of the blackbox approach is found in the elimination of the internal details of the system, content and structure during analysis and design; this is based on rule of thumb that the principle solution should follow the requirements imposed by the external relationships. The delivery time of standard products can be acceptable in the example, the lead-time for special products well above competitive standards: to analyse the order processing we need only to examine the processing of orders for special products and might skip all what involves processes for standard products. Hence, the example demonstrates the advantage of the blackbox approach for investigating systems: the analysis of behaviour without being burdened with details about the internal structure and the creation of more focus during next stages of analysis.

Aggregation Strata

A second distinctive principle for examining systems from a holistic perspective is by using aggregation strata. During an analysis the details may overwhelm and, therefore, there is the need to create an overview by models and by arranging data, whether of qualitative or of quantitative nature. Particularly, when we examine the internal structure of systems, the distinction between systems, subsystems (or aspectsystems) and elements provides a model for allocating related observations to different levels of aggregation (aggregation strata); see Figure S.5. Naturally, we do so already by creating hierarchy in organisational structures, and by the design of units or modules within products and machines. The use of aggregation strata for analysis and design aims at arranging data and information in such a way that the perception of the problem and the systems clarifies the causes for underlying poor performance; this includes deficiencies for improving behaviour to meeting newly set criteria and the possible options for structuring a system. This also means that when a system is analysed or a

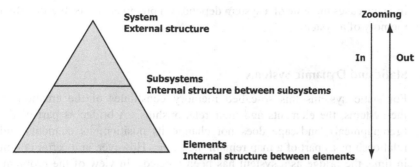

Figure S.5 *Aggregation strata and zooming in and out applied to systems. In this drawing the levels of system, subsystems and elements represent levels of detail for investigating a system. By zooming in more details (i.e. subsystems and elements of the system) become visible. Zooming out results in distinguishing (emergent) properties of the system as a whole and mostly makes it possible to examine the external structure of a system more purposefully.*

problem is resolved, going into more detail does not necessarily contribute to better understanding. For example, if the profit and loss statement of a company shows that the expenditures outstrip the revenues, then in the first instance going into more detail about the revenue generation, such as market segmentation, will not contribute to solving the problem of profitability; however, many are inclined to provide as much as detail as possible to solve a problem, relevant or irrelevant. Therefore, the distinction of aggregation strata assists in avoiding detail when not necessary. However, a higher level within the aggregation strata 'absorbs' the details of the lower level, resulting in loss of accuracy but at the same time gaining overview. Aggregation strata accommodate a better grip on the problem by defining levels of detail for searching and analysing data derived from a problem definition (for both analysis and design).

III   Processes

Going back to the temporal aspect of looking at systems, often a study does not look at systems from a static point of view, as a snapshot, but how they evolve over time. For example, an oven for baking cakes and pastries, viewed as a system, has achieved a certain temperature at a given time. This temperature reflects the state of the system – the values of relevant parameters at a certain point of time (note that the problem definition prescribes which relationships are relevant). The change of temperature is an event, a change of relationship, caused by another event, e.g. the setting of the temperature by the cook or baker; an activity indicates an event induced by another event and activities generally consume time, that means that it takes a while before the changes in relationships take effect. For the given example it means that the oven will reach the set temperature after a certain time has lapsed. Therefore, in many cases the state of a system depends on previous events, the so-called memory of a system.

Static and Dynamic Systems

For static systems this so-called memory constitutes of the creation of the systems, the elements and their relationships. A bridge as part of the (geographical) landscape does not change its position; its elements and relationships as part of a map remain the same. However, at a certain point in time, the bridge as a system has been created. In view of the problem, the position of the bridge in the geographical landscape, no changes take effect during the period of observation because of its purpose (nevertheless, one might come up with events that change the position of the bridge in the landscape). However, for organisations and increasingly for technical systems, responses to changes in external relationships (events) determine the potential behaviour under varying conditions. How do companies react to the dynamics of the markets? How does a computer network react to loads of request for web-sites? Dynamic systems may imply solely changes in the values of relationships, external changes in relationships leading to internal changes, or modifications of the structure of a system. When restructurings companies that is mostly done by adapting the external structure, for example how it communicates with (potential) customers, and the internal structure, for instance by changing the business processes. Whereas static systems are 'created' once and have a fixed state, dynamic systems go through different states (values of relationships and properties or changes in structure) induced by events.

Processes: Change of State

In the case of dynamic systems processes happen when events lead to activities that act on the system and eventually these activities may lead to changes in the external relationships or internal relationships of the systems. For example, a piece of equipment is assembled by putting components together that were initially not connected to each other; hence, the assembly process creates relationships between the components that did not exist before. In other words, the initial event – called the input – leads to output. To establish

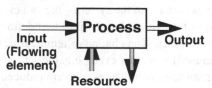

Figure S.6 *Process as interaction between flowing element and resource. The transformation of input into output requires the presence of resources. The changes of the state of the flowing element correspond the changes of state of the resource.*

these changes we need elements or systems to interact with the flowing elements as a process. In the case of assembly the process needs labour or equipment to establish changes in properties of the flowing elements; these are called systems of resources, see Figure S.6. When analysing the processes of an organisation, it may also be necessary to consider how the resources are structures (in groups and departments). Generally speaking, for the analysis and design of dynamic systems, such as machines and organisations, we focus on processes rather than on the elements and we have interest in the processes for displaying behaviour that the environment 'requires'; in addition to the analysis and design of processes, the structure of the system of resources may be subject of investigations.

Function

The execution of a process delivers output, a flowing element or flowing elements, to the environment that will fulfil a need, called the function. The function is an aggregate of the flowing element(s) allowing us to look for more principal solutions to fulfil the need. For example, take electricity; the function of electricity is energy. However, energy can also be delivered by mechanical processes, such as a watermill, and by radiation, such as a nuclear reactor; each solution has specific advantages and drawbacks, which will determine the feasibility. For particular cases that means more sources of energy can be considered than just electricity for solving problems and for creating designs. Thus, the primary objective of denoting functions is not to get 'trapped' by particular solutions but increasing the scope by looking at the essence of the output.

IV Control of Processes

However, processes, such as manufacturing and agriculture, do not always produce consistently the same output when fulfilling their function within the environment. For example, during manufacturing irregularities in supply or production processes may cause disruptions. The same goes for the growth of agricultural products, when weather conditions determine the quantity and quality of the output of farms. Eventually, often we want to achieve predictable behaviour despite the irregularities that occur when conducting processes. That stresses the need for controlling the primary process, albeit that three conditions need to fulfilled to make that possible:

- The existence of a target state. If no target state exists, the control mechanisms will not exert effective interventions.
- The capability to measure a parameter relevant to the target state.
- The capability to influence the outcome of a process through an intervention.

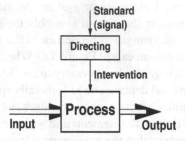

Figure S.7 *Directing. A control signal, the standard, is converted into interventions for the process (or input). Observe that no measurement takes place, the control process relies on the adequate translation of the standard into an one-time intervention (or directives).*

The comparison of the target state with the measurement of relevant parameters represents monitoring, which in case of deviations will lead to an intervention. Thus, control depends entirely on the capability to exert influence on the primary processes to achieve a pre-determined target state.

Directing

As the first of four principal control mechanisms, directing means only generating a one-time intervention for control of the system. Such an activity may consist of setting the value of a parameter or introducing a structure. After this activity the system should produce the desired output without further intervention, e.g. setting the temperature of a house; in general, such an activity more or less generates 'norms' for processes seen as blackbox no matter their internal structure. Whatever behaviour of the primary process will occur after setting the signal or standard, no correction will take place (see Figure S.7). An example is setting the amount of electric power to be generated based on past patterns for demand, the day in the year and the actual time of the day. To this purpose, the controller must know exactly which specific intervention produces those results. However, most processes do not comply with this prerequisite due to disturbances in input, resources and throughput beyond the capability of directing; therefore, this control mechanism has a limited range of applications.

Feedback

To correct for disturbances in input, resources and throughput, often feedback is used; this second principal control mechanism measures parameters of the output or process parameters and intervenes upstream (see Figure S.8). The intervention upstream corrects the input of flowing elements, the parameters of the process or the system of resources. An example of feedback is when the operations of a company fail to reach pre-determined output levels and

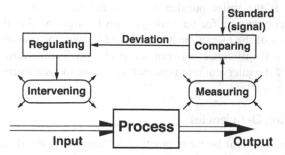

Figure S.8 *Feedback. Deviations in the measurement of parameters of the output's state lead to interventions in either the input or the process' parameters. The comparison might include calculations to make it possible to compare the standard with the measurement. The intervention depends on a model to convert deviations into interventions.*

then employing more workers increases the level of production. Or it might be that feedback is provided on coursework to a student who then sources more suitable textbooks for the subject of study. Generically speaking, the feedback mechanism responds to deviations no matter their cause.

Feedforward

Whereas feedback measures parameters of the transformation process or its outcome and intervenes in the process, resources or input, in the case of feedforward – the third principal control mechanism – parameters are measured upstream of the intervention. The intervention by a feedforward control mechanism (see Figure S.9) could be directed at either the influx of flowing elements or parameters of the process. Feedforward happens when a company has to process unexpected rush orders and decides to increase its capacity to fulfil the overall demand. Another example is that there is leakage in the case of water supply to water purification plants, and the distribution to users by water utility providers are temporarily decreased or suspended;

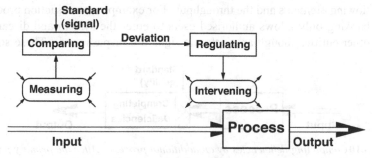

Figure S.9 *Feedforward. Generic representation showing that a measurement taken from the properties of the input results in an intervention downstream. The regulatory mechanism depends on a model connecting the deviation to the intervention.*

however, if the water purification plant decides to increase its influx of water to compensate for the leakage, then it might be that the intervention is upstream of the measurement and hence it could be considered feedback. The essential difference between feedback and feedforward is the relative position of the intervention in comparison to the measurement.

Completing Deficiencies

Sometimes it might be more practical to 'complete' the deficiencies in the output of flowing elements rather than intervening in the parameters of the process, the system of resources or the input of flowing elements; to recover from the deficiencies in output, there are two different manifestations of the fourth principal control mechanism. The first one is by adding missing elements in the output or by correcting the output (the additional process after the quality filter in Figure S.10). Such might be the case when at the end of an assembly line, cars are checked and missing components added and faulty components replaced; a car manufacturer would do that because discarding the entire vehicle would be more expensive. Alternatively, after the quality check, the 'faulty' output might be fed back into the transformation process by a loop; sometimes that requires an additional process to dismantle the product partially or entirely to make it suitable for the transformation process again (see Figure S.11). An example of the latter mechanism for feedback is when kids are building a house from toy bricks and come to the conclusion that it does not look as good as they had in mind, they take the bricks apart and have a go at it again. Therefore, completing deficiencies intervenes directly in the output of a process itself rather than adjusting the input of flowing elements, the system of resources or the process parameters.

V   Steady-State Model

Processes and systems of resources only operate within certain the capabilities of the control mechanisms and within limits for variations in the input of flowing elements and the throughput. For example, a transaction process for banking only allows authorised users to enter the system and discards any other entities though an electronic signature or password. These so-called

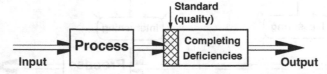

Figure S.10 *Completing deficiencies by an additional process. After the primary process, a check on properties will reveal deficiencies, which will be completed; note that this is an additional process to the primary process, thus requiring its own system of resources.*

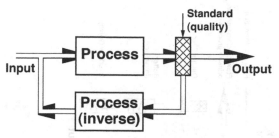

Figure S.11 *Completing deficiencies through a feedback loop in the primary process. After a check on the properties, the output of flowing elements (the defected ones) is returned to the input of the process. Please note that in practice this requires an additional process to convert the flowing elements to a state that processing becomes possible again (as indicated in the figure by the inverse process).*

boundary control mechanisms act on the flowing elements of the primary process itself and on the internal control processes. These boundary control mechanisms are found in three zones: the input boundary zone, the output boundary zone and the regulatory boundary zone; see Figure S.12. Because these zones are interrelated and have some common elements, the common features elements are explained in the next subsections as well as the specific features of each of the three boundary zones.

Coding and Decoding

Generically speaking the input of flowing elements needs to be made suitable for the primary process and the output of flowing elements for the environment; these processes are called coding and decoding. For the process itself the input needs to be coded before it can be used in the primary process. Take for example, the chewing of food as a coding process for the digestion in the human body; chewing properly breaks food down into smaller pieces that allow more effective decomposition during the processes in the stomach and intestines. Whereas coding occurs before the primary process, decoding happens after the completion of the process and adopts the flowing elements to the environment. For example exhaust gasses of a car are processed through a catalyser before streaming into the environment, that way reducing the output of CO and NO_2. Coding and decoding ensure that the interaction with the environment through the flowing elements stays within the capabilities for the primary process and the capability of absorbing output by the environment.

Quality Filters

After coding and before decoding, the flowing elements pass through a filter that compares the quality of the flowing elements with pre-set standards. If these standards are not met, then principally the flowing elements are

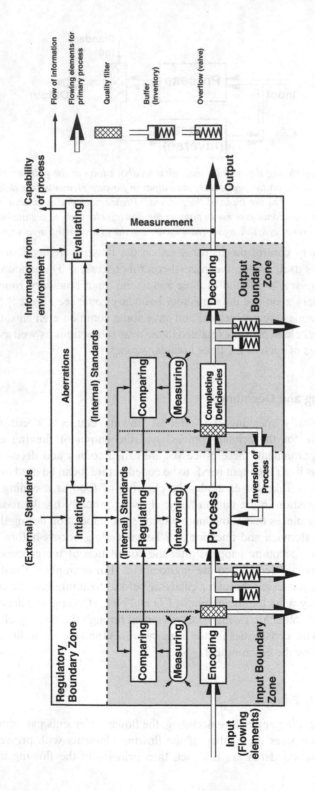

Figure S.12 *Steady-state model. This generic model provides a complete overview of all processes in the boundary zones, the regulatory activities and the control mechanisms. For reasons of simplification the resources needed for all processes (primary process and control processes) have been omitted.*

discarded. Using an example again, the coding of authorised users of the banking system for transactions happens through connecting to a (debit or credit) card, a signature, a fingerprint or username and password. The quality filter determines of a user appears on the authorised list of users. If not, then they are discarded by the system, but if they do they can conduct and authorise transactions as part of the actual banking process. On the output side, similar processes happen. However, if necessary, output that does not comply with standards for authorisation is rejected; for example, incomplete details for a bank transfer. Then it may be recycled through the process for acquiring the required quality or deficiencies are added or replaced (see control mechanisms for completing deficiencies); if the user does not succeed in providing all correct details, the transaction is cancelled entirely (and exited from the primary process). The example also implies that the quality control in the output boundary zone has more options at its disposal than the quality filter in the input boundary zone.

Buffers and Overflow Valves

The synchronising of the influx of flowing elements with the capability of the process and the transfer of flowing elements into the environment means that buffers need to absorb the differences. For instance, companies do so by holding inventory of materials and parts they need for manufacturing products; sometimes this type of inventory is held for economic reasons (cost savings by ordering large batches) or for uncertainty (irregularity of supply). Conversely, at the output side they might store finished products in warehouses to be distributed according to actual demand by consumers. For example, this is the case for products for festive seasons, such as Christmas puddings, which are made months before their actual purchase by consumers. Hence, buffers smooth irregularities between the primary process and input and output.

In actual situations, the capability to absorb the differences between process and input or output may be limited. Might the input exceed the capability or can the output not be distributed to the environment and can the buffers absorb no more flowing elements, then the flowing elements are discarded into the environment through valves as overflow. Take the case of a power plant based on traditional fossil fuel; generally, the level of production can only be limitedly altered on the very short term (it might take days before a power plant reaches its maximum capacity). For this reason, the excess energy needs to be discarded. But also the flow of water to a watermill might exceed the capacity of the watermill itself and might bypass the mill because it cannot be stored or used. Hence, the buffers and overflow valves act to align the inflow and output with the capability of the process, albeit in different and complementary ways.

Initiation and Evaluation

In addition to the boundary zones for the input and output, the regulatory boundary zone interacts with the environment about the standards for the control mechanisms for the primary process. One of its mechanisms is converting standards of the environment into operational, normative standards within the system. This so-called initiating process might transform the deadline for an order into deadlines for individual steps in the production process and capacity requirements for specific activities. As a second mechanism, the actual performance of a system might call for the need to adapt the standards. Such a situation occurs when the processing of orders in a firm does take structurally two weeks rather than the standard of one week. To perform the evaluation process, information from the environment helps to assess the standards or creates the need for adaptation. Also, market growth might end up in increasing the levels of inventory to allow the same service degree for delivery of products to customers. If the change of standards affects the performance to the environment, a signal will be generated to inform the environment about the changing capability of the transformation process. Hence, the regulatory boundary zone consists of (i) the initiation process, which is the conversion of external standards into operational standards for the control processes, (ii) the evaluation of the performance of the transformation process to revise standards, if necessary, and (iii) the transfer of information to the environment about the actual of capability of the system of resources to perform the primary process and to maintain standards.

Limitations of the Steady-State Model

The primary process, and possibly secondary processes, together with the control processes and the three boundary zones form the steady-state model (see Figure S.12); however, this steady-state model only applies to recurring processes. Through the control processes it will adapt to changes in its environment and the adaptation is limited to the capabilities of the system of resources and the limitations embedded in the control processes for dealing with variations in the influx of flowing elements and the system of resources. Does one want to go beyond the existing capabilities of a process than a new internal (and, if necessary, an external) structure should make that happen; this redesign is not covered by the steady-state model but by the breakthrough model (see Section VIII of this synopsis).

Note also that a steady-state model, consisting out of a primary process, control processes and boundary zones, applies to one aspect only. So far, there is no process model that captures multiple aspects, mostly because of the subjective weighting of individual aspects. Each person attributes different values to certain aspects, e.g. environmental impact versus financial results. If we want to apply the steady-state models for different aspects, we need to construct separate models for each aspect.

VI Autopoiesis

This makes the steady-state model a representation of repetitive processes that also occur in so-called autopoietic systems. Autopoiesis is a process whereby a structure, i.e. a system of resources, reproduces itself. An autopoietic system is an autonomous and self-maintaining system that has processes in place for producing the elements and subsystems it consists of. By doing so, the elements and subsystems, through their interaction, generate recursively the same structure of processes which produced them. Cells and organisms are examples of autopoietic systems; generally speaking they produce their own offspring as a 'copy' of their own contents and structure. This principle has profound implications since the generation of contents and structure of offspring depends on the original state of the autopoietic system and, therefore, any mutation can be traced back to former states.

Structurally Closed and Self-Referential

This also means that the state of an autopoietic system is determined by the processes for generating offspring, which is differing from its (internal) primary processes; this also called being structurally closed. Take a human being as a simplified example of an autopoietic system. The primary processes of maintaining its steady state – breathing in air, drinking water and consuming food – differ from the process of creating children, although the primary processes is needed to achieve that offspring is born. The intake for the primary process can be changed relatively quickly to the environment. However, the capabilities of the offspring are building on the capabilities of the parents. For instance, it takes people starting to live at the high altitude of the Andes mountain range a few generations before their respiratory system is fully adapted to the conditions of 'thin air', air with less oxygen. Therefore, the concept of autopoiesis explains processes of mutation and how these are linked to previous states.

In this respect, the theory of autopoiesis adds further insight to the more cybernetic views that have dominated the previous sections in this synopsis. Principally, it tells that next generations of autopoietic systems build on the elements and structures of previous generations. Such is the case in evolutionary biology for offspring. Autopoiesis implies also that these systems are self-referential in their interaction with the environment; only that what can be perceived acts as stimulus for activities in the system and for the next generation. A case in point is the vision of human beings; generically speaking, the vision is limited to certain frequencies of light, and, therefore, the capability of observation is limited in comparison the vision of birds and insects. Furthermore, the concept of autopoiesis is also a very difficult theory to apply to systems because the observers have cognitive limitations as well. Think about the observations of stars that use equipment that goes beyond the frequencies of the naked eye as an enhancement of the visual capabilities of human beings. Therefore, the principles of autopoietic theory serve as explanation for systems with a high degree of complexity but should be applied with reservations.

Allopoietic Systems

A special class of autopoiesis systems are so-called allopoietic systems. These systems do not self-produce through 'offspring' as autopoietic systems do but are 'created' or emerge from systems external to them. The dependency on the external systems for justifying its existence means also that it depends on the perceived need of the output or function by the external entities (or systems; these are also called actors or agents). Thus, mutations are changes in structures and content of these systems. An example of allopoietic systems is organisational systems; these are created by some entities labelled as stakeholders (owners, shareholders, and, in case of cooperative arrangements, employees) for the sake of other stakeholders (customers, employees, society, etc.). Both internal and external actors influence mutations of organisation, such as changes in organisational structures; a case in point is the implementation of safety and health regulations triggered by legislative requirements. Therefore, adaptations of an allopoietic system are driven by external enactment, while at the same time building on the self-referential aspects of cognition within the system. A conversion of an allopoietic system relies on its extant subsystems and elements and in that sense follows principles of autopoiesis (many companies integrated the safety and health systems into the existing structures for quality systems). Only when further adaptations are not possible the allopoietic systems become extinct and, in the best case, elements or subsystems are re-used by external (perhaps, different) entities.

VII   Complex Adaptive Systems

In addition to the concepts of autopoiesis, theories about complex adaptive systems state describe systems that have non-linear behaviour; this type of behaviour is more difficult to predict as opposed to deterministic behaviour. These systems have many autonomous entities, that they are able to respond to external changes and that they form self-maintaining systems with internal pathways for feedback. The concepts related to complex adaptive systems aim at explaining the non-linear behaviour of such systems that cannot directly be explained as a result of the behaviour of individual entities of these systems.

Simple Rules

One of the mechanisms to explain non-linear behaviour is that simple rules for the interaction between entities in a complex adaptive system could lead to complex patterns. A famous example is the flocking of birds; computerised simulations that use rules such as maintaining a certain distance to the nearest-by entity result in patterns that look very similar to the patterns of flight by a large group of birds. However, such patterns might only appear in homogeneous and regular environments.

Fitness Landscapes

Another mechanism for explaining the non-linear behaviour is the search by complex adaptive systems for optimal points on a fitness landscapes. These landscapes can be imagined as real-life landscapes with rugged areas, hilly areas and flats. Complex adaptive systems seek out peaks on these landscapes, which might be either local peaks or global peaks. Moving from one peak to an other (higher) one follows pathways that will lead to passing through sub-optimal points, such depressions and valleys in the landscape. These pathways can be circumvented if a complex adaptive system takes larger steps (for mutations), i.e. long jumps, however these steps also increase the chance of missing out on reaching other peaks. Note that complex adaptive systems consist out of more entities and its thoughts need to be applied to groups of entities, for example, species, economics sectors and national systems rather than a single specimen or firm.

VIII Breakthrough Model

However, the changing environment and the changes induced by organisations themselves are often put into a coherent approach called a strategy. For for the formation of strategy organisations can deploy dynamic strategies, forecasting and scenario planning. The use of each of these methods depends on how organisations perceive their environment, since organisations can be considered allopoietic systems.

The breakthrough model (Figure S.13) offers an overview of the necessary internal processes for the formation of strategies and the implementation of changes, called breakthrough. Related allopoietic principles for self-cognition and self-learning, in the breakthrough model an organisation sets out a strategy by scanning the environment; such a strategy might have new or adapted goals and also includes the capabilities of the resources that are accessible to the organisation as a system. Thus, the process of 'strategy formation' informs tactical and operational decisions and also forms the base for reviewing decisions. As a next step, the process of 'confrontation and tuning' leads to more specific decisions on the utilisation of resources

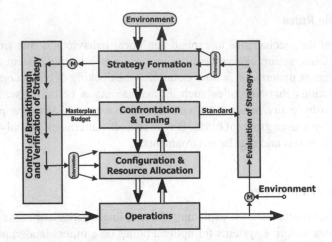

Figure S.13 *Breakthrough model. By scanning the environment new or adapted goals are set and the derived policy acts as a reference for the review of tactical and operational decisions. The process of confrontation and tuning takes the possibilities into account leading to specific decisions on the utilisation of resources and structures for operations. Through the configuration and resource allocation process the actual implementation of the structural changes in operations takes place. The evaluation of strategies may create new input for the breakthrough processes. The verification enables companies to follow the progress of the breakthrough processes.*

in the context of achieving the objectives of the organisation. Through the configuration and resource allocation process the actual implementation of the structural changes in operations takes place; this means that new subsystems and elements are introduced (or replacing previous ones) together with a new structure (that might cover both internal and external structures) for the recurrent processes in operations. The revised or new structure for operational processes is principally a steady-state model. Also note that all these steps are iteratively linked. Hence, the breakthrough model is a reference model for the iterative (internal) processes that link the observations of change in the environment to actual structural changes in the operational (steady-state) processes.

In addition to the iterative process linking strategy formation to operational processes, there are two specific control processes. Note that these control processes have similarities to the control mechanisms in Section IV and V of this synopsis, but are also unique for the breakthrough model. The first mechanism is the evaluation process for the strategies based on the actual performance of the primary (operational) processes; it is found on the right-hand side of the breakthrough model in Figure S.12. Note that the performance evaluation of the steady-state model only related to the capability of the primary process; the feedback mechanism for the achievement of strategy looks also at to what extent objectives are met. The evaluation of strategies may create new input for the breakthrough processes. As the second mechanism, the verification process enables companies to follow the progress of the implementation of strategies and resource allocation. This process differs substantially from the feedforward control process in Section IV; whereas feedforward focuses on the performance against set standards (and, therefore, a target state that was already achieved or the process has the capability of meeting it), the verification process measures against a future state that has not been reached, yet. Therefore, it does not only measure against the master plan derived from the strategy formation process, but also reviews whether the capabilities of resources (and resource allocation) meet the 'requirements' of the future state for the primary process. Both verification and evaluation are essential parts of the breakthrough model to enable organisations to achieve future states.

IX Model for the Dynamic Adaptation Capability

For assessing strategic renewal, this breakthrough model is augmented with learning processes with the breakthrough model; see Figure S.14. The so-called innovation impact point serves as an indicator to determine the effect of innovation process on firms and to determine the involvement of management in decision-making. When the innovation impact point moves at lower levels in the breakthrough model, potential innovations have a limited effect on the market position; at higher levels the innovations have a potential impact on the development of the organisations and the market position (incl. the business model).

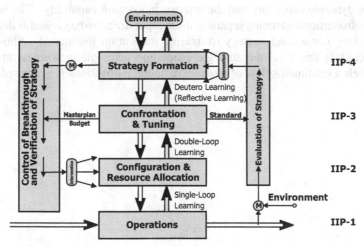

Figure S.14 *Model for the innovation impact point. The breakthrough model shows the learning modes and the identified innovation impact points (IIPs). The higher the impact point, the more changes and innovations from lower levels affect organisational decision-making. Architectural, and often radical, changes and innovations come about through accumulation of minor changes and innovations.*

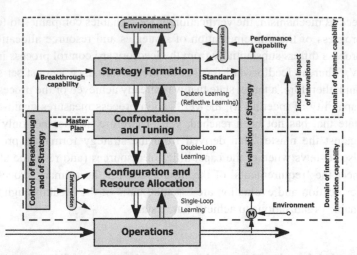

Figure S.15 *Model for the dynamic adaptation capability. Expanding on the model of the innovation impact points, this particular model distinguishes the internal innovation capability and the dynamic capability with its external orientation.*

Additionally, the central role of confrontation and tuning in the breakthrough model points to the competency of organisations to adapt strategies and internal processes to the dynamics of the environment; this leads to the model for the dynamic adaptation capability. Fed by bottom-up innovations through learning cycles and technological improvements in addition to the dynamics of the market itself, continuous reflection on possibilities and opportunities leads to a continuous stream of project and programmes for innovation. This model in Figure S.15 has two components: the dynamic capability and the internal innovation capability. The level of confrontation and tuning separates these capabilities (to be viewed as decision-making connecting strategy to internal innovation initiatives). Above and including this level the strategic adaptation takes place, whereas at lower levels a continuous flow of more incremental innovations is generated.

Contents

Foreword to the Second Edition

During my studies I got involved with systems thinking; the quest for its applications and its meaning continued during my years in practice and in academia. Right away from the beginning the attention was drawn to how this thinking emerged as an interdisciplinary approach, building bridges between sciences and fuelling cross-fertilisation to take advantage of conceptualisations and applications.

In this edition of the book I have extended the research to find out more about the history of systems thinking. During this search I found some originators of systems thinking, who were forgotten. Reading their writings made me realise how much we are indebted to these early thinkers. Combining the past with the present led to adjustment of inconsistencies in systems thinking and to furthered understanding how this thinking can be applied to both practice and research itself.

Scotland, March 2017.

1 Introduction

Thinking in systems goes back to both reasoning about physical objects and philosophising about the interrelationships between (scientific) disciplines. Over time, the thinking in systems has elevated systems theories from a theoretical framework to a practical methodology for application in a wide variety of disciplines, for example engineering, management and psychology. No matter their origins and applications, system theories have deepened our understanding of the complexity of real-life systems and also contributed to finding solutions for the challenges in dealing with systems.

This book describes that particular way of looking at the world, thinking in wholes rather than thinking only in the properties of individual elements. Before elaborating the concepts of Applied Systems Theory in the chapters that follow, this introduction will briefly describe the origins of systems theories, a discipline of science that dates back to the 1940s and even before. The next section indicates the areas of science, especially engineering and management, which use the theories. The origins of Applied Systems Theory, as recorded in this book, are elaborated in Section 1.3; this section will mention the main sources and how these were used. Over the course of time, systems theories have diverged in various kinds of strands. Section 1.4 will present the distinction between hard and soft systems methodologies, as two main strands of systems theories that have emerged. This section will also explain why this book about Applied Systems Theory represents a so-called hard systems theory with characteristics of soft systems theory. Finally, Section 1.5 outlines the scope of this book and how the book might be used.

1.1 Concise History of Systems Theories

Systems theories and their foundations are neither new nor recent. They are rooted in early scientific traditions. Peter Checkland [1981, p. 75], known as the founder of soft systems methodology, mentions that Aristotle argued that a whole was more than the sum of its parts, one of the most fundamental principles for systems theories. It was common that academics in ancient times would study multiple disciplines in science. From literature it is known that Faust [von Goethe, 1808, 1832], a character in a play, became the last single person to encompass all knowledge existing that time; although he embraced this thought of mastery at the expense of his soul. That exemplifies that Aristotle's holistic view was overthrown by the scientific revolution of the 17th century. Since this revolution, a reductionist approach prevailed in which decomposition of problems into smaller ones constituted the main methodology, following principles outlined by Descartes [Wilson, 1998, p. 28]. This reductionist approach allowed individual sciences to progress

© Springer International Publishing AG 2017
R. Dekkers, *Applied Systems Theory*,
DOI 10.1007/978-3-319-57526-1_1

and develop their own concepts, theories and methodologies. Consequently, individual sciences developed in different directions and displayed unawareness of each other's progress and insight into the nature of phenomena. Thus, sciences turned into isolated communities of practice, which yielded tremendous progress in individual fields. However, already in the 18th century an exploratory discourse about systems was written by de Condillac [1749]. After that, in the beginning of the 20th century Bogdanov's interpretation of tektology [Gorelik, 1975, pp. 348, 351; 1987] describes a discipline with the aim of unifying all social, biological and physical sciences by considering them as relationships and by seeking the organisational principles that underlie all systems. This must be seen as the precursor of systems theories, although he referred to systems also as complexes [Bogdanow[1], 1926, p. 23 ff.; 1928]. Furthermore, he refers to the concept of a regulator [Bogdanow, 1928, p. 102 ff.], to be seen as feedback. The idea of control and feedback also appeared in the work of Ştefan Odobleja produced between 1929 and 1937, according to Iancu [2009]. Similar ideas were popularised later by von Bertalanffy [1973] as general systems theory and Wiener [1948] for cybernetics, though Bogdanov's legacy may have had little impact on their writings after all [Rowley, 2000, p. 5]. It looks like that systems theories took off when halfway through the 20th century the need for integrative approaches appeared due to the increased complexity of engineering systems, particularly in control engineering and aerospace engineering (for example, airplanes and spacecraft), and because of advances in evolutionary biology that pointed to complex interactions in ecosystems. Cleland and King [1969, p. 3] trace the germ of the idea for integrative approaches back to a writing by Henry P. Kendall. Therefore, the emergence of systems theories is embedded in holistic thinking that emerged in the first half of the 20th century, possibly hastened by the need to develop more complex systems in a coherent manner.

Originally, these integrative approaches – the system approaches – with their focus on the interaction between an entity and its environment, originated for a large part in biological perspectives on entities. The aim was twofold: to understand the complexity of reality for solving intricate, layered challenges and to stimulate multi-disciplinary thinking for enriching disciplines. The communication and exchange between academics, leading thinkers and practitioners allowed the emergence of new concepts through cross-fertilisation [Boulding, 1956, p. 201], similar to the ancient cities of Cordoba, Spain, and Baghdad, Iraq, that represent two historical sites at crossroads of cultures that boosted advances in science in their times of prosperity. That interaction between different disciplines and the search for cross-fertilisation inspired the series of Macy Conferences (1946–1953), where specialists from different areas of expertise met to discuss inter-disciplinary work and to lay the foundations for the general systems theory (see Box 1.1). Von Bertalanffy's drive to establish a discipline to enhance communication between experts, created ground for cross-fertilisation and

---

[1]   Note that Bogdanov's name is spelled 'Bogdanow' in German.

Box 1.1: Macy Conferences

The ten Macy Conferences between 1946 and 1953 were the first organised approach to interdisciplinarity, spawning breakthroughs in systems theory and leading to the foundation of what later was to be known as cybernetics. They were organised by the Josiah Macy, Jr. Foundation.

The participants were leading scientists from a wide range of fields. Casual recollections of several participants stress the communicative difficulties in the beginning, giving way to the gradual establishment of a common language powerful enough to communicate the intricacies of the various fields of expertise present. The scientists that participated in all or most of the conferences are known as the *core group*. They include:

- Gregory Bateson – anthropologist
- Julian Bigelow – electro technician
- Heinz von Foerster – biophysicist
- Lawrence K. Frank – social scientist
- Ralph W. Gerard – neuro physiologist
- Molly Harrower – psychologist
- Lawrence Kubie – psychiatrist
- Paul Lazarsfeld – sociologist
- Kurt Lewin – psychologist
- Warren McCulloch (chair) – psychiatrist
- Margaret Mead – anthropologist
- John von Neumann – mathematician
- Walter Pitts – mathematician
- Arturo Rosenblueth – physiologist
- Leonard J. Savage – mathematician
- Norbert Wiener – mathematician.

In addition to the core group several invited guests participated in the conferences; amongst many others: Claude Shannon (information theorist) and Max Delbrück (geneticist and biophysicist).

Source: Wikipedia [2007].

---

initiated the search for the so-called general systems theory [von Bertalanffy, 1973; Kline, 1995; Klir, 1969], as universal language between sciences (see Box 1.2 for its founders). In the spirit of his thinking, the interaction between the many different fields did hold the promise of new perspectives. Nowadays, the notion prevails that advances in theory and practice should be drafted from not only a specific domain of research but also give way to

---

Box 1.2:   Founders of General Systems Theory

William Ross Ashby (1903–1972)

William Rosh Ashby was an English psychiatrist and a pioneer in the study of the organisation and control of complex systems. He created the concept of a 'homeostatic machine', which proved to be a fundamental concept for the development of mathematical models of cybernetics.

Gregory Bateson (1904–1980)

Gregory Bateson was a British anthropologist, social scientist, linguist and cyberneticist whose work intersected that of many other fields. He strongly opposed those scientists who attempted to 'reduce' everything to mere matter and was intent upon the task of re-introducing 'Mind' into the scientific equation. Gregory Bateson and his colleagues developed the double bind theory (about communicative situations where a person receives different or contradictory messages). He helped to elaborate the science of cybernetics with colleagues, such as William Ross Ashby, Heinz von Foerster and Norbert Wiener.

Ludwig von Bertalanffy (1901–1972)

Karl Ludwig von Bertalanffy was a biologist. Already in the 1930s, Bertalanffy formulated the organismic system theory that later became the kernel of the General Systems Theory. His starting point was to deduce the phenomena of life from a spontaneous grouping of system forces, comparable to contemporary system developmental biology. Von Bertalanffy introduced the General Systems Theory as a new paradigm for model construction in all sciences. As opposed to the mathematical system theory, it describes its models in a qualitative and non-formalised language. Thus, its task was a very broad one, namely, to deduce universal principles that are valid for systems in general.

Kenneth E. Boulding (1910–1993)

Kenneth E. Boulding, also known as one of the founders of General Systems Theory, emphasised that human economic and other behaviour is embedded in a larger interconnected system: to understand the results of our behaviour, economic or otherwise, we must first research and develop a scientific understanding of the eco-dynamics of the general system, the global society in which we live. Boulding believed that in the absence of a strong commitment to the right kind of social science research and understanding, the human species might well be doomed to extinction. But he died optimistic, believing our evolutionary journey had just begun.

BOX 1.2   (CONTINUED)

MARGARET MEAD (1901–1978)

Margaret Mead was an American cultural anthropologist. It was through her work that many people learned about anthropology and its holistic vision of the human species.

Sources: Brauckmann [1999], Heylighen [2004], Wikipedia [2009]

---

the spirit of interdisciplinary thinking, as the founders of systems theories envisioned.

In parallel to the development of science and the rise of systems theories, the view on systems has changed over the years. McCarthy [2004, pp. 125–126] points to the following perspectives that have influenced thinking in systems (note that the first four existed before systems thinking surfaced as a scientific discipline):

- the Aristotelian view (the perception of systems as organic, living and spiritual);
- the Cartesian view (the observation of systems as mechanistic and as resulting from reductionism);
- the Newtonian view (the examination of systems as obeying principles of mechanics);
- the romantic view (the concept of systems as self-organising wholes);
- the general systems science view (the notion of systems as consisting of elements and their relationships to the whole, and open versus closed systems);
- the cybernetic view (the examiniation of systems with feedback, and the capability for self-balancing, self-regulating and self-organisation);
- the soft systems view (the consideration of systems as mental constructs);
- the complex systems view (the perspective of systems as an expression of non-linearity, self-organisation and emergence).

Despite their differences in points of departure, among these perspectives a common and binding theme is the understanding of complicated entities in many fields by:

- determining the system boundary, the constituent elements, the relationships between elements, the attributes of elements and the input and output of the system;
- supporting the integration of views and knowledge to study the total system and how it interacts with its environment.

Thus, systems theories essentially constitute a multi-disciplinary perspective for studying entities, or objects, as some call them (for example, Kline [1969, p. 36]), with the ultimate purpose of providing a better understanding of objects and constructs of the mind, paving the way for more adequate descriptions, purposeful analysis and design.

1.2   Application of Systems Theories

As a result of the multi-disciplinary approach systems theories for studying complicated entities have found their way back to many fields of science [Heylighen and Joslyn, 1992], such as theoretical development and conceptual foundations (e.g. the philosophies of Bahm [1981], Bunge [1979] and Laszlo [1972]), and applications ranging from mathematical modelling and information theory (e.g. the work of Mesarović et al. [1970] and Klir [1969]) to practical applications for decision-making in organisations (e.g. Checkland [1981]). As one of two strands of applications, the mathematical systems theories arose from the development of analogies between electrical circuits and other systems; this is often called hard systems thinking. Applications of this strand of systems theories include engineering, computing, ecology, management and, to a certain extent, family psychotherapy. The second strand of applications, that of systems analysis, which developed independently of systems theory, applies systems principles to aid a decision-maker with problems of identifying, reconstructing, optimising and controlling a system (usually a socio-technical organisation), while taking into account multiple objectives, constraints and resources. This strand combines systems theories as aid to decision-making and soft systems thinking (the latter viewing systems often as mental models). It aims to specify possible courses of action, together with their risks, costs and benefits. Systems theory is closely connected to cybernetics, and also to system dynamics, which models changes in a network of coupled variables (e.g. the 'world dynamics' models of Forrester [1968] and the Club of Rome). Related ideas are used in the emerging 'sciences of complexity', studying self-organisation and heterogeneous networks of interacting actors, and associated domains such as far-from-equilibrium thermodynamics, chaotic dynamics, artificial life, artificial intelligence, neural networks and computer modelling and simulation. Systems analysis not only covers mathematical models for decision-making but also qualitative models for the study of information systems and organisational design; Applied Systems Theory constitutes one of the methods for these domains. These applications demonstrate the wide array of sciences that systems theories have boosted.

During the 1980s cybernetic approaches found their ways to many disciplines as basic approaches or methodologies. These approaches became so common that they seemed unique to many disciplines rather than presenting an interdisciplinary perspective. Wilson [1998] revived the discussion about interdisciplinary thinking by pointing to the concept of consilience as transferring knowledge from one domain to another, particularly the connection between biological phenomena and social sciences (akin the intent of the founders of general systems theory). He assumes that all phenomena are based on material processes that are causal and, however long and tortuous the sequences, ultimately reducible to the laws of physics [ibid., p. 266]. For example, consilience of knowledge about the management of organisations would demand a vision capable of sweeping from whole societies to an individual human brain. It would involve reduction – decomposition of events and phenomena – and synthesis – the integration of knowledge. To dissect something into its elements is consilience by reduction, and to reconstitute it is consilience by synthesis [ibid., p. 68]. Wilson [ibid., pp. 70–71] offers an example of consilience in practice from his early research on ants. To explain communication within an ant colony (e.g. an internal alarm alerting an entire colony to an attack by a predator), Wilson and his associates studied an ant colony across four levels of organisation, from the whole colony, then reductively to the organism (individual ants), to glands and sense organs, and finally to molecules (pheromones). He also worked in the opposite direction (synthesis) when he predicted the meanings of signals observed in the colony (for example, alarm, danger versus food, follow me) by linking various signals to matching changes in the molecular composition and concentration of individual ant pheromones; he used simulation models to compare theoretical findings with actual behaviour of ants. The result was a comprehensive or holistic study of ant communication [Peroff, 1999, p. 98]. The value of Wilson's thinking is that we have not reached yet the limits of interdisciplinary research and that at least some phenomena might be better explained by explicitly deploying principles of consilience to enhance cross-fertilisation among disciplines; such advances can be made by interdisciplinary studies, supported by systems theories.

1.3   Foundations of Applied Systems Theory

The striving for cross-fertilisation implies that any systems theory is interdisciplinary in nature and constitutes a blend of concepts originating in the rise of systems theories and later developments. The text of this book has its origins in the rise of general systems theory and cybernetics during the 1950s, 1960s and 1970s; the book pays tribute to these by referencing to the early writings about systems thinking. A lot of interaction happened in the spirit of interdisciplinary exchange of thoughts (as also evident from Box 1.1), which makes it difficult at times to pinpoint the emergence of ideas and concepts and to trace it back to specific authors. Therefore, generally accepted terms and definitions in systems theories have not been traced back to specific literature; however, as much as possible, specific concepts and thoughts have been marked by reference to the originator as identified during the research for this book. That means that works of others are only cited when they are not considered part of the generic and interdisciplinary knowledge about systems theories.

The book also includes references to more recent developments, found in the science dealing with complex systems and the science of complexity (especially in Chapters 8–10). Inspired by the Zeitgeist of the late 1980s, the trend of decentralisation and the postulation of non-hierarchical, participative and distributed control in society and organisations also penetrated complexity science [Malik, 1992]. Starting with the works of the Santa Fe Institute in the early 1980s, the paradigm of self-organisation emerged and opened a new strand in the explanation and control of complexity [Jost, 2004]. With the increasing number of elements in artificial systems – turning them into networked entities – their control became increasingly complex [Tucker et al., 2003]. This made the deterministic, top-down approach to systems control inefficient, if not impossible, especially against the background of a highly dynamic environment. Thus, Applied Systems Theory – as presented in this book – has embraced the recent advances in the science of complexity.

For the actual text many sources have been used. First, it builds on the systems theories that emerged during the 1950s, 1960s and 1970s and the general systems theory manifesting itself during the same period. Although homage is paid as much as possible to original authors, at the same time, that is hindered by the interdisciplinary approach (see the description of the Macy Conferences in Box 1.1). The theories from that period have found their way into other writings as well, such as the socio-technical approach of Miller and Rice [1967]. Second, Beer [1959; 1966; 1972] and Checkland [1981] have made major contributions to systems thinking and some of their ideas have been incorporated in this book. Third, additional sources have been consulted, e.g. the web-sites from Principia Cybernetica and Wikipedia, to clarify terms and journal papers to obtain additional information. By using a wide range of sources the text has moved away from its origins in general systems theory to a concise and consistent systems theory for practical use in engineering sciences, and business and management science (and other sciences as demonstrated by the examples throughout).

Although widely accepted in the scientific and engineering community, the principles of systems theories have limitedly reached the business and management science community. Most theories developed for that domain constitute a generic approach, which hardly combines with the drive to arrive at specific solutions as necessary for adoption in business and managerial applications; for example, see for the latter Sagasti and Mittroff [1973], who integrate problem-solving for real-life problems with systems theories for the domain of operations management. For operations management, most presented systems approaches limited themselves to relatively simple representations of operations management; the notes by Hill and Hill [2011, pp. 12–14] and Slack et al. [2010, pp. 11–13] are cases in point. Hence, operations management could benefit from approaches derived from systems theories, such as the analysis and design of organisational structures (for instance, Dekkers [2008]). Such approaches follow a rationalist view for analysis and decision making. However, the search for optimal solutions that is at the core of the application of systems theories seems far off from the practical decision-making by managers; think about the principle of satisficing of Herbert Simon [1959, pp. 262–264]: accepting the first available solution that meets criteria, disregarding whether it is marginally or substantially better than other solutions. This means that Applied Systems Theory aims at offering a more applied approach than other systems theories from a rational perspective; but it will take time before the systems theories catch on in managerial practice and management science sufficiently to turn it into a basic tool for resolving problems, since they are used more as a backdrop than as an essential skill for managers.

1.4 Hard Systems Approach vs. Soft Systems Approach

Using problem-solving applications of systems thinking to real-world problems is characteristic for Applied Systems Theory. According to Laszlo and Krippner [1998], this strand of the study of systems – problem solving – can be divided into three strands: (a) work in 'hard systems', e.g. the development and use of systems engineering methodology, (b) aid to decision-making, for example operations research and management science methodologies and (c) work in 'soft systems', such as the development and use of social systems design methodologies. Applied Systems Theory, as presented in this work, covers mostly (a) and (c), being a methodology for qualitative analysis and design. Its application resembles methods used in engineering for analysis and design, though these are not explicitly covered in the book, whereas its attention goes to social organisations in later chapters. However, Applied Systems Theory is complementary to other methods and methodologies, as used in operations research and Soft Systems Methodology. Depending on the type of problem, one might choose the most appropriate methodology to resolve the problem (see Section 11.4). In thaat sense, Applied Systems Theory is only one set of approaches and heuristics as a methodologyto tackle real-world problems, may be a comprehensive one.

The combined hard systems and aid to decision-making approach of Applied Systems Theory contrasts with soft systems methodology of Checkland [1981] – another popular method for social systems –, although both aim at resolving real-world problems. Both have their own domain of application within managerial science. The focus of Soft Systems Methodology is on the analysis of systems by a process of inquiry and involvement of stakeholders. This concerns mostly unstructured problems, which require iteration and questioning, whereas structured problems, especially in engineering, do not; during design stages, the latter calls for the understanding of structures and patterns. In this sense, Applied Systems Theory centres on a more formal way of modelling, which is complementary to Soft Systems Methodology.

1.5   Who Might Benefit from Applied Systems Theory and How?

In a more generic sense, systems theories help to understand the complexity of the world we are in better; this can be used to describe real-world problems, to analyse these problems and to find solutions for them. Particularly for complex situations, interrelationships and understanding key mechanisms takes centre-stage during analysis and synthesis. Systems theories aim at identifying relevant interrelationships from the perspective of entities and processes. For those purposes, systems theories are of interest to managers, consultants, engineers, students, researchers and stakeholders in processes of change.

For *managers* and *consultants*, the first generic group that will directly benefit, systems theories help in multiple ways. First, by applying the concepts in this book, they will be able to develop an adequate holistic picture. This may lead to decisions that are based on considering principle solutions rather than concentrating on optimisation of current structures and processes; too much emphasis on optimisation, when not appropriate, will result in sustaining solutions that are not feasible anymore, therefore reducing the productivity of an organisation. Moreover, Applied Systems Theory will aid in finding root causes and bottle-necks, and finding solutions for processes and structures for the domains of operations management, new product and service development, logistics and supply chain management, quality management, health and safety systems, maintenance, etc. To that end, Chapter 4 presents the generic steps for analysis and solving problems. Furthermore, Applied Systems Theory can be used for designing or re-designing organisational structures (see Section 11.3). And, finally, the concepts can be applied to strategic renewal (see Chapter 10). Therefore, one might even conclude that a working knowledge of system theories should belong to the 'basic toolbox' of managers and consultants.

For *engineers*, including software engineers, system theories support analysing and designing technological systems. Building on the basic concepts of systems, Chapter 3 contains principles for analysis of those systems; some of these principles can also be used for design purposes. Chapter 4 extends these principles to a generic process for analysis and design of systems. Chapter 5 concentrates on modelling of (business) processes. Furthermore, Chapter 6 presents the principle control mechanisms and Chapter 7 extends these to the steady-state model. Particularly, the distinction between primary processes, secondary processes and control processes in these chapters will be helpful for designing control and information systems. That is further supported by some basic principles of systems engineering that are found in Section 11.1. Since technological systems might have a huge impact on stakeholders, that perspective is part of Section 8.6 and critical systems thinking in Section 11.4. Hence, Applied Systems Theory offers a wide range of concepts and tools for engineers in a broad range of domains.

For *students*, not only restricted to those that are studying engineering and business and management, the concepts and the applications of Applied Systems Theory provide foundations for processes of analysis and design (for artefacts, methods and structures). Those foundations are an addition to the support it gives managers, consultants and engineers as described in the previous two paragraphs. By relating basic skills of analysis and synthesis to examples, some of these principles are highlighted. Furthermore, the concepts of systems theories have found their way into biology, psychology, etc. That means that students can gain transferable skills from this book and have an entry to understanding multiple disciplines better.

For *researchers* in a broad scope of domains Applied Systems Theory offers some basic principles for analysis and synthesis. This is particularly the case for Chapter 3 that presents principles of modelling and reasoning that reach beyond the use of systems theories. Furthermore, the descriptions of systems in Chapter 2, processes in Chapter 5 and cybernetic mechanisms may be of help for more specific strands of research, such as management and design of organisations. Furthermore, Chapters 8 and 9 provide a brief introduction into the principles of autopoiesis and complex adaptive systems. Those researchers that are interested in foresight and strategic renewal will find concepts in Chapter 10.

For *stakeholders in process of change* the book is of interest if they want to know more about some projects that use the principles of systems theories. For those Chapter 2 offers some basic explanations, whereas in other chapters some features of so-called critical systems thinking are discussed. However, this is an addition to the text and not the mainstay of the book.

Whereas the book discusses the wide-ranging concepts and applications of Applied Systems Theory, it goes into limited detail about the methods and techniques used for analysis and synthesis. Even though the concepts are presented in a holistic way and always linked to a purpose, be it describing systems, analysis of systems and related processes or finding and detailing solutions, the specific methods and techniques are touched on. Readers that want to know more about problem solving in addition to Chapter 4 are referred to other works.

1.6 Outline of Book

Given the focus on solving real-world problems, one might wonder what writing a book about Applied Systems Theory rather than focusing only on the practical applications will bring to the table. Indeed, without the application in any field this theory would become a framework without meaning and lead to philosophical discussions on what systems represent or not (following the spirit of Hamlet's famous words). The analysis and design of systems are specific issues and vary for different fields of study, although present in the background and descriptions provided. Throughout the whole book, practical examples are given to relate the theory to the daily practice of engineering and organisations; from other fields than these two some examples are given throughout in order not to restrict the application of the theory.

The writings on Applied Systems Theory originated in describing adequately both technical systems and organisations. In later chapters it will appear that the theory might also be applied to biological systems. In those sections of the text, it becomes apparent that concepts have been borrowed from evolutionary biology to explain phenomena of complex networks of technical systems and organisations. Furthermore, this writing focuses on general systems concepts, cybernetics and complex (adaptive) systems in an attempt to provide the reader with a coherent approach to analysis and design of systems. The book does not describe the design methodologies, these can be found in other books on engineering topics, information systems, and business and management science.

Right after this introductory chapter, the second chapter of this book introduces the basic concepts of Applied Systems Theory. Through understanding these concepts it becomes possible to view systems and their relation to the environment. At the heart of studies into specific systems, we often aim at understanding their interaction with the environment and at inducement of internal changes caused by external events. The so-called design approach for systems, partly an engineering perspective, builds on that insight. To that purpose, how to study systems constitutes Chapter 3, especially paying attention to why a system is more than its elements in the context of analysis and design. Additionally, the way to study systems in their environment becomes more defined. Chapter 4 builds on the concepts of the previous chapter by presenting a generic template for analysing and resolving problems. In the next chapter the concept of processes is worked out in more detail along with ways to look at systems and dynamic conditions. Chapter 5 also pays attention to alternative modelling of processes. Chapter 6 expands on the control of primary processes. It introduces the basic cybernetic concepts that link to primary processes and presents the mechanisms for exerting effective control. Extending the control mechanisms to the boundary zone for the primary process, Chapter 7 will introduce the steady-state model. Describing living systems and organisations requires the understanding of the principles of autopoiesis as outlined in Chapter 8. Further building on non-linear behaviour, the topic of complex adaptive systems constitutes Chapter 9. Thereafter, Chapter 10 describes the breakthrough model as a specific model for adaptation by organisations. Finally, Chapter 11 deals briefly with the application of the theories to engineering, biology and management science, even though other fields in social sciences have benefited from system theories as well.

References

Bahm, A. J. (1981). Five Types of Systems Philosophy. *International Journal of General Systems*, 6(4), 233–238.

Beer, S. (1959). Cybernetics and Management. New York: Wiley.

Beer, S. (1966). Decision and control; the meaning of operational research and management cybernetics. London: Wiley.

Beer, S. (1972). Brain of the Firm - the Managerial Cybernetics of Organization. Chichester: John Wiley & Sons.

Bertalanffy, L. von (1973). General System Theory. New York: George Braziller.

Bogdanow, A. (1926). Allgemeine Organisationslehre Tektologie (Vol. I). Berlin: Organisation Verlagsgeschellshaft (S. Hirzel).

Bogdanow, A. (1928). Allgemeine Organisationslehre Tektologie (Vol. II). Berlin: Organisation Verlagsgeschellshaft (S. Hirzel).

Boulding, K. E. (1956). General Systems Theory. Management Science, 2(3), 197-208.

Brauckmann, S. (1999, January). Ludwig von Bertalanffy (1901–1972). Retrieved from http://www.isss.org/lumLVB.htm

Bunge, M. A. (1979). A world of systems. Dordrecht: Reidel.

Checkland, P. (1981). Systems Thinking, Systems Practice. Chichester: John Wiley & Sons.

Cleland, D. I., & King, W. R. (1969). Systems, organizations, analysis, management: A book of readings. New York: McGraw-Hill.

de Condillac, E. B. (1749). Traité des Systêmes, Où l'on en démêle les inconvéniens et les avantages (Vol. II). Paris: Libraires Associés.

Dekkers, R. (2008). Adapting Organizations: The Instance of Business Process Re-Engineering. Systems Research and Behavioral Science, 25(1), 45–66. doi: 10.1002/sres.857

Forrester, J. W. (1968). Principles of Systems. Boston, MA: MIT Press.

Goethe, J. W. v. (1808). Faust: Eine Tragödie. Tübingen: Cotta.

Goethe, J. W. v. (1832). Faust: Eine Tragödie, zweiter Teil. München: Wilhelm Goldman Verlag.

Gorelik, G. (1975). Reemergence of Bogdanov's Tektology in Soviet Studies of Organization. Academy of Management Journal, 18(2), 345–357. doi: 10.2307/255536

Gorelik, G. (1987). Bogdanov's Tektologia, general systems theory, and cybernetics. Cybernetics and Systems: An International Journal, 18(2), 157–175. doi: 10.1080/01969728708902134

Heylighen, F. (2004). Cybernetics and Systems Thinkers. Principia Cybernetica Web. Retrieved from http://pespmc1.vub.ac.be/csthink.html

Heylighen, F., & Joslyn, C. (1992). What is Systems Theory? Principia Cybernetica. Retrieved from http://pespmc1.vub.ac.be/SYSTHEOR.html

Hill, A., & Hill, T. (2011). Essential Operations Management. Basingstoke: Palgrave Macmillan.

Iancu, Ş. (2009). Ştefan Odobleja – The Main Romanian Forerunner of the Cybernetics. Annals of the Academy of Romanian Scientists – Science and Technology of Information, 2(1), 41–58.

Jost, J. (2004). External and internal complexity of complex adaptive systems. Theory in Biosciences, 123(1), 69-88.

Kline, S. J. (1995). Conceptual Foundations of Multi-Disciplinary Thinking. Stanford: Stanford University Press.

Klir, G. J. (1969). An Approach To General Systems Theory. New York: Van Nostrand Reinhold.

Laszlo, A., & Krippner, S. (1998). Systems Theories: Their Origins, Foundations, and Development. In J. S. Jordan (Ed.), Systems Theories and A Priori Aspects of Perception (pp. 47-74). Amsterdam: Elsevier Science.

Laszlo, E. (1972). The Systems View of the World: The Natural Philosophy of the New Development in the Sciences. New York: George Braziller.

Malik, F. (1992). Strategie des Managements komplexer Systeme. Bern: Haupt.

McCarthy, I. P. (2004). Manufacturing strategy: understanding the fitness landscape. International Journal of Operations & Production Management, 24(2), 124-150.

Mesarović, M. D., Macko, D., & Takahara, Y. (1970). Theory of hierarchical, multilevel systems. New York: Academic Press.

Miller, E. J., & Rice, A. K. (1967). Systems of Organization: The control of task and sentient boundaries. London: Tavistock.

Peroff, N. C. (1999). Is Management an Art or Science? A Clue in Consilience. Emergence, 1(1), 92–109.

Rowley, D. G. (2000). How Important Was Alexander Bogdanov? H-Net Reviews in the Humanities & Social Sciences. Retrieved from https://www.h-net.org/reviews/showpdf.php?id=4154

Sagasti, F. R., & Mittrof, I. I. (1973). Operations Research from the Viewpoint of General Systems Theory. Omega, 1(6), 695–709. doi: 10.1016/0305-0483(73)90087-X

Simon, H. (1959). Theories of Decision-Making in Economics and Behavioral Science. The American Economic Review, 49(3), 253–283.

Slack, N., Chambers, S., & Johnston, R. (2010). Operations Management. Harlow: Prentice Hall.

Tucker, B., Furness, C., Olsen, J., McGuirl, J., Oztas, N., & Millhiser, W. (2003). Complex Social Systems: Rising Complexity in Business Environments. Cambridge, MA.

Wiener, N. (1948). Cybernetics: Or Control and Communication in the Animal and the Machine. Paris: Hermann.

Wikipedia. (2007, 2 February). Macy Conferences. Retrieved from http://en.wikipedia.org/wiki/Macy_conferences

Wikipedia. (2009, 1 August). Margaret Mead. Retrieved from http://en.wikipedia.org/wiki/Margaret_Mead

Wilson, E. O. (1998). Consilience: the unity of knowledge. New York: Alfred A. Knopf.

2 Basic Concepts of Systems Theories

When defining what systems are, the first thing that comes to mind is how often people use the word: system. All kind of professions and knowledge domains denote different meanings to this concept used in daily language. Engineers frequently talk about systems when they review designs or analyse technical equipment, e.g. the propulsion system of a ship. Computer experts point to information and communication systems. Biologists see the oceans as ecological systems. In addition, many consider organisations as systems. Thus, the word systems refers to objects (discrete systems) as well as purposeful constructs of the mind that are abstract in exchange between people, such as the conceptualisation of an organisation as a system. Distinguishing systems within reality helps to describe, to analyse and to create. To support the analysis of problems and to generate solutions, this chapter will define systems, discuss their properties and expand on their application in the domain of technical design, biology and organisations, while keeping in mind that the principles are applicable to many (scientific) disciplines.

The use of systems theories as a methodology of description and analysis originates from the drive to simplify reality and to comprehend natural events. The interpretation of reality has fascinated mankind since long and many have tried to explain phenomena that we experience daily, to understand patterns and to predict what will happen. The complexity surrounding us has forced investigators to look at interrelationships between objects and events. How does a propulsion system of a ship react to changes in forces? How does an information and communication system react to a cyber attack? How does an ocean as a system react to pollution? And how does an ecological system react to human interventions? Putting it all together, we are looking for approaches and methodologies to understand what is going on and how to solve a wide range of problems presented to us. As mentioned in Section 1.2, in the spirit of generating knowledge and solving problems, systems theories attempt to bridge different disciplines by their range of applications and at the same time act as a platform for multi-disciplinary perspectives. Applied Systems Theory, as one of the systems theories (later, Chapter 11 will introduce briefly a few other theories), provides such an opportunity to describe and analyse problems, mainly due to its holistic approach.

This chapter starts by looking what the concept of systems means and by defining them in Section 2.1; some of the key concepts for the definition will be elaborated on. The section thereafter discusses the properties of systems, needed for further analysis of a system. The chapter continues by looking at subsystems and aspectsystems appear in Section 2.3 and 2.4, as specific ways of examining systems, in more detail. The state of systems, related to their properties, is the topic of Section 2.5, and pertains to events and activities

© Springer International Publishing AG 2017
R. Dekkers, *Applied Systems Theory*,
DOI 10.1007/978-3-319-57526-1_2

that happen in the environment of a system. Since systems may respond to changes in their environment, Section 2.6 introduces the various concepts of systems' behaviour. Finally, Section 2.7 addresses the system boundary.

2.1   Systems

Although incorporated in daily language, when we talk about systems, each of us might attribute total different meanings to this comprehensive word, the key to any systems theory (see examples of definitions in Box 2.1). What we intend telling is that we separate elements from a total reality to study the nature of the system driven by the purpose of a particular study. This will enable the investigator to analyse and to predict the behaviour of such

---

BOX 2.1:   DEFINITION OF SYSTEMS

This box provides definitions of systems to show similarities and differences in what a system is according to different authors.

APPLIED SYSTEMS THEORY

A system consists of elements discernible within the total reality (universe), defined by the aims of the investigator. All these elements have at least one relationship with another element within the system and may have relationships with other elements within total reality.

ALTERNATIVE DEFINITIONS

... any entity, conceptual or physical, which consists of interdependent parts. *[Ackoff, 1969, p. 332]*

... sets of elements standing in interrelation ... *[von Bertalanffy, 1973, p. 38]*

... the word "system" has been defined in many ways, all definers will agree that a system is a set of parts coordinated to accomplish a set of goals. *[West Churchman, 1979, p. 29]*

... system is defined as a set of concepts and/or elements used to satisfy a need or requirement. *[Miles Jr., 1973, p. 2]*

System ... is a set of entities, real or abstract, comprising a whole where each component interacts with or is related to at least one other component and they all serve a common objective. *[Wikipedia, 2007]*

a system, for example an organisation. Anybody wanting to describe or analyse an organisation does not start by enumerating everything outside the organisation or by defining all small objects within a company, for example individual employees and forms in use. The simplification starts by defining objects and entities of interest given the problem statement. That means that the elements of study may quite differ when we perform a analysis of a quality system or a logistics system even when it concerns the same company. Once the entities have been defined, the investigator will examine the relationships enabling the understanding of the behaviour of a system.

Defining Systems

That means that a system is more than just listing its elements. Think about a watch; all separate elements (parts) of a watch do not make it work and indicate the time; however, when the parts are put together and an energy source activated (manual winding, automatic winding, battery, etc.), then the watch starts showing the time. For the purpose of analysis and design, the separation of a system from its surroundings helps understanding the relationship between the system and its environment, the relationships between its elements and elements in that environment and the interaction between elements within the system (see definition in Box 2.1). Cutting the relationships of the system, better those of its elements, from the environment will result in limiting any study to the optimisation of the system itself; it will not lead to embedding in its environment or to adapting the system to its environment from which it makes part. Which interaction to review, within the system and with the environment, depends entirely on the nature of the study and the analysis. As Checkland [1993, p. 101] notes: 'the observer will, *for his own purposes*, use systems thinking as a means for arriving at his description'.

Some examples will illustrate this definition of systems. Box 2.A shows a map of the Galápagos Islands and demonstrates that the view on the system will differ when considering it from a geographical perspective, from a socio-economic view or from an evolutionary perspective (the Galápagos Islands appear in the work of Charles Darwin [1859]). Another example is the service and overhaul of airplanes by an airline that may serve as an element of the airline when exploring the adherence to flight schedules. When looking at the way that interaction takes place between workers within the Technical Service Department to optimise co-operation, people will serve as the elements of the study. However, if we want to observe the maintenance and overhaul of the airplanes themselves, only the steps necessary for this process constitute the elements of study. The interaction with the environment will differ as well. We might consider the propulsion of a car as a system. The propulsion system then includes all elements related to moving a car, e.g. engine, transmission and tyres, but other elements of the car, such as the dashboard, will not be part of the system. The two examples merely demonstrate the notion that the nature of a study entirely determines how to look at any system or even how

Box 2.A: Introducing the Galápagos Islands

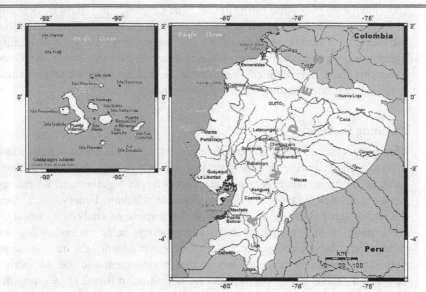

Source: Wikipedia [2009]

The Galápagos Islands have become famous through the work of Charles Darwin (1809–1882); they are an archipelago of volcanic islands distributed around the equator in the Pacific Ocean, 972 km west of continental Ecuador (South America), see figure above. The Galápagos islands and its surrounding waters form an Ecuadorian province, a national park and a biological marine reserve. The islands' population (ca. 23,000) lives mostly of tourism, farming and fishing.

When examining the geographical position of these islands, the only interest is into the shape and the position of islands relative to continents, countries or islands (for example, the relative position to the rest of Ecuador). However, if an analysis would concentrate on the social-economic conditions, the elements and relationships to consider are social-economic entities, such as fisherman, fleets, food processing companies, traders, tourism agencies and their collaborations. Although the geographical location might be to the advantage of social-economic prosperity, it is not the prime concern. As a third case, Darwin's study focused on the fauna, particularly, the populations of finches that he studied and that allowed him to verify the theory of natural selection that was simultaneously proposed by Alfred Russel Wallace (1823–1913) [Darwin and Wallace, 1858]. Again, the geographical location might favour the study of natural selection but does not include it in the first instance. Hence, these three examples of investigations into a system show that the elements and relationships to consider might differ substantially from study to study.

to define a system (and therefore, it depends even on the perception of the observer).

Modelling systems by using Applied Systems Theory starts out with analysis of the elements and their relationships, and the interaction of a system with its environment. Figure 2.1 shows also that you can distinguish a system within total reality, but not separate it from that same total reality. This points to the need to consider organisations as open systems rather than closed systems. The definition, the one of Applied Systems Theory, mentions several key concepts (see Box 2.2) that need elaboration before moving on to discussing the properties of systems and closed versus open systems.

Elements

The elements constitute the smallest parts needed for the purposeful analysis of a system within a specific study. In Figure 2.1 all elements, except G, are part of the analysis undertaken; the systems itself consists of the elements A, B, C, I and J. To understand the purpose of any system, you need to look at the relationships between the external elements and the internal elements. For example, an element of the propulsion system of a car is the engine. The engine converts thermal energy (through the combustion of fuel) into mechanical energy and transfers that energy through the drive shaft to the gearbox, another element of the propulsion system, to the axles that are attached to the wheels; finally, it creates the driving force through the contact with the road surface (this contact constitutes the relationship with the environment). As another instance, within the logistics system of a factory, the department responsible for the supply of materials may be seen as an element of the system when analysing the flow of goods. Both the propulsion system and the logistics system have relations with the environment, which affect the performance of the system. For example, the propulsion system is linked to the driver as an element from the environment; actions generated by the driver influence the behaviour of the propulsion system. And conversely,

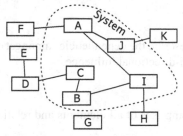

Figure 2.1 *System with its elements and relationships. Each of the internal elements has at least one internal relationship to other elements within the boundary (A, B, C, I and J). The environment consists of those external elements that have direct relationships with internal elements (D, F, H and K). Some elements outside the system boundary have no or no direct relationship with internal elements (E and G) and should not be considered part of the system's environment.*

the reaction of the propulsion system to external circumstances, think about slippery road conditions, determines how the driver has to adjust the speed of the car. In the case of the logistics system of a company, the department responsible for material supply within the logistics system connects to suppliers as external elements. Thus, the logistics performance of any company depends not only on the internal elements but on the performance of suppliers as well. Both examples show that the environment has a strong impact on the performance of systems. For that reason, the examination of the interaction of a system (through its elements) with the environment often constitutes the first step of analysis.

Elements may range from physical objects to constructs of the mind, depending on the study's objectives. When examining the material flow as such within a company, the flowing elements of the system consist of the materials and parts transformed into products. In the case of information systems, the elements also depend on the problem definition. The microprocessor within a computer or server handles bits or bytes, the elements that make up a system through batch-jobs or files that pass through that processor. However, in the case that the investigator wants to analyse the infrastructure of the information system, the servers, computers and cabling are the elements of the system. In another case when we are examining the interaction between the organisation and the information system, the information is considered the element (information is then the combination of bytes into data with an attributed meaning; information is a construct of the mind). As might become clear, the problem definition has a strong influence

---

Box 2.2: Key Concepts of Systems

Elements

The elements constitute the smallest parts needed for the purposeful analysis of a system within a specific study.

Relationships

Relationships describe the dependencies amongst elements, whether it be a mono- or bi-directional influence.

Universe

The universe comprises of all elements and relationships, known and unknown.

Environment

The environment is that part of the universe that has any (known) direct relationships with the elements of the system.

on what to consider as elements and whether these have a discrete, physical nature or have abstract meanings within a particular study.

Relationships

The elements in the environment and the internal elements have relationships that describe the dependencies amongst elements, whether it be a mono- or bi-directional influence. This influence reflects the change(s) in values of properties of systems. Between elements, different relationships might exist. For example, in the propulsion system of a car the engine and the gearbox share both an energetic relationship as well as a geometric relationship. Note that within a system interrelationships exist between elements, which implies that all elements are connected by relationships and no isolated elements are present. When examining the logistic relationship between a supplier and the customer aimed at the physical goods flow, the human interaction is of no interest for the study at that point of time; therefore the directors of the company are not part of the system being studied, but part of total reality. Hence, the aims of a study determine the relationships, both internal and external, to be considered for analysis.

Universe

The total reality points to the universe comprising of all elements and their relationships, known and unknown. Depending on the nature of the study, we will consider only a partial set of elements and specific kinds of relationships within the total reality as identified by a problem definition. This implies that not all elements and relationships bear any weight for a specific study. Besides, no one can be aware of all elements and relationships; the regular discovery of stellar systems, planets, etc. demonstrate this notion. In most cases it is possible to distinguish the elements in the universe that have an impact on the system under investigation. For the propulsion system of a car, the universe consists of other systems from the car, e.g. the suspension system, as well as other systems, such as the weather system, which do not directly influence the behaviour of its propulsion system. Even so, this applies equally well to the human resource management when studying the optimisation of the logistics system of a personal computer manufacturer for deliveries to customers; at first sight human resource management is not dorectly related to this system, unless, for example, their skills are inadequate for tasks or training is needed. The concept of the universe as a total reality beyond full comprehension points to the limitation of any study: taking only a part of total reality into account.

Therefore, the view on a system might totally differ when considered from distinct disciplines and sciences each having their own objectives as well. Considering the definition of a system, each study requires emphasising specific elements and relationships within total reality. This notion indicates that each of the different disciplines working together in a context of solving a problem within a specific study should generate an unique focus on the elements and relationships within the universe. Take for example, the customer service department of a bank. A computer specialist may look at it from the perspective of hardware, integration with communication technologies and software applications that control workflow. However, a marketeer will approach the same department from the frame of reference for communication with customers, offering of products and resolution of complaints. In other words, no system will be the same for investigators from different backgrounds; it will even vary from study to study how a system is defined.

Environment

When analysing a system, in the first instance, we tend to restrict ourselves to that part of the universe that has any (known) direct relationships with the system. The environment consists of the elements that have any relationship with the system but are not part of the system and for that matter are part of total reality (universe). In Figure 2.1 element G is no part of the environment, whereas the environment itself is part of the universe; even element E should not be considered part of the environment, because it does not have a direct relationship with any internal element. West Churchman [1979, pp. 35–37] notes that those elements that are outside the system's control but relevant to its objectives constitute the environment. The examples as mentioned above identify the driver as part of the environment for the propulsion system of a car and the supplier as environment for the logistic system for deliveries to customers. Again, the objective of the study determines which elements outside the system make up the environment.

The environment exerts a strong influence on systems, even beyond what is visible by the eye. For example, the ancient Egyptians did cut trees and papyrus under the moonlight during the full moon, which was for a long time considered superstitious. As it later turned out, this timing for cutting papyrus would enhance its durability due more being saps present in the logs. Nowadays, we would not consider the timing of harvesting wood in relation to the lunar cycle. From the point of view of the increasing importance of durability in our age, this superstition turns into expansion of our view on sustainable production of wood. The example shows that we need to consider carefully what constitutes the environment for a specific problem and how it affects the behaviour of the system, because this possibly influences the effectiveness of an intervention or solution.

2.2 Properties of Systems

Once you have defined a system related to the scope of the specific study, the need to describe the system emerges for the purpose of further analysis and later for generating solutions. The description of a system allows further

---

BOX 2.3:   GENERIC PROPERTIES OF SYSTEMS

CONTENTS

The contents of a system represents all elements that constitute the smallest parts needed for the purposeful analysis of a system.

STRUCTURE

The structure consists of all interrelationships that describe the dependencies amongst elements, whether it be a mono- or bi-directional influence. The structure is consisting of both an internal structure (relationships between elements of a system) and external structure (the relationships between external elements and relevant internal elements).

ATTRIBUTES

The attributes consist of the properties of the system or the properties of its constituent elements.

EMERGENCE

Emergence refers to properties of the whole that cannot be solely explained by the properties of the constituent elements.

---

analysis by pointing out which properties it possesses in relation to the problem definition. The generic properties of a system (see Box 2.3) allow doing so and are divided into the content, the structure and the attributes. In addition to content, structure and attributes, the emergent properties of systems demonstrate the principle of systems that they are more than their elements. And finally, the degree of interconnectedness between elements is expressed by the dimension of wholeness and independence. These five properties help to understand the behaviour of the system.

Content

The content refers to the listing of all internal elements of a system; for example, in Figure 2.1 the content of the system is: A, B, C, I and J. The concept of content compares to a list of parts on a technical drawing, the bill of materials used for logistics management and a directory of files on a computer. The content does not describe the relations between the elements and between the system and the environment. However, it does separate those elements that belong to the system from those that do not within the universe. It is simply a list of elements and the level of detail for the elements may vary depending on the problem definition. For example, if somebody wants to know if all meetings have been documented a list of minutes of

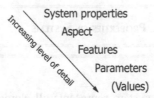

System properties
Aspect
Features
Parameters
(Values)

Increasing level of detail

Figure 2.2   *Interrelation between properties, aspects, features and parameters.   The system properties may be broken down into aspects, which reflect the types of relationships that are investigated.   When decomposing aspects, the investigator of a system considers features and parameters; to parameter values can be attributed.   That means that the figure shows that by going into more detail, sometimes the relationship with the original property of the system might become less clear.*

meetings (content of a file) will suffice. If somebody else wants to know whether a specific issue has been addressed, the contents of the minutes of the meeting will need to be examined (that means a lower level of detail). Henceforth, the distinction of the level of detail of elements depends on the aims of the investigator and the content merely enumerates the internal elements of a system.

Structure

To understand the properties of a system, the investigator needs to examine the structure of the system, i.e. the listing of all interrelationships between elements; please note that it always concerns the relationships of interest to the study undertaken. Relationships imply that elements do have a mutual influence on each other that stretches beyond the fact that each of these elements is present within the system. It approaches the concept of connectivity as described by Hitchins [1992, pp. 79–80]. He notes that only when elements have some influence on each other, an interrelationship exists, changing at least one of the properties under consideration. For example, when a user saves a document on the hard disk of a computer and the programme saves data and settings in separate files. For retrieving documents, the specific application will have to use the interrelationship between the data and settings to make an adequate representation of the document in the user interface (for example, the display of the computer or tablet). Thus, the examination of the structure clarifies the influence elements may have on each other.

To distinguish the relationships within the system from the relationships with the environment, a division exists between its internal structure and its external structure. The internal structure records all the relationships between the elements within the system (internal relationships). For example, Wilson [1990, p. 70] mentions that physical layout, power hierarchy, formal and informal communications reflect the structure of the organisation as a system; most of these descriptions are internally oriented. The relationships with the environment, so-called external relationships, are the domain of the external structure, which means that exploring these requires crossing the system's boundary. For instance, the relationships with suppliers are part of an organisation's external structure. Generically speaking, the external structure has a strong affect on the internal structure.

Attributes

Elements do have attributes that we commonly describe by using features. The shape and performance represent attributes of a delivery van in general, the dimensions and transport capabilities features belonging to them (attributes carry similarity to aspects which Section 2.4 will introduce). A feature may be classified as either determinate or determinable. A determinable feature is one that can get more specific. For example, colour is a determinable property because it can be restricted to redness, blueness, etc. A determinate feature is one that cannot become more specific. These features may be described by parameters that in turn may have values. Features do not have necessarily a quantitative value, for example the colour is also a parameter that does not have directly a numerical value (although physicists use wavelengths of light as numerical value and the painting industry a standardised coding to describe colours). Instead you may describe features with meaningful adjectives, such as blue for the parameter colour. Figure 2.2 depicts the relations between the system properties, aspects, features and parameters.

Emergence

Especially, when describing complex systems, the whole may have properties that refer to the whole and are meaningless in terms of the parts that make up the whole [Checkland and Scholes, 1990, pp. 18–19]; these we call emergent properties of the whole system. This notion becomes increasingly important when systems consist of many elements and numerous types of relationships. This is apparent for a car: all individual parts cannot provide its transport function, however when put together it is capable of transporting passengers and goods. Conversely, when we strive for reducing systems by distinguishing elements, the emergent properties might be lost. Organisations often achieve performances that exceed the sum of the individual capabilities (often referred to as synergy). These performances elevate the organisation from being a collection of elements to a level of self-being. Thus, the performance of an organisation cannot be traced back to an individual even though that person might have had a strong influence. In that respect, emergence is a property of the whole system more than of individual elements.

For organisational, biological as well as technological systems this points to an integration step when discussing the properties of a system. When looking at the whole system we might attribute different properties then when reviewing the elements themselves. For the purpose of analysis, we may loose perspective moving from the system level to the level of elements. Conversely, when shifting attention from elements to the system as a whole, we will discover properties that were not noticeable before. These phenomena might explain why ecosystems show resilience when a species (as an elements of it) becomes extinct because of self-regulating mechanisms at the level at the whole ecosystem; for example, the bird called dodo (raphus cucullatus) died out at the Indian Ocean island of Mauritius during the mid-to-late 17th century, however, Mauritius' ecosystem has continued to flourish (that is until recently). Whereas for some events biological systems show resilience, ecosystems can also collapse because of changes at the level of the elements. For example, that happened when the sea lamprey (Petromyzon marinus) – a marine invader from the Atlantic Ocean that entered the Great Lakes (between Canada and the U.S.A.) through the ship canals and locks built to bypass obstacles, such as the Niagara Falls – outcompeted smaller, native lampreys and devastated the fish communities of these lakes from the 1930s on. These examples show that systems theories are by definition not reductionist in the sense of Descartes' view but provide a balancing insight between properties that can be attributed to the system as a whole and properties that are an extension of the properties of the elements.

BOX 2.4: DEFINITION OF SUBSYSTEMS AND ASPECTSYSTEMS

SUBSYSTEMS

A subsystem is a subset of elements within the system, while retaining all original relationships between these elements.

ASPECTSYSTEMS

An aspect system is a subset of relationships within the system, while retaining the original elements on the condition that all remaining elements have mutual relationships within the system.

In contrast to denotation of subsystems, there are many definitions on how to call a system with a subset of relationships under consideration. For example, aspectsystems are also known as partial systems. Calling them partial systems might cause confusion because some authors [e.g. de Leeuw, 1979, p. 97] follow the definition of aspectsystem from Applied Systems Theory and some use the mathematical sense where it means a subset of equations. Further adding to the confusion, some denote a partial system as a subsystem. And a subset of relationships has been called a functional system as well ([Gershenson and Heylighen, 2003, p. 608]. Finally, aspects are equated to subsystems ([van der Zwaan, 1975, pp. 150, 153]. In the definitions of Applied Systems Theory, there is a strong distinction between looking at specific elements (subsystem) and at specific relationships (aspectsystem).

Wholeness and Independence

This especially holds true when elements have many interrelationships. Wholeness indicates that all elements have relationships with all other elements within the system whatever these might be. In such a case changes in any relationship will affect all its elements and in practice lead to instability within the system and towards the environment. Some extended operating systems for computers, such as versions of the Windows® operating systems, and large software applications tend to possess this characteristic and within the community of information technology have a name that adaptations have unpredictable outcomes. At the other side of this spectrum is independence when elements within the system have no interrelationships at all. In fact, we cannot call this a system since it does not comply with the definition in Box 2.1, which presumes the presence of relationships between elements. Hence, the degree of interconnectivity within the system also indicates how difficult it might be to intervene. For systems that are gravitating towards wholeness, there are possible ways to counter that; for example, modular design of products and services aims at achieving a higher degree of independence that way creating more flexibility and less dependence of production control to market demands. Practically, systems span a wide range of connectivity between elements ranging from wholeness to near independence; however, the higher the degree of interrelationships between elements, the more complex it is to understand, to describe and to analyse the system (note that Chapter 8 will expand on complex adaptive systems, partially addressing this type of complexity).

2.3   Subsystems

When conducting a study of a particular system, the need to examine specific parts of the system might emerge. In the case of evaluating the performance of an organisation we may need to analyse the purchasing system as part of the logistics system. While designing a windmill, we might need to look at the energy conversion. Both examples show an expansion of details, while ignoring other elements or relationships. Since a system consists of elements and relationships, we might as well distinguish two ways of breaking down a system into 'partial' systems (that follows from Figure 2.1 that depicts the key elements of a system: elements and relationships). First, we might look at specific set of elements contained within the total system, then we speak about a subsystem (the purchasing system) and, second, we might examine certain types of relationships by distinguishing an aspectsystem (the energy conversion).

Looking at clusters of elements, subsystems, helps defining main parts within a system without describing endless lists of elements (see the definition in Box 2.4), especially when the original system contains a large number of elements. Imagine a listing of all the parts of an airplane, the individual organisms of an ecosystem (such as the rainforest) or all the personnel working for a company, such as Shell or Philips, through which the investigator has to find his way to find a certain type of parts, species or personnel, e.g. database specialists. Subsystems define sets of elements as purposeful entities within the system. Doing so, the relationships within the system, with the environment and, therefore, the relationships between the subsystem and the other elements remain the same. The original system becomes now part of the environment of the subsystem. When looking at a system, an investigator might view it as a set of interrelated subsystems.

The application of dividing systems into subsystems strongly relates to simplifying the structure of a system to purposeful entities without losing overview. During the study of these systems and their elements, subsystems serve as intermediates between the system as a whole and the elements. The airplane has different subsystems, e.g. the electrical system, the fuselage of the plane, the wings, etc. An organisation will have subsystems, too. When studying the logistic system, the purchasing system and the warehouse system are subsystems. Within the purchasing system, goods receipt is one of its subsystems. All these subsystems have interrelations. In a house, one might have a subsystem for water supply and a subsystem for supply of electricity; these are interconnected by the geometrical position in the building. Thus, subsystems may have various levels, depending on the depth of the study, but the different type of subsystems also have interrelationships to each other.

The definition in Box 2.4 also reflects that a subsystem itself is a system (see Figure 2.3). Defining subsystems results in studying the smaller parts of a total system without isolating them from the system and its properties. It implies as well that a specific study into a subsystem requires an adapted problem definition for the subsystem. The original objective of the study results in distinguishing that particular set of elements and relationships while the need for exploring a subsystem has narrowed down the focus caused by further analysis at system level. This further analysis informs limiting the scope to these particular elements, i.e. the subsystem. Thus, the need for investigating a subsystem has a strong link to the progress of the analysis and detailing as part of a study.

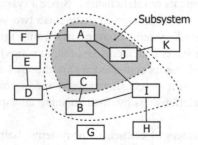

Figure 2.3 *A subsystem within a system. Some elements are not looked at as part of the particular subsystem (elements B and I in this example); they become part of the environment of the subsystem.*

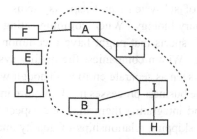

Figure 2.4 *Aspectsystem. While principally retaining all elements, the specific relationships studied result in discarding some elements as being part of the system under review. A comparison with Figure 2.1 shows that element C is no longer part of the aspectsystem and that elements D, E and K are no longer part of the environment, because they have no specific relationships of this type with elements of the aspectsystem.*

2.4 Aspectsystems

The second principle for having a more detailed look at a system focuses on which relationships in particular draw interest in the perspective of the problem definition. An example in economics illustrates this. When a cost-price analysis results in the quest to find data on prices of parts or on units of labour, the physical characteristics of the product are of no interest except for obtaining the proper data for the cost-price calculation. Other relationships than those related to the objectives of the study have no impact on the results, for example, the aesthetic aspect of the product. This way the number of relationships under examination is reduced to the necessary ones according to the nature of the study. The relationships subject to closer study are called the aspect or aspects and an aspect always concerns a subset of the relationships present in the system and its relationship with the environment.

An aspectsystem reflects the choice for particular relationships as the area of interest. Basically, we eliminate all the relations except the ones we choose to explore (see the definition in Box 2.4). If it occurs that some of the remaining elements do not have any more relationships with other elements in the system, these elements need to be removed, leaving the aspectsystem always with elements that have mutual relationships within the system or with elements outside the system; see Figure 2.4 in relation to Figure 2.1, where elements D and E have a relationship of the aspect but no relation to elements within the system for which these should be discarded as part of the study. An example might illustrate this; the fuel consumption of a jet engine has no direct relationship to the use of lubricants for rolling parts within the total system of the airplane. Hence, the focus on a specific type of relationships not only reduces the number of relationships to consider but may also affect the number of elements in a more detailed study.

No predefined aspectsystems do exist since the aspect under review, a particular set of relationships, finds its origin in the specific problem definition. The example of the Galápagos Islands in Box 2.B demonstrates how the concept of subsystems and aspectsystems can be applied to the study of evolutionary biology. When communicating people often point to general classes of aspects that might have a common meaning for all, e.g. the energy system. Within companies, the quality system and the logistics system are mostly seen as separate entities. Though when quality is a must, improvement of the business processes might require investigating both these general aspects and integrating them into one aspect for the study at hand (and may be even skipping relationships of quality and logistics that do not relate to the focus of the study). This example underlines the necessity to articulate in each situation the aspect for further evaluation.

For illustration of the concept of aspects, two examples in addition to Box 2.5 will follow with generic classifications of relationships. To describe an office building and an organisation an architect may distinguish several aspects (the first example):

BOX 2.B: GALÁPAGOS ISLANDS: SUBSYSTEMS AND ASPECTSYSTEMS

The Galápagos Islands have become most famous through the work of Charles Darwin (1809–1882) when he studied populations of endemic species. From the perspective of systems theories, he has applied thinking in subsystems and aspectsystems. Consider all fauna in these islands as a system. By taking a specific species within the fauna, a subsystem is created. A specific population of one the species on one of the islands should then be called a subsystem of a subsystem. By comparing subsystems of subsystems, most notably the finches, Darwin did find the evidence for the theory of natural selection and adaptive radiation. However, by concentrating on the anatomical appearance of species, he has focused on only one aspect; for example, he did not consider predatory relationships. Nowadays, biologists would rely on a number of aspects before concluding on relationships between populations of species, DNA samples being one of them.

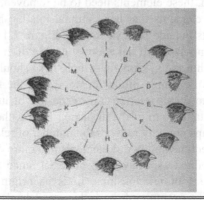

- the geometrical aspect. This includes the dimensions of the structure of the building, the size of the offices, the lay-out, the position of the building in the environment, etc.;
- the functional aspect. The functional aspect describes the use of the building, the flow of people through the building, the goods entering the building, the catering facilities and so on;
- the energetic aspect. In present times, the energy consumption plays an important role in the design and construction of buildings;
- the utilities aspect. This aspect consists of the power supply throughout the building, the information and communication infrastructure, the water supply and piping, the drainage system, the illumination and so forth;
- the aesthetic aspect. Office buildings should be a pleasant place to work in and might have to leave an impression in people's mind;
- the structural aspect. Buildings have to withstand external influences, weather and earth movements, and display internal strength during the time of occupancy;
- the maintenance aspect. The building has to be kept in a working state due to the deterioration (as wear and tear) appearing in the course of time.

Each of these aspects describes particular sets of relationships of the building, which may have little or limited interrelations. Eventually, the problem definition will define which particular sets of relationships are of interest and this way what the aspect compromises. Generic classifications of aspects have little meaning for specific problems except that they may be helpful for generating theories for generic aspects, such as in the case of an organisation, the second example:

- the logistic aspect. This aspect consists of the flow of materials and goods through the company and to the customers. It also includes planning of production, storage and movements;
- the quality aspect. This aspect entails meeting the customers' requirements and maintaining the standards for products and processes;
- the technology aspect. The deployment of skills and knowledge to expand the product range and to improve primary processes are the domain of technology;
- the human aspect. It addresses the way people within the company communicate and collaborate either with others in the company and with persons outside the organisation;
- the information aspect. This entails the flow of information through the company and the processing of data;
- the financial-economic aspect which compromises the cash-flow, the budgeting, the decision-making on investments, etc.

These aspects of an organisation follow more or less the division of (scientific) disciplines, neither one describing the system in its full extent. When choosing for a specific aspect, the study limits itself to a partial description of the object under review. However, such a description might go beyond a single aspect as a generic classification. For example, in the case of the firm, if a study is undertaken into handling of complaints, only a part of the quality aspectsystem and part of the logistics aspectsystem will be of interest. Thus, for specific problems the aspects under consideration may be unique and not following canonical divides and that means that what is considered a system and aspectsystem is contingent on the problem definition.

Describing a system in fuller detail requires the comprehension of interrelationships between aspects that may exist, though at a specific point in time little might be known about these. These interrelationships come into the picture during evaluation, appraisal and decision-making. Managers and engineers take decisions regarding trade-offs between quality and cost-price but each person attributes different values to the two aspects. Even persons fulfilling similar jobs will have different opinions. Therefore, the trade-off between aspects is subjective and may differ from one occasion to an other, mainly because little is known about the interrelationships between aspects.

2.5   State of Systems

At a certain moment in time, a system might have defined properties, the content, the structure and the attributes, the so-called state of a system (see definition in Box 2.5). For example, a company has a set of elements, an organisational structure and has certain values for the financial and logistic

---

BOX 2.5:   DEFINITIONS OF STATE AND BEHAVIOUR OF SYSTEMS

STATE

The state of a system describes its content, its structure and (the values of) its attributes at a given moment in time.

BEHAVIOUR

Behaviour is the capability of a system to respond to variations in external relationships and modifications of the external structure, either through changes in attributes, adjustments of the structure or adaptation of the external structure.

One could say that the state of a system is related to a specific point in time and behaviour is considered during a certain period of time.

Behaviour can be static or dynamic; behaviour is called dynamics when properties or relationships change. The properties can change deterministic or probabilistic. If the properties remain stable within a specific time-frame, the system is in a steady state; however, when the outcomes depends on the memory of the system, the behaviour is transient.

---

aspect, all representing the state of the system at a given moment. When the state of the system does not change in view of the problem definition, the entity is a static system, the properties remain the same within the given time-frame. The position of a bridge in a landscape on a map marks a static system. In the case that as a result of an event any of the system's properties changes, whether it concerns the content, the structure or attributes, then we call the system dynamic. A capital expansion of a firm represents such an event, some of the features concerning the financial aspect do change with the intent to strengthen the financial-economic position of the company. Activities, events leading to other events, take time in general, creating interdependencies between several states of a system. To summarise, when the state of a system remains unchanged, the system is called static; when activities cause any type of changes in the state, the system is regarded as dynamic.

Therefore, the state of a system is dependent on previous events and states; all these successive stages, the history of all previous states and events, correspond to the memory of the system. In the case of companies, the state of an organisation has roots in previous organisational structures, the intake from orders and the knowledge gained by people working in the organisation. If the elements and relationships remain unchanged over a period of time and only the attributes change, that means that the scope of a system is limited to its present capabilities for dealing with events and perturbations. The memory will tell us about the adaptations taking place in response to changes in the environment because they are embedded in the current state of a system.

When modifications occur in the interrelations, within the system or those with the environment, or elements, this implies directly also alterations in relationships; the system has an altering structure. Such an altering structure might display a repeating pattern; in general it is assumed that these variations are irreversible due to the memory. Managers and engineers exert a similar characteristic in this view when both look for interventions to enhance performance of either organisations or technical objects through structural changes. These structural changes always concern changes in elements and relationships; for example, changes in the organisational structure or redesign of equipment. The observation that the structure does not change may entirely depend on the interval between the monitoring moments, pointing out the caution to drawing early inferences on the dynamic capabilities of a system. To summarise, the behaviour of a system tells about the ability to undergo changes in state, whereas the changes in the structure also reflect only on the internal and external relationships among the elements (also called the dynamic capability).

Based on the previous typologies for changes in the state of systems, examples of the various possibilities are:

- bridge on map (position in the landscape): static system;
- car engine (delivering power for propulsion): dynamic system, permanent structure;

- company (delivering products to customers): dynamic system, altering structure.

Generally speaking, systems either have a permanent structure that we intend to change in a revised or new permanent structure or have a changing structure that we influence as participants in the system.

2.6 Behaviour of Systems

A dynamic system will display specific behaviour during the time of the study depending on the nature of the objectives either through variation in attributes or by modifications of the internal structure. The time frame may influence the outcomes of the study depending on its horizon: how did the system respond to changes in the external structure during different periods. Take a company as an example, on the short-term substitutes for products from competitors might lead to direct changes in price and delivery time by a company, whereas on the long run it should develop a new product range to divert the threat of this unexpected event. Therefore, the events in the external structure always lead to an internal response by the system. In addition to the internal response, in many cases the internal activities also result in changes towards the external relationships again causing a reaction by the environment, as seen from the example of the company (for instance, the new products may lead to different customers to trade with it). Behaviour denotes the capability of a system to respond to variations in external relationships and modifications of the external structure (see definition in Box 2.5).

During the studies, the investigator might encounter one of the two typical cases of a system's behaviour:

- static system behaviour. The properties of external relationships depend only on the specific values of events acting on the system and the timing of these values;
- dynamic system behaviour. The properties of external relationships depend also on the history of events over time.

For example, a company processes orders for standardised products from a wide variety of customers. If the lead-time remains the same no matter how many orders it accepts, the company displays static behaviour in terms of the lead-time; to achieve this, it should be possible to tune the capacity to the order flow, which implies that the company should have an infinite capacity. When the capacity has limitations, the actual lead time of the company will also depend on the intake of orders during previous periods. Whether the behaviour of a system is static or dynamic does not only depend on the time-frame but also on constraints embedded in the system's properties.

When we can predict the behaviour of a system entirely, then the behaviour is deterministic whether the nature of the system behaviour is static or dynamic. The responses of control systems in petrochemical plants possess this characteristic by reacting on deviations in the chemical processes. In contrast to deterministic system behaviour, the system might also display

---

BOX 2.C: GALÁPAGOS ISLANDS: BEHAVIOUR

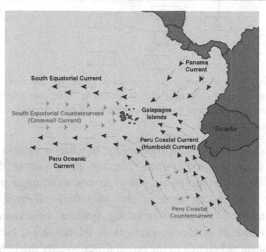

The unique, relatively stable subtropical climate at the Galápagos Islands has contributed to the study of endemic species. The climate is determined almost exclusively by ocean currents, which are themselves influenced by the trade winds that push them. The marine biota are also affected by these currents. The Galápagos Islands are situated at a major intersection of several ocean currents, the cold Humboldt current (which predominantly influences the climate), the cold Cromwell current (also known as the Equatorial Countercurrent, which is responsible for much of the unique marine life around the Galápagos) and the warm Panama current, see figure above. The unique mixture of relatively cool waters, tropical latitudes and islands with different altitudes produces an ever changing environment that has resulted in flora and fauna found nowhere else on earth.

From a system's perspective, the climate is relatively constant but its behaviour is dynamic and stochastic when predicting the weather for a relatively short period of time, say from days to weeks; the weather can be predicted but there is uncertainty about the exact conditions. This caused by the memory of the system (today's weather conditions depend on yesterday's ones). Taking a time horizon of years, the climate is fairly constant with predictable cycles, in systems theory's terms: the climate system of the islands is in a steady state. Even the El Niño, occurring every four to seven years, has a fairly predictable impact on the climate of the islands. The recent trends in climate change can be labelled as causing a transition; for example, at least 45 Galápagos species have now disappeared or are facing extinction. For the study of endemic species at the Galápagos Islands, the climate constitutes the environment.

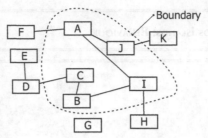

Figure 2.5 *Boundary of a system. The boundary separates the internal elements from the elements that constitute the environment. The relationships that cross this boundary, i.e. relationships between internal and external elements, are called the external structure.*

behaviour with a degree of probability, stochastic system behaviour. For example, fuzzy control systems coming about during the 1990s found their way in home appliances; they do not exerting a predefined action but adjust interventions more or less on a trial-and-error basis depending on the outcomes. Although capturing systems' behaviour becomes more difficult in case of stochastic changes in relationships, tuning of attributes and relationships belongs to the possibilities to alter the behaviour. Another example is given in Box 2.C for the Galápagos Islands for the their climate. In all three cases, deterministic or stochastic behaviour, the outcomes of changes in the relationships are predictable, albeit to a varying degree.

In case of recurrent behaviour, either deterministic or stochastic, the system is in a steady-state. The system repeats the same changes in relationships and attributes, mostly related to the fact that similar events act on the elements which requires no adaptations in relationships and elements. When events cause changing the behaviour in course of time, the systems is called a transient system. The memory may prevent that particular behaviour appears again as happens during the growth stages of a human being. A steady-state becomes only possible when a homeostasis, a balance, occurs in the relation to the environment (see Section 3.1, and Chapters 5 and 6).

2.7   Systems Boundary

Around a system the investigator will draw a system boundary, separating the elements from the environment (see Figure 2.5). The purpose of study will determine this separation to examine the specific elements (system) within the universe, and the external structure and internal structure. As a way of illustration might serve the study of whales in their habitat in New Zealand. Although this might help to study behaviour of local populations of sperm whales, some whales also follow migratory routes from the Antarctic to the tropics. If the study wants to understand, the behaviour of all whales, it might be necessary to include the migratory routes, resulting in an increase of geographical spread of the system studied. That means that the problem definition determines mostly the system boundary. To that end, there are a few practical guidelines for setting the system boundary:

- the exchange with the environment concentrates on a few elements. The internal structure has in this case a more dominant role in determining the system behaviour than the external structure. This might result in the practical guideline that the number of internal relationships equals or exceeds the number of relationships in the external structure;
- the exchange with the environment might require more effort then maintaining the internal structure. This indicates the capability of the system to maintain itself within its environment;
- the capability of the system to serve a purpose within its environment. Again, this refers to the capability of the system to maintain itself within its environment but directed at its purposefulness.

Might a system experience difficulties in maintaining itself in the environment than the it has to adapt its behaviour to the events taking place in the external structure or has to adapt its structure matching the (external) event or has to dissolve itself. Such situations might arise from the diffusion of the system boundary, a problem encountered by many companies through the increased capabilities of information and communication technology where customers have a stronger influence on the behaviour of a system; the permeable boundary leads to customers having more impact of the structure of the system, even though the customers may not notice the change. Whether a static or a dynamic systems, or whether it displays static behaviour or dynamic behaviour, the internal structure should match with the performance requirements imposed on it through the external structure.

The interaction with the environment points to so-called open systems. In the case of closed systems, the interaction with the environment is not considered. It is hard to imagine that to be the case, any system has a position in the universe and is interrelated. Nevertheless, if that occurs, the only consideration is the internal structure; the system boundary merely serves as a separator of the internal structure and content from the universe (not just the environment, following the definition of Applied Systems Theory).

2.8 Summary

Looking at systems means purposeful distinction of elements and relationships within the universe (therefore, systems are always part of the universe). The separation should serve the nature of the study and an investigation will take only those elements and relationships within the system into account plus the relationships with its environment, i.e. those elements in the universe with which the internal elements have direct relationships. By describing a system by its contents, its structure and its attributes, it becomes possible to define the state of a system and its behaviour in view of the nature of the study undertaken.

Subsystems and aspectsystems represent two different ways of examining a system in more detail (see Figure 2.6). Subsystems leave the relationships intact in favour of looking at a subset of elements, whereas aspectsystems concentrate on certain type of relationships within the system. Defining an aspectsystem means eliminating elements that have no interrelations of a specific type anymore with any other element present in the system. Practically, it means that a study always considers an aspect, or perhaps some, while at the same time the investigation concentrates at subsystems of a larger set.

Ultimately, most studies look for ways to modify the behaviour of a system, which is the change of the state of a system by events happening in the external structure. The modification of behaviour of technical or organisational systems results either from optimisation of attributes (of elements) or from altering the structure of the system. When the behaviour repeats over time the system has achieved a steady-state. Especially organisational systems show transient behaviour due to the memory caused by earlier events that led to adjustments especially in the structure of the system and, therefore, will hardly reach a steady state.

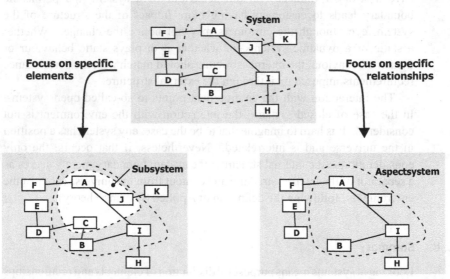

Figure 2.6 *Two principles for investigating a system in more detail. When focusing on specific elements a subsystem may be distinguished; in this case, this subsystem is consisting of elements A, C and J. When focusing on a specific relationship only, the elements that have interrelationships of this type are looked at; in the figure these elements of the aspectsystem are A, B, I and J. Note that elements in the system that do not have this particular type of relationship with other elements are omitted; by doing so, an aspectsystem will fulfil the definition of a system. This also means that element D is no part of the environment of the aspectsystem in this case.*

References

Ackoff, R. L. (1969). Systems, Organizations, and Interdisciplinary Research. In F. E. Emery (Ed.), Systems Thinking (pp. 330–347). Harmondsworth: Penguin Books.

Bertalanffy, L. v. (1973). General System Theory. New York: George Braziller.

Checkland, P. (1981). Systems Thinking, Systems Practice. Chichester: John Wiley & Sons.

Checkland, P., & Scholes, J. (1990). Soft Systems Methodology in Action. Chichester: John Wiley & Sons.

Darwin, C. (1859). On the Origin of Species by Means of Natural Selection or, The Preservation of Favoured Races in the Struggle for Life. London: John Murray.

Darwin, C., & Wallace, A. R. (1858). On the Tendency of Species to form Varieties; and on the Perpetuation of Varieties and Species by Natural Means of Selection. Journal of the Proceedings of the Linnean Society of London. Zoology, 3, 46–50.

de Leeuw, A. C. J. (1979). The control paradigm as an aid for understanding and designing organizations Progress in cybernetics and systems research (Vol. 5, pp. 93–100). London: Hemisphere.

Gershenson, C., & Heylighen, F. (2003, 14–17 Sept.). When Can we Call a System Self-organizing? Paper presented at the 7th European Conferene on Advances in Artificial Life, Dortmund.

Hitchins, D. K. (1992). Putting Systems to Work. New York: John Wiley & Sons.

Miles Jr., R. F. (1973). Systems Concepts: Lectures on Contemporary Approaches to Systems. New York: John Wiley & Sons.

van der Zwaan, A. H. (1975). The sociotechnical systems approach: A critical evaluation. International Journal of Production Research, 13(2), 149–163.

West Churchman, C. (1979). The Systems Approach: Revised and Updated. New York: Dell.

Wikipedia. (2007, 14 March). System. Retrieved from http://en.wikipedia.org/wiki/System

Wilson, B. (1990). Systems: Concepts, Methodologies, and Applications. Chichester: John Wiley and Sons.

3 System Approaches

Using the concepts of Chapter 2 and supposing that a problem has been defined in conjunction with an associated system: how to start investigating this system to solve the problem? Does solving this problem mean that it has to be taken apart, all individual elements examined, deficient ones replaced and the elements put back together to see if the system works? Examining all single elements could mean looking at many details, omitting relevant relationships between the constituent elements and ignoring the interactions between the system and its environment; particularly, getting lost in detail may happen when a systems has many elements and many relationships. Or, alternatively should we look at the system from the outside, occasionally opening a suspect subsystem for closer investigation? Starting from the whole to investigate the systems carries the risk of not gaining enough depth to identify the source of the problem. Either way might end up in addressing symptoms instead of finding the root causes; reason why this chapter looks into more detail about how to deploy systems theories and related concepts for resolving problems.

In any case, an adequate problem definition constitutes the first step for the investigation of a system; see Section 4.2 for a more detailed description of problem definitions as part of the steps of solving problems. For example, if a car does not move after starting its engine and pushing the accelerator (gas pedal), the reason could be a (mechanical) fault somewhere inside the system *car*. However, a redefined problem statement '*the engine hums, the wheels spin, but it does not move*' indicates that the engine supplies power to the wheels, but the turning of the wheels does not lead to motion of the car. This could lead to a wider systems boundary that includes the icy road conditions or the mud pool the car is actually in; in this way, the investigator takes into account that the car has relationships with its environment. Therefore, the problem description should pinpoint which system to be considered and also which aspectsystem will be examined in relation to its environment. In this particular example, the aspect *motion* could quickly be narrowed down to the feature *traction* (see Section 2.1 for the relationship between aspects and features), leaving out the need to examine other features, such as the *conversion* of potential energy in the form of fuel into kinetic energy (which is done by the subsystems engine and gearbox). This process of turning a problem description into a way of looking at the system and its relationship with the environment is called modelling.

This chapter explores modelling and associated system approaches by building on the concepts introduced in Chapter 2; an approach to problem analysis and solving will be presented in Chapter 4. First, Section 3.1 will discuss three methods of abstraction associated with modelling; abstraction constitutes one of the basic steps during the creation of a model needed for

R. Dekkers, *Applied Systems Theory*,
DOI 10.1007/978-3-319-57526-1_3

analysis and synthesis (design). Thereafter, Section 3.2 will describe the blackbox approach as a specific tool for system analysis. That leads to the distinction between inductive, deductive and abductive reasoning in Section 3.3, which is closely related to analysing a system from the perspective of elements or as a whole. Section 3.4 presents a classification of models, partially based on systems theories. Consecutively, Section 3.5 will pay attention to a hierarchy of systems, which helps to understand the validity of models for different types of systems. Therefore, reasoning goes hand in hand with modelling to resolve problems.

3.1 Modelling and Abstraction

This means that it is first necessary to understand what modelling is and how it is positioned within approaches for problem solving. In terms of systems theories, a model is a simplified system to study another system from the perspective of a given problem definition. This goes back to Rosenblueth and Wiener [1945, p. 316] who state that abstraction consists of replacing the subset of the universe that is under consideration by a model of similar but simpler structure. Manifestations of the simpler structures are concepts, drawings, equations, formulae, graphs, laws, rules, etc.; these model reality by simplifying that reality into elements, relationships and properties or parameters that matter. They reflect on elements to study with specific relationships or aspects in mind (see Figure 3.1). Hence, a model is always an aspectsystem, pointing out the elements for closer observation. Beware; a model does not equal reality. For example, the social-economical reality of a country, such as the Netherlands, has a higher complexity than even captured in the statistics, through its Statistics Netherlands (in Dutch: Centraal Bureau voor de Statistiek), even though the data are complex.

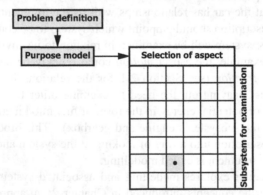

Figure 3.1 *Models as choice of aspects and subsystems. The purpose of the model, often captured by a problem definition, prescribes the primary areas of interest, both for the type of relationships to be examined and the particular subsystems as objects of study. The environment (as part of the universe) denotes an object as subsystem.*

Thus, looking at the complexity of real systems leaves often no choice but simplification. In addition to modelling leading to simplification, de Rosnay [1979, p. 141] states that the building of a model implicitly relies on the stock of existing knowledge and also indicates which knowledge has not yet been obtained. The existing knowledge is embedded in concepts, drawings, equations, formulae, graphs, laws, rules, etc., whereas parts of the problem definition may not yet be captured by the stock of existing knowledge. That means that modelling implies both the purposeful simplification of reality to understand systems and their behaviour, and the exploration of the limitations of contemporary knowledge for these systems.

Moreoever, the modelling of a system is related to its position in the universe. In this sense, West Churchman [1979, p. 76] remarks: *More generally, it shows there is a fundamental limitation of any modelling of a system, that a system is always embedded in a larger system.* Again, this statement signifies the selection of the aspect(s) and subsystem(s) to consider for the investigation and that way defining the environment of the selected subsystem(s). For example, looking at the earth's ecosystem means understanding its connection to the solar system it is part of; changes in the radiation of the sun influence temperatures and the position of the moon influences tides. But the remark by West Churchman also means that aspectsystems and subsystems that do not contribute to the (solution of the) problem are best left out. Continuing with the Earth's ecosystem as part of the solar system, Jupiter does not have any effect on the radiation from the sun received on Earth and consequently variations in temperatures; hence, it is best left out for the study of that particular aspect of the earth. That makes it all more important for any analysis to consider explicitly the boundaries of a system, which entails its environment, and the type of relationships with the environment the analysis focuses on.

This search for understanding how to take what into account and what not continues in the concept of abstraction. According to Timpf [1999, p. 126]:

An abstraction denotes the essential characteristics of an object that distinguishes it from all other kind of objects and thus provide crisply defined conceptual boundaries, relative to the perspective of the user.

Consequently, this statement by Timpf implies that modelling constitutes an abstraction process that has the onus on essential characteristics of a system for a given problem definition. She recognises three types of abstraction processes (although slightly modified to link these to Applied Systems Theory):

- Classification: a form of abstraction in which an object type is defined by a specific set of observed properties; note that observation precedes classification. A famous example of classification is the practice of taxonomy by biologists for species, following the tradition set by Carl Linnaeus (1707–1778). Such a classification makes it possible to compare objects and to relate them to each other; however, to classify, it is necessary to observe the individual elements first. At the same

BOX 3.1:   APPLICATION OF THE THREE TYPES OF ABSTRACTION

CLASSIFICATION

Classification takes place when properties of systems or elements call for grouping to decrease the number of variables. By way of illustration, houses, schools, industrial buildings could all belong to the generic class *building*, or local roads, city streets, highways could all belong to the generic class *roads*. By putting together similar entities, their differences are selectively ignored.

AGGREGATION

Aggregation is the putting together of different entities to form a coherent whole (in terms of Applied Systems Theory: either relationships or elements). For example, houses, apartment buildings, offices and amenities could form a community area; in that case, the focus is not only on the individual buildings but the total offering that community area provides. Aggregation purposefully leads to loss of detail (the perspective of the investigation determines whether that loss of detail leads to ignoring relevant properties of the system).

GENERALISATION

Generalisation is the application of behaviour derived from one set of entities to another set of entities, by the existence of similar relationships and elements but not necessarily all. For instance, after investigating the application of solar energy in houses, an inference is drawn for potential savings in offices, purely based on the fact that both consume energy; however, their energy consumption patterns might differ substantially. Therefore, a key question is not whether to generalise at all, but to what extent.

time, the classification decreases the number of elements, relationships and properties of elements to consider. Taking a viewing point from the elements, based on similar properties, grouping elements and relationships into subsystems and aspectsystems could be considered classification.

- Aggregation: a form of abstraction in which a relationship between similar or related objects is considered as a higher-level aggregate object for fulfilling an objective. An illustration of aggregation is to move from looking at an individual colouring pencil to considering a pencil set. Again, taking a viewing point from the elements, subsystems and aspectsystems could also be regarded as abstraction if they are related to fulfilling an objective.

- Generalisation: a form of abstraction in which similar objects are thought to be related based on having a limited range of similar elements and relationships. For example, in economics some scientists use selection models derived from evolutionary biological models to describe market mechanisms for companies, while understanding that organisms in evolutionary biology do not equal organisations as entities. As seen from the example, generalisation also occurs when transferring concepts (knowledge, theories and models) from one domain to another one.

These three types of abstraction have their own specific applications when modelling (see Box 3.1). Classification takes place when properties of systems or elements call for grouping with the explicit purpose of reducing the number of 'elements' considered. Aggregation acts as a measure of composition to decrease the relevant information known about a system. And generalisation helps us to transfer explore the validity of mental constructs to a larger set of entities (or domains of knowledge); mental constructs are ideas, theories, models, frameworks, methods, etc. All these three abstraction mechanisms assist an investigator in gaining a better overview to resolve problems.

Classification

The use of classification as a principle for abstraction typically arises during the early stages of observation, particularly when little is known about relevant properties and patterns of behaviour. Sometimes, when little knowledge exists about certain objects, comparison might lead to understanding the nature and behaviour of objects. It can be seen as a start to gain knowledge. For example, the Linnaean taxonomy in biology describing species was performed as comparative anatomy of animals and plants; Carl Linnaeus (1707–1778) formalised this taxonomy based on binomial nomenclature (naming of species composed of two parts) without necessarily understanding the underlying concepts and mechanisms that directed biological evolution as we perceive today. So it happened in biology when Darwin and Wallace's [1858] notion of natural selection made it possible to reach beyond classification and to understand evolution as a pattern of behaviour (in the sense of Applied Systems Theory, see Section 2.6). Even though this notion of natural selection did shed a different light on how species are related, the search for purposeful classification and how species are actually related still continues. At the end, through classification and successive stages of observation the governing principles should be detected, which leads to exposing laws and the nature of change.

Classification has also limitations to its application as a mechanism for abstraction, especially when the initial set of properties used has been proven irrelevant during later stages. An example from (evolutionary) biology serves as case in point; the single-celled organism *euglena* has properties from both plants and animals: it is green and processes sunlight for energy, as plants do, and it has a tail like structure to propel itself, akin animals have.

For many years, it has been classified as a plant. Later, a new classification was created, called *Protista*, for organisms that are neither plant nor animal (note that the term was introduced by Ernst Haeckel in 1866, though preceded by attempts to find the correct classification by other scientists). This example demonstrates that classification resides in the potential of the investigator to differentiate between observable characteristics to those that matter and those that do not contribute to understanding a specific phenomenon; it also indicates that a classification is subject to knowledge and the relevant classifications may change over time.

Instantiation (see Box. 3.2), the opposite of classification, is describing an event, activity, element or system by looking at an individual specimen or phenomenon and setting it apart from the other entities in a class. Hence,

---

BOX 3.2:   APPLICATION OF THE OPPOSITE PRINCIPLES OF ABSTRACTION

INSTANTIATION

Instantiation, as being the opposite of classification, defines properties of either elements or relationships, that makes these observably different from other elements or relationships in relation to the problem statement. In the case of a generic class of buildings, offices are a distinct instance of buildings when examining energy consumption. Instantiation only contributes to understanding when the observably different properties have an impact on solving problems.

DECOMPOSITION

When applying decomposition, the opposite of aggregation, to a system, it is separated into subsystems or aspectsystems of a single kind. When applied to a house, it could be split into living quarters, areas for food preparation and consumption, relaxing spaces and so forth. Or alternatively, the relationships of the house could be divided into geometry of the house, aesthetics, use of utilities, interaction of users with environment, etc. By nature, decomposition leads to distinguishing more detail driven by uncovering relevant subsystems and aspectsystems.

SPECIALISATION

When using specialisation, as the opposite of generalisation, the emphasis is on those relationships and elements that distinguish systems from each other. For example, a meeting room is not commonly found in houses but in offices. Therefore, the design of an office building includes the integration of meeting rooms, or better meeting spaces. Specialisation is also an important way to generate more specific knowledge by applying general knowledge to specific instances.

enumerating the elements of a class is instantiation, which means that the class is not considered a sufficient description for problems at hand. Through listing and specifying elements (or subsystems) the search starts either for a more accurate description of the class to suit the objectives of the study or for an adequate distinction between subsystems. Listing more details does not always succeed in the resolution of a problem; particularly in management (studies), more details are often sought than necessary by lack of understanding of the governing principles for solving problems. In these cases, a check proves to be necessary whether the newly introduced properties relate to the problem, and whether these properties indeed resolve the problem; on hindsight it might turn out that the identified properties do not at all address the symptoms found in the problem definition. By applying instantiation, an investigator could identify if properties of individual elements will contribute to comprehending a problem statement better.

The same dilemma exists in Applied Systems Theory when deciding which elements or (sub)systems and which aspects will classify as pertaining to the problem statement. Timpf [1999, p. 131] explicitly states that classification is a prerequisite for all other abstraction mechanisms, even though it might be less useful during later stages of an investigation. This means that a classification of the elements and aspects relevant to the problem definition constitutes a first step. However, at the beginning of an analysis there is often simply not enough understanding about the problem and not enough relevant information to make a sensible choice with regard to relevant elements, subsystems and aspects. Unfortunately, no methodology exists (yet) that enables to draw the system boundary in predefined steps (see Section 2.7 and the strand of critical systems thinking in Section 11.4). But aggregation, as discussed in the next subsection, gives the opportunity to review the sensibility of the choice made in the first instance and to adapt the problem definition. In this way of thinking, classification precedes aggregation.

Aggregation

Once a classification has been made, aggregation becomes possible; aggregation can be described as the combining of (different) elements, etc. into a single group or whole. An example is regarding all storage, transportation and distribution as parts of a (total) delivery system for a company. Timpf [1999, p. 131] notes that only members of the same class can be aggregated, because classification precedes aggregation. In practice, that means that elements or subsystems with similar relevant properties are aggregated into a single group or whole. In the case of a logistic system the resources used for storage (for instance, warehouses), transportations (trucks) and distribution (vans) are quite different in their appearance and even use; however, they all hold products with the purpose of transferring goods to customers and, therefore, constitute the delivery system. Aggregation is justifiable whenever units are sufficiently independent and similar, for instance expressing political opinions through voting or market preferences through individual purchases.

Aggregation leads to misleading indicators and theories whenever the whole system or aspect of a system exhibits a behaviour not expressed in a mere summation. Aggregation should not lead to loss of erratic detail but consider purposefully the relevant properties at the level of the grouping or whole.

This implies that aggregation describes at which level of detail a system will be investigated. For taking a close look at something, we can use a microscope; thus, we get a lot more detail, but only from a small piece of the original. Moving the microscope at the same level means another detailed view, but losing the previous one. However, looking at a big, complex system would require a different device; something that reduces unnecessary details and clutter, but amplifies the essential relationships and relevant (sub) systems or elements. That is why de Rosnay [1979, Introduction] presents the concept of the macroscope, see Figure 3.2: *The roles are reversed: it is no longer the biologist who observes a living cell through a microscope; it is the cell itself that observes in the macroscope the organism that shelters it.* Note that this macroscope represents a symbolic instrument, a way of viewing and understanding; it is not a piece of hardware. Symbolically or not, the microscope and the macroscope represent two ways of viewing, either in more detail or from an overview.

Obviously, the analysis and related modelling requires shifting between different levels of detail and overview, depending on the state of the analysis. Mesarovic et al. [1970, p. 37] mention these different levels of detail as levels of aggregation or aggregation strata. In system theories, the process of

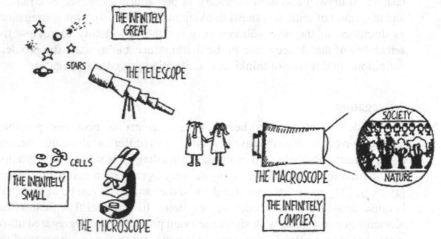

Figure 3.2: *Methods for exploring the unknown [de Rosnay, 1975, p. 10]. The microscope has permitted a dizzying plunge into the depths of the living matter, the discovery of the cell, microbes, and viruses. The telescope has opened the mind to the immensity of the cosmos, tracing the path of the planets and the stars. The macroscope symbolises the study of the infinitely complex, especially the interdependence and the dynamism of systems, transforming at the moment we study them.*

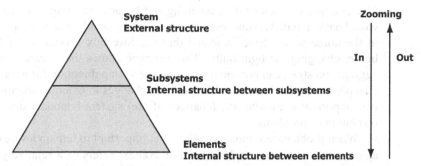

Figure 3.3: *Aggregation strata and zooming in and out applied to systems. In this drawing the levels of system, subsystems and elements represent levels of detail for investigating a system. By zooming in more details become visible (i.e. elements of the system). Zooming out results in distinguishing properties and mostly makes it better possible to examine the external structure of a system.*

going to a level with more detail is called zooming in or moving to a lower aggregation stratum (see Figure 3.3); it only results in losing overview. For planning a trip it best first to examine a general map to find the way from one city to another; later the inspection of a detailed city map is necessary to find out how to arrive at the exact location. The reverse process is called zooming out and means gaining overview, but loosing detail (aka helicopter view). Looking at a map when lost in a city is an example of zooming out. Looking into more detail, zooming in, and gaining more overview, zooming out, helps solving problems at different stages of the problem analysis.

Many may have experienced the following well-known example of implicit use of aggregation strata (a popular illustration for teaching how to solve a problem). Suppose somebody comes home and switches on the lights, but notices that there is no light. Fortunately, a spare light bulb is at hand, so changing the bulbs should normally solve the problem. Considered from the perspective of Applied Systems Theory when there is a problem with a lighting system, zooming in on the most probable cause based on knowledge and experience – in this case the light bulb – and taking action – replacing it – should resolve the situation. When after changing the bulb, there is still no light, it may be noticed that the digital clock of the music player does not light up either. Perhaps a fuse blew? This further analysis, zooming out to the system level of which the first lighting system is a subsystem, could explain the failure. This, with all fuses intact, still no light and other electrical equipment in the house not working, the resident decides to check with the neighbours. They answer the door carrying a candle. Therefore, another step of zooming out tells that the problem is beyond the control of an individual household and that the utility company providing electricity should be notified. Without stating it explicitly, the system boundary has shifted along with the aggregation stratum from light bulb via the electrical system of the house to the provision of electricity in the area; each stage for shifting depended on the information available at the specific stratum.

The example also shows that zooming out before changing the light bulb would have avoided an unnecessary step; looking at all electrical equipment in the home would have indicated that the fuse box should be looked at before changing the light bulb. This example shows that a careful use of aggregation strata can prevent a search in the wrong direction for too long; it also points out that because of the possible changes in external relationships it is important to monitor the (changes of the) system boundary during the resolution of problems.

When problems are more complex, it is important to remain aware of the aggregation stratum; complexity in this context refers to a relatively large number of elements and interrelationships. For example, society is faced with an ever-increasing consumption of electric energy (see Fourastié [1949] already making this case); therefore, the choice for more environmentally friendly energy sources, such as solar or wind energy, could be made instead of relying on traditional sources of energy, such as fossil fuels. At another aggregation stratum however, one could say that the only green energy is the energy neither produced nor consumed by actors in the system. It is up to the reader to argue whether this is an example of zooming in or out. This example demonstrates that the aggregation stratum at which reasoning takes place also influences which inferences will be drawn.

The opposite of aggregation is decomposition (see Box 3.2). Particularly, decomposition becomes necessary when behaviour of a system cannot be explained any more or no further inferences drawn at the current level of observation. An example is decomposing a personal audio system into a data-reading subsystem, an amplifying subsystem and a sound rendering subsystem to identify the subsystem at fault when the quality of music heard does not match expectations. However, decomposition does not help when examining emergent properties that cannot be clarified by looking at a lower level of abstraction. A case in point is an airplane; its potential to fly is difficultly understood by just looking at the 'nuts and bolts' it is made of. When designing or creating a system, decomposition might assist in enumerating or specifying elements at a lower level of detail so that it becomes possible to produce or buy these elements. Hence, the need for decomposition is associated with lower levels of abstraction if that helps to understand better the problem at hand.

Generalisation

Generalisation is another abstraction mechanism through which we realise that elements, subsystems or even systems have a limited number of characteristics in common with other elements, subsystem and systems; these common characteristics lead to a more generic object at a higher level of abstraction or knowledge that has a wider validity than the original application. Note that in terms of creating knowledge generalisation extends to building theories, too. However, generalisation does not require all properties of the elements to be (sufficiently) similar. An example of generalisation is the application of theories and models from evolutionary biology to companies and market mechanisms. When talking about the functioning of markets, people often refer to competitive forces between companies as being selective towards the fitter (note that for the sake of argumentation, natural selection aims at weeding the least fit rather than selecting the fittest). However, companies and organisms or species appear quite distinct and share only few properties from the perspective of an observer as relevant for applying this analogy for natural selection. As another example, recognising that both a car engine and a watermill exploit the aspect rotation is generalisation for transferring one form of potential energy into another form of energy. But a car engine and a watermill do so in a quite different fashion and to a very different purpose. Hence, generalisation might mean that a phenomenon that is studied can be taken easily out of context and that generalisation has a limited application defined only by what it examines.

In everyday language, the difference between generalisation and aggregation may be blurred. However, modelling requires a clear understanding that generalisation focuses on a common (well known and understood) aspect of (entirely) different entities. Aggregation is merely a method for composing similar elements that constitute part of the same system. Again, this depends on the problem definition. For generalisation, no direct connection between systems considered is necessary. By way of illustration, cybernetics applied to management is generalisation (as happened during the 1950s and the 1960s); cybernetic theories arrived from control theories for technical systems and were used to advance control of production but also to introduce objective-oriented management. Principles of cybernetics could be applied to management of organisations because some aspects are similar (the control of primary processes, see Section 5.3 for its definition) but not all. It were these similarities that allowed this generalisation, whereas for differences this generalisation did not work out, a case in point being the interaction between employees in an organisation. However, for both generalisation and aggregation, classification that deals with combining entities at the researcher's choice, building a system (named *class*) from elements and subsystems is necessary. The difference between generalisation and abstraction resides in the problem definition. Abstraction is based on the same types of elements and similar properties; generalisation is about one aspect or some aspects at best.

Specialisation (see Box 3.2) is the opposite of generalisation as abstraction mechanism in the context of system approaches. Please note that specialisation may refer to quite different concepts in daily communication. Specialisation as an abstraction mechanism occurs when a generic phenomenon or knowledge insufficiently explains necessary details. Again, using the metaphor for natural selection for companies, the survival of the fitter might include systems, elements, relationships (or aspects) that are hardly relevant to organisms; take financial contracts, transactions and instruments (loans, shares, etc.) that have no direct equivalent in evolutionary biology. For the other case, *transforming energy* specialisation could be a gearbox, an electric transformer, a hydraulic torque converter, a watermill, etc. However, for understanding these, quite different expertise is needed although at a generic level they achieve the same: converting one energy source into another type of energy. Specialisation leads to inclusion of elements and relationships that have not been considered in the generalisation of knowledge or the generic object.

3.2 Blackbox Approach

Very different from the three abstraction mechanisms, a way of examining a system is by only looking at its external relationships. This type of investigation means momentarily forgetting about the content and internal structure of the system and only observing the changes in external relationships; this system approach is called the *blackbox approach* (see Figure 3.4). When all is in order, denoted as no problem has been defined, there is no need to look into the details of a particular system, being its elements or subsystems and relationships or aspectsystem. Only when there seems to be a problem, the blackbox should be opened (note the parallel with decomposition). A common example is that when a car does not start at the moment the ignition key is turned, the driver can open the hood, and check all kind of parts, connections, and other intricacies with the purpose of finding out how to get the engine to work again. However, a simple examination of the external relationships might have revealed the absence of fuel (refilling taken as supplying fuel being an external relationship to the car as a system). Hence, the blackbox approach supports the examination of a system as a whole and avoids getting lost in (unnecessary) details.

The blackbox approach, typical for systems theories, investigates the external structure of a system without identifying any of the internal

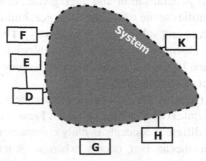

Figure 3.4: *System as blackbox. The blackbox approach allows examining the external structure and behaviour of a system. The elements and the internal structure are not looked at. The blackbox approach for analysing systems supports deductive reasoning by examining the behaviour of the system in response to external stimuli.*

Box 3.A: BUILDING AS BLACKBOX

Imagine that a building is blackbox. Everyday, an analyst observes which people are going in and out of the building and at what hour. That way the observer could find out what type of building it is without knowing the shape, the size and more details.

For example, when people arrive and they are all adults, dressed in formal outfits, carrying briefcases and heading straightforward into the building as a blackbox, an inference could be that it is an office building. What type of an office building would depend on a closer inspection of those entering.

Alternatively, a blackbox where every day two middle-aged people are entering and leaving five days a week at regular times the building wearing formal clothing, going in and out infrequently during the weekend dressed in casual clothes, five days a week, two kids with schoolbags are leaving and entering the house, in weekends accompanied by the two elder persons carrying sports gear, etc., it would be reasonable to assume a family was living there.

If these recordings of entrances and departures in the blackbox were combined with consumption levels of utilities, like electricity, that would even increase the probability of the inferences. An office building will utilise more energy during office hours, while a house would most likely consume more energy during the early morning, evening and early night.

---

elements. This approach supports a study by not looking at the elements of the system and the internal relationships and so creating space to focus on the behaviour of the system as if it was one element. In that sense, a blackbox is equivalent to a system [see Beer, 1959, p. 49]; for example, the human body or a house. At this level, the need arises to identify the relevant properties and the relevant (external) relationships for the problem definition. A remark should be made: to a certain extent considering the blackbox as one element eliminates the notion of external relationships, it strengthens the inevitability of linking a system to its environment. When at a later stage of the study it becomes necessary to open the blackbox, subsystems and elements may serve again as blackbox, akin the description of aggregation strata.

When applying the blackbox approach a study will aim at relating changes in one or more external relationships to other relationships in the external structure. A change in a specific relationship might cause changes in another relationship. It might also occur that multiple relationships become affected. Or that multiple changes in relationships produce changes in one relationship. Consequently, observing a system as a blackbox requires an understanding of the mutual influence of the external relationships. For example, practising physicians deploy this method by deducting from the behaviour of the human being as a system, e.g. temperature, pain, coughs, what internal causes bear relevance to the well-being. Through purposeful dosing of medicine and assessing their efficacy for a particular case, doctors draw conclusions about their earlier findings (or rather their presumptions). Therefore, the response to medicines generates indications about the internal structure of the system (see Box 3.A for an other case of applying this method). More generically speaking, there is a strong relation between the exertion of stimuli on a system and its externally oriented behaviour (and response).

However, the higher the number of external relationships and the related possibilities for exerting stimuli, the harder it gets arriving at inferences. Even to the extent, that it might become impossible to determine with certainty how the system responds to these changes in external relationships. Consider a system with only two external relationships. Even when the behaviour of the system is stochastic it becomes possible to describe the influence of one on the other. Increase the number of relationships to four, eight and so on. Then you will notice that it becomes harder to detect changes in relationships in comparison to the variation in only two relationships. The complexity of such an exercise increases certainly when multiple relationships affect multiple others. It seems reasonable that the blackbox approach becomes most effective when the modelling has reduced the external relationships that are considered to an acceptable number.

3.3   Deductive, Inductive and Abductive Reasoning

In addition to the blackbox approach, the behaviour of a system might be analysed by three different ways of reasoning: deductive, inductive and abductive reasoning. These three types of reasoning have different starting points, inductive reasoning starts by looking at the internal elements and relationships and their influence on the behaviour of a system and deductive reasoning works the other way around, from the outside to the inside of a system; typically, abductive reasoning begins with an incomplete set of observations and proceeds to the likeliest possible explanation. Each of these three ways has its specific applications.

Deductive Reasoning

Deductive reasoning looks at the behaviour of a system and tries to arrive at underlying causes for problems or what needs to be resolved for future states of a system. To apply deductive reasoning to a system, one needs a description of the expected state of a system. An example of deductive reasoning might be the case when running out of cash unexpectedly; there are reasons why which expenditures did lead to the current deficit, esp. by comparing these with a budget. The deficit might have been triggered by factors that could have been controlled; for example, there might have been spent more on entertainment than planned or more exclusive food items bought with the purpose of preparing a meal for a partner. Or alternatively, the deficit resulted from a rise in food prices, fuel, etc. Only by this comparison between what 'ought-to-be' and 'as-is'[1] inferences become possible. In quite of number of cases this type of reasoning involves a number of these steps for analysis of the gap between 'ought-to-be' and 'as-is' before more definite conclusions can be drawn and underlying causes appear. The application of deductive reasoning is found in the analysis of current systems and the evaluation of performance during 'design and engineering' of systems.

Deductive reasoning has a close connection with the blackbox approach. When applying the blackbox approach, the investigator tries to understand the system without considering the internal structure. The changes in the external relationships after an event that is induced externally to the system determine how it will behave as response to external stimuli. The response to the external stimuli also indicates what possible weak spots in the internal structure exist. Hackers use this technique amongst others for understanding how computer systems operate before entering them. For this reason, the blackbox approach uses the principles of deductive reasoning for evaluating the behaviour of a system by comparing its performance against 'objectives'.

Equifinality, Homeostasis and Deductive Reasoning

The principle of equifinality presents another challenge to determine, or for that matter predict, a system's behaviour. Equifinality refers to achieving set objectives through a dynamic balance during changing circumstances [Baker, 1973, p. 9]. Von Bertalanffy [1968, p. 40] writes about equifinality, when introducing the concept as part of the general systems theory:

In any closed system, the final state is unequivocally determined by the initial conditions... If either the initial conditions or the process is altered, the final state will also be changed. This is not so in open systems. Here, the final state may be reached from different initial conditions and in different ways. This is what is called equifinality, and it has a significant meaning for the phenomena of biological regulation... It can be shown, however, that open systems, insofar as they attain a steady state, must show equifinality, ...

An example of equifinality is provided by Feiring and Lewis [1987]: a sample of children was observed on interactions with their mothers during 3 and 24 months after birth; initially, at 3 months the group differed on several social behaviours and after 24 months only on one. This example of equifinality shows that different initial conditions might lead to the same outcome.

If a system reaches a final state – the state describes the properties of elements and relationships (see Section 2.5) – that might be robust to exactly how it has reached that state and therefore for disruptions by the environment, henceforth, it will tend to maintain a homeostasis with the environment.

[1] These terms 'ought-to-be' and 'as-is' are related to the German words 'Soll' and 'Ist'; these words are often associated with the so-called gap model.

This concept of homeostasis was formulated by Cannon [1932, p. 22] for processes of interaction or mechanisms that balance various influences and effects such that a stable state or a stable behaviour is maintained; in turn, his writings on this matter were based on the thoughts of Claude Bernard, a French physiologist living in the 19th century, on *milieu intérieur*. Because of reaching a final state, in a sense independent from the initial state, equifinality directly connects to homeostasis [von Bertalanffy, 1968, p. 46]:

> ... *equifinality, the tendency towards a characteristic final state from different initial states and in different ways, based on dynamic interactions in an open system attaining a steady state; the second, feedback, the homeostatic maintenance of a characteristic state or the seeking of a goal, based upon circular causal chains and mechanisms monitoring back information on deviations from the state to be maintained or the goal to be reached.*

Later scientists [Kast and Rozenzweig, 1970, p. 467; Hagen, 1973, p. 79] further developing the general systems theory affirmed this principle. As a trivial example, the human body maintains itself at a certain temperature, about 37° C. Disruptions by viral infections and other diseases will be counteracted by perspirations, etc., aiming at restoring the normative body temperature for the human body. Note that this principle of homeostasis applies to a wide range of domains, not only to biological systems but also to self-stabilising mechanical systems and organisations among others.

Furthermore, Hagen [1973, pp. 79–80] denotes that in case of homeostasis one property could return to its old value only if another one changed permanently in magnitude. When standing still, humans use muscles to maintain the standing position; imagine to be without muscles and to hold a standing position (relying only a skeleton). Hence the movements of muscles, though minute and hardly visible, make humans stand upright, seemingly without effort. This indicates sometimes as a signal of weakness for the effect of one-time interventions for systems; these interventions must draw on the resources at the disposal of a system before it can reach its equilibrium again.

Another overlooked phenomenon, called heterostasis, makes it possible that systems operate at multiple points of equilibrium, even though a limited number exist in practice [Selye, 1973]. For example, when having fever, patients are at equilibrium with their environment, even though they are maintaining a higher body temperature. This temporary equilibrium is less stable and requires the consumption of additional resources to maintain it in comparison the normative homeostatic state; in the case of the human body that is energy. Hence, the concepts of equifinality, homeostasis and heterostasis imply a link between the variable kept at a constant level and the resources needed for achieving that (this comes together in the concepts for processes in Chapter 4).

In addition, there should be processes for beneficial mutations (adaptive systems), which move an entity along its life cycle in response to external and internal stimuli [Kast and Rozenzweig, 1970, p. 467] for which maintaining the homeostasis does not suffice. Rather, in such cases, the system moves or searches for a new equilibrium. Later chapters about autopoietic systems and complex adaptive systems will elaborate on these adaptive processes for mostly biological systems and organisations. Also, the model for breakthrough processes, as part of Applied Systems Theory (see Chapter 10), demonstrates the related steps to identify new needs and changed requirements and to transfer these into a new structure for steady-state process, particularly for the case of organisations. Although the principle of equifinality assumes that different internal structures may produce the same or similar results and performance levels of systems and processes, the growth of biological systems and social organisations may require the adaptation of internal structures for future fits.

Multifinality is the opposite developmental principle to equifinality, whereby similar initial conditions lead to dissimilar outcomes. This indicates that for the investigator either the mechanisms are not understood or the relevant aspects have been left out; the first explanation is also related to the phenomenon of the impact of tiny variations on the behaviour of complex systems and that will appear in Chapter 8. An example of multifinality is given by Feiring and Lewis [1987] in the same article about equifinality. They demonstrate the principle of multifinality by referring to the study of children assessed for attachment classifications at the age of 1 and for emotional functioning at the age of 6. The study revealed that 6-year old boys who were securely attached to their mothers exhibited fewer behavioural problems, but that female behaviour at the age of 6 did not significantly differ among attachment groups. Hence, multifinality indicates that similar starting conditions do not always result in similar outcomes. This implies that by applying deductive reasoning seemingly different phenomena or outcomes might be traced back to similar root causes when multifinality comes into play.

Thus, the principles of equifinality, multifinality, homeostasis and heterostasis have far-stretching implications for the application of deductive reasoning. The paradox of equifinality and multifinality means that when observing the behaviour of a system, it might be moving towards a final state irrespective of the initial state or moving away from an initial state without being able to predict the final outcome. That implies that the observer should select the appropriate period for observation of a system to know whether the system moves to a stable state. Also, variations in initial state and the response of the system might indicate whether the entity is subject to equifinality and multifinality. Similarly for homeostasis and heterostasis, the period of observation might affect the inferences by the observer about the behaviour of the system. Generically speaking, the poised state of heterostasis is more difficult to maintain over longer periods of time and has side effects by straining the system, for example the use of resources. Therefore, the state of heterostasis will be punctuated by longer intervals of homeostasis; this phenomenon is called punctuated equilibrium [see Eldredge and Gould,

1972] in evolutionary biology. That means in all cases that dynamic of behaviour interpreted through deductive reasoning should be interpreted with care; however, as shown in Box 3.B, the understanding of equifinality, homeostasis and equifinality is very helpful for solving practical problems in conjunction with deductive reasoning.

Inductive Reasoning

As opposed to deductive reasoning, inductive reasoning looks at elements or subsystems rather than the performance of the whole in the context of systems thinking. In the case of Applied Systems Theory, inductive reasoning refers to understanding the impact of changes in properties of elements (or even discarding, replacing or adding elements) on the state and the behaviour of the whole. An illustration of this is the replacement of handwritten documents by electronics ones. Although many of the steps for the procedures might remain the same or are similar, the implementation of this electronic documentation system will affect the performance of an organisation and how people work with it. Within the domain of synthesis as integrating components into a

---

Box 3.B: Equifinality, Homeostasis and Heterostasis

When looking at a house, it might be maintaining a stable temperature (this can be measured externally), certainly if there is a thermostat in the house for the heating or a temperature control for the air-conditioning). At the same time, the observer might notice that on hotter days the house consumes more electric energy; colder days result in the use of more gas for the heater. According to the principles of homeostasis, maintaining the stable temperature, one variable, requires variations in the usage of energy, in the form of electricity and gas, the other variable(s).

Also, after the house has been empty for a while, the temperature of the house will return to its set temperature. That final state will be reached, no matter the original external temperature, which the house has reached after being inhabited for a while. This is the principle of equifinality; the stable state of the house will be reached from any initial external temperature.

When for some reason or another, the temperature of the house will be maintained at an extreme high temperature during harsh winter conditions that will come along with a higher consumption of gas. The internal system for converting the gas in heat will be used longer and more intense, not only increasing the fuel consumption but also reducing the life span of the heater. This is called heterostasis, maintaining an extra-ordinary state during a brief period of time, albeit at the expense of durability of the system.

whole (covering a wide range of applications from engineering itself to design of organisations), inductive reasoning entails if another one replaces one element or subsystem, how the system will modify its behaviour. The foremost application in engineering and design constitutes the introduction of novel solutions as part of a total system, for example a new type of storage device in a computer. Think about the use of flash drives instead of hard disks in laptop computers; although flash drives perform the same tasks as hard disks, they consume less energy and are more compact, thus influencing the design and use of electronic devices. Hence, inductive reasoning takes the elements or subsystems as starting point for studying their effects on the whole.

Inductive reasoning as such is applied to situations that are complicated or ill-defined, when we look for patterns and simplify the problem or situation by constructing temporarily hypotheses or schemata to work with. This is often called the hypothetico-deductive method and became later better known as 'induction logic' (made popular by Popper [1966, pp. 98–99]; his thoughts seem to be rooted in the research of Selz [1913, p. 97]). An example of this are chess players who form hypotheses or schemata about the opponents' intentions by studying their moves; note that in the case of chess the ultimate objective is known: defeating the other player, but the objectives are not always known. Based on those hypotheses or schemata, localised deductions serve as specific conjecture for observations. In the case of chess, the player assumes a certain attack and deducts from that pattern which move the opponent most likely will make. The cycle becomes complete when feedback from the environment about the state or the behaviour of a system is considered that might strengthen or weaken beliefs in current hypotheses, discarding some when they cease to perform, and replacing them as needed with new ones [Arthur, 1994, p. 407]. In the case of chess, not only the moves but also the facial expressions might serve as indicator about the match develops; in terms of systems theories, this is zooming out and extending the boundary of the system considered. In other words, where we cannot fully reason or lack a full definition of the problem, we use simply models that bridge cause and effect for filling the gaps in the understanding of reality.

One of the most paramount constraints for applying inductive reasoning is caused by limitations in knowledge and experience. In most situations, it is presumed that decision-makers or those that analyse problems have access to and gather all relevant information, and act based on full rationality. But in reality this is hardly the case. In terms of Applied Systems Theory, any observer of a system has limited knowledge of the universe (see Section 2.1 for the definition of the universe); those elements and relationships known to the observer are called the *real-life system*. That implies that for each observer the real-life system will differ; in terms of Applied Systems Theory, the environment of each observer is different from all other observers (this theme will return in Chapter 8 about so-called autopoietic systems). Building on this notion, the concept of bounded rationality states that rationality of individuals is restrained by the real-life system, cognitive limitations and finite amount of time for decision-making [Simon, 1947, 1959]. The cognitive limitations indicate that not only the real-life system influences thinking but also the capability of processing information and deriving useful hypotheses or schemata. Hence, bounded rationality embedded in the real-life system of the observer or decision-maker restricts in practice the application of inductive reasoning.

However, inductive reasoning should not be confused with the approach of 'trial-and-error' to analyse and solve problems. In its most extreme form, by using trial-and-error, an investigator changes randomly any relationship or element to find out whether the intended effect will be achieved. This is particularly useful when no apparent hypotheses or schemata apply to the problem; the search for some new drugs is often characterised by this trial-and-error method. This does not mean that the approach needs to be careless; for an individual, it can be methodical in manipulating the variables in an attempt to sort through possibilities that may result in success. It is possible to use trial-and-error to find all solutions or the best solution, when a finite number of possible and testable solutions exist. To find all solutions, one simply makes notes about observations and continues, rather than ending the process, when a solution is found, until all solutions have been tried. To find the best solution, one finds all solutions by the method just described and then comparatively evaluates them based upon some predefined set of criteria, the existence of which is a condition for the possibility of finding a best solution. Nevertheless, people who have little knowledge about a problem domain often use this method. Hence, trial-and-error has many characteristics of the search processes related to inductive reasoning but lacks the formulation of hypotheses or schemata in advance.

Even though inductive reasoning may be limited by bounded rationality and by search processes akin the trial-and-error method, it has its advantages as a complementary mechanism for the complexity of reality and those problem situations that are ill-defined. Particularly, it enables dealing with complexity of reality through constructing plausible, simpler models that we can cope with. An example of simplification by modelling is the drawing of a house by architects showing the four faces without going into much detail about how the actual construction should take place. The drawings are sufficient for a customer to understand how the house will look like once being built. Additionally, inductive reasoning makes it possible coping with situations that ill-defined: where we have insufficient definition of a problem definition, the working models or hypotheses fill the gap. This is similar to the stance taken by the soft systems methodology of Checkland [1981, pp. 169–177] when talking about constructing conceptual models (to be seen as hypotheses); although his approach is more a means to involve stakeholders rather than a formal modelling technique (for more on Soft Systems Methodology, see Section 10.4). Modelling in his view is seen as understanding reality so that meaningful actions can be undertaken that lead to improving a 'problem situation'; the extent of resolution determines whether another iterative cycle of problem solving should be evoked. This only indicates that inductive reasoning may imply that investigations of system are going through a number of cycles before a plausible model is found, particularly in ill-defined and complex problem situations.

That premise of iteration is quite similar to scientific approaches but that comparison introduces further limitations to inductive reasoning. For scientific research inductive reasoning requires the following steps: the observation of all relevant facts (or properties), the classification of these facts, the generalisation and the testing of assumptions. The four-stage model of scientific discovery or inductive logic [Popper, 1999, p. 14], as an example of inductive reasoning, not only follows the principle of inductive reasoning but should also lead to caution with respect to hasty conclusions. The four stages of inductive logic consist of: (a) the definition of the old problem, (b) the formation of tentative theories about the phenomenon, (c) the attempts at elimination of at least some of these tentative theories, (d) the uncovering of new problems that gives reasons to repeat this cycle again. According to Popper [1999, p. 10], tentative theories (or hypotheses) should be falsifiable. Further investigation of theory and empirical data should reveal whether a refinement becomes possible such that testing will allow the assessment of theories [Popper, 1966, pp. 52–55]. For testing of hypotheses, it is necessary that the class of opportunities to falsify the theories should not be empty on beforehand. The generic character of theories allows no verification, but they could be open to falsification [Nola and Sankey, 2000, p. 18]. The second point of Popper's philosophy towards scientific discovery, inductive logic, warns for drawing generalisations where possibly inappropriate [Popper, 1966, pp. 98–99]. It shows that inductive reasoning is open to interpretation whereas deductive reasoning is closed and more directed at fact-finding within a given framework.

Abductive Reasoning

The fact that not always all facts for deductive reasoning are available and that inductive reasoning is open to interpretation leads to so-called abductive reasoning. This type of reasoning, which is also called abduction, abductive inference or retroduction, is a form of logical inference that goes from an observation to ideally seeking the simplest and most likely explanation. A well-known example is the observation that a lawn is found wet in the morning. The knowledge that it has been raining during the night leads to the inference that the rain has caused the wetness of the lawn. However, there are also other causes for a wet lawn in the morning; take for example dew and lawn sprinklers. Additional facts might provide better substantiations whether the rain was the cause or not; such facts include the exact time in the morning, the degree of wetness and meteorological conditions, such as temperature, strength of the winds and cloudiness. These facts will increase the certainty that the rain has caused the wetness of the lawn or that other causes are also feasible. Such searching for further facts to corroborate a likely cause also depends on the knowledge of the investigator and the willingness to consider triangulation; the term triangulation refers to indicate that two (or more) methods are used in a study in order to verify the results of the other one for the same system or phenomenon. Thus, unlike in deductive reasoning, in abductive reasoning, the premises do not guarantee the conclusion. Henceforth, abductive reasoning should be understood as the 'inference to the best explanation'; that is why the philosopher Charles Sanders Peirce (1839–1914) first introduced the term as 'guessing' (or later known as 'making an educated guess' [1901, par. 219].

However, there is a very thin line to be crossed before abductive crossing turns into so-called 'justified true belief'. In the case of justified true belief, someone mentally assents to some proposition ('belief'); if this belief is 'true', then there is some fact about reality that makes the proposition true; and then

---

BOX 3.C: FAMOUS STATEMENTS ABOUT MODELS AND MODELLING

Essentially, all models are wrong, but some are useful. [Box and Draper, 1987, p. 424]

A theory has only the alternative of being right or wrong. A model has a third possibility: it may be right, but irrelevant. [Eigen, 1973, p. 618]

The purpose of models is not to fit the data but to sharpen the questions. [Samual Karlin, at the Eleventh R. A. Fisher Memorial Lecture, Royal Society (20 April 1983)]

There are many specific techniques that modellers use, which enable us to discover aspects of reality that may not be obvious to everyone ... [Silvert, 2001, p. 261]

Models are of central importance in many scientific contexts. The centrality of models such as the billiard ball model of a gas, the Bohr model of the atom, the MIT bag model of the nucleon, the Gaussian-chain model of a polymer, the Lorenz model of the atmosphere, the Lotka-Volterra model of predator-prey interaction, the double helix model of DNA, agent-based and evolutionary models in the social sciences, or general equilibrium models of markets in their respective domains are cases in point. Scientists spend a great deal of time building, testing, comparing and revising models, and much journal space is dedicated to introducing, applying and interpreting these valuable tools. In short, models are one of the principal instruments of modern science. [Frigg and Hartmann, 2006]

if the belief is 'justified', it means that the believer has some evidence or good reason for the belief. See the case of the wet lawn in the previous paragraph. These cases for reasoning where not necessarily all facts are considered and where an observer beliefs something is true are called Gettier problems; this term is derived from his writing [Gettier, 1963]. These types of 'false belief' can be circumvented by falsification (see previous Subsection 'Inductive Reasoning'), by searching for more facts to increase certainty about possible inferences, by considering other beliefs and by model-based reasoning.

3.4 Types of Models

Because any type of reasoning deploys models as a tool to solve a problem, it is important to understand the relation between models and the system to be studied beyond the notion of simplification introduced in Section 3.1. *The best material model of a cat is another, or preferably the same, cat*', said Norbert Wiener, one of the founders of cybernetics, together with Rosenblueth [1945, p. 320]; see Box 3.C for other famous statements about models and modelling. Such a statement does not help very much to reduce the complexity of the system to gain overview. Rephrasing the notion of a model as a system with a simplified structure we might say: a model is an aspectsystem of a (sub)system on a higher level of abstraction – any of the three types in Section 3.2 – for the purpose of studying an other system. That introduces the concepts of isomorphism and homomorphism that will be elaborated in the next subsections, followed by analogies and metaphors. After the discussion of these four concepts a classification of models is presented that should support the selection of models and understanding their limitations; this classification builds on Figure 3.1.

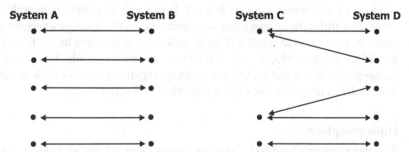

Figure 3.5: *Isomorphism and homomorphism for systems. Isomorphism means a one-on-one relationship between the two systems studied (left picture with System A and System B). When distinguishing one system in reality and one artificial one to be studied as an abstraction of reality, it becomes hard to achieve isomorphism; System C and D are homomorphic (depending on the problem definition).*

Isomorphism

When there are as many elements and relationships in the model as in the original system, we speak about isomorphism (see Figure 3.5). 'Isomorphic' (in Greek ισοσ = equal and μορπηε = shape) means 'having a similar form' and one system is said to be isomorphic with another when, in formal terms at least, they could be interchanged [Beer, 1959, p. 42]. The terms goes back to Eilhard Mitscherlich (1974-1863), who introduced the law of isomorphism, which states that compounds crystallising together probably have similar structures and compositions. In mathematics, the word *isomorphism* applies when two complex objects can be mapped onto each other in such a way that to each part there is a corresponding part in the other object. Hofstadter [2000, p. 49] expands this definition:

The word 'isomorphism' applies when two complex structures can be mapped onto each other, in such a way that to each part of one structure there is a corresponding part in the other structure, where 'corresponding' means that the two parts play similar roles in their respective structures.

Isomorphic structures are 'the same' at some level of abstraction; that implies ignoring the specific elements and relationships at lower level of detail. Here are some everyday examples of isomorphic structures:

- A solid cube made of wood and a solid cube made of lead are both solid cubes from a geometric perspective, although their materials differ entirely;
- The Clock Tower in London (that contains the Big Ben) and a mechanical wristwatch when looking at their mechanisms for reckoning time are similar, even though both devices vary greatly in size;
- A six-sided die and a bag, from which a number 1 through 6 is chosen, have random number generating abilities that are isomorphic, despite the method of obtaining a number being completely different.

From these examples it also becomes clear that isomorphic model building always includes a perspective for observation. Consequently, the building of models for complex systems is unlikely to result in perfect isomorphism (even cats differ from specimen to specimen); in the case of exceedingly complex systems, that result will be by definition impossible to verify, if not to achieve [Beer, 1959, p. 42]. Ultimately, the extent to which a model is isomorphic with the real system at a given aggregation stratum will determine its usefulness for predictions given a problem definition.

Homomorphism

The imperfection of model building expresses itself in homomorphism. This occurs when the model has fewer elements than the original, but with all relevant relations intact from a given perspective; note the similarity with defining an aspect system (see Section 2.4). Thus, homomorphism from one system to another of the same kind or to a model is a mapping that is compatible with all relevant structures, given the problem definition. When structuring systems for a given problem definition we are looking for a homomorphic model, because a model is always a simplification, with isomorphism for the aspect under consideration; and if not, as much isomorphic as possible. Homomorphism is important in establishing whether one system is a model of another and which properties of the original the model retains for the specific purpose of investigation. Some examples are:

- The comparison of the English language and the Sami language (the Sami are Europe's most northern indigenous people living in Finland, Norway, Russia and Sweden) when using the word snow. The Sami language has hundred words for snow while in English language that is limited to a few words, such as pack, powder, sleet and snow.
- The similarity between natural ecosystems and economic ecosystems for interactions among its constituent actors. A natural ecosystem might have a wider variety than an economic ecosystem and an economic system is a construct of the human mind. These differences will make it difficult to make a direct comparison.

For each system one can construct a lattice of homomorphic simplifications. In that sense, the three abstraction mechanisms – classification, aggregation and generalisation – are forms of homomorphism.

Analogies and Metaphors

In addition to isomorphism and homomorphism, analogies and metaphors are distinguished as archetypes for how similar distinct systems might be. In that sense, an analogy is a comparison between two different things, in order to highlight some form of similarity. Analogies are often used to explain new or complex concepts by showing the similarities between these and familiar concepts. Some types of analogies can have a precise mathematical formulation through the concept of isomorphism. A famous example, in engineering is the mathematical description of a simple mechanical oscillating system and a simple electronic oscillating system. Whereas their components are entirely different in their appearance, the behaviour of these two systems expressed in equations is exactly identical. Within the domain of logic reasoning, an analogy focuses on similarities in known respects to similarities in other respects. Ultimately, an analogy is a kind of generalisation. Very differently, metaphors address a figure of speech, which is not literally applicable to the object of comparison. For instance, an organisation and a human body are compared with each other, leading to the proposition that an organisation needs a head (note that this concept has even penetrated daily language). But there is no way that an organisation functions like a human body; the metaphor acts symbolically to underline the importance of somebody in charge of an organisation. In this perspective, an analogy might be considered a kind of extended metaphor or long simile in which an explicit comparison is made between two things (events, ideas, people, etc.) for the purpose of furthering a line of reasoning or drawing an inference; it is a form of reasoning that employs comparative or parallel cases.

For applying analogies or metaphors in the context of systems theories, a few words are in order. First, the application of concepts in use in other disciplines does not deny the reality or concreteness of phenomena in the domain of application. By way of illustration, the process of creation within the arts might be compared with the process of design in engineering-based sciences, albeit that the context of creation is very different for both of them. Second, if some fields of science are further along in their understanding of reality, related fields should make sure that their explanations are consistent with the latest discoveries and insights – which is the notion of consilience [Wilson, 1998, p. 8]. Current crossovers of biological concepts to all other kinds of sciences, such as information and communication technologies, urban planning, material sciences, etc. serve as an example. Third, using metaphors and analogies is sometimes judged negatively. However, any conceptual or formal model can be said to involve metaphors and analogies to some extent; the systems theories themselves are a case in point. According to Hodgson [1993, pp. 18-19], analogies and metaphors should not be regarded just as literary ornaments that hide the core of a theory or model. Playing with metaphors means approaching reality from various perspectives, and recognising that concepts have subjective interpretations as well as inevitably a social and academic history and context.

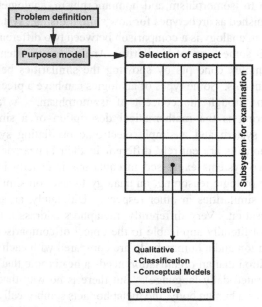

Figure 3.6: *Expansion of models (building on Figure 2). The problem definition points to the aspects and subsystems to be considered. Additionally, models have the dimension of either being qualitative or quantitative. Qualitative models are further divided into classifications and conceptual models.*

Of course, one needs at all times to be conscious that analogies and metaphors can be misleading, because they are incomplete or inaccurate or inappropriate in the worst case [Morgan, 1997, p. 5]. Kickert [1993, p. 262] asserts that caution is particularly appropriate when applying theories derived from natural sciences to social sciences. The same holds of course for formal models, which can be considered a particular type (subcategory) of metaphor. Many models in economics can be traced back to modelling traditions in natural sciences, such as biology, physics and engineering. For example, an analogy that is persistent in economic theory is that firms, industries or even countries as a whole behave like individuals (at the level of firms perhaps enforced by the acknowledgement of a corporation as a legal entity [Bakan, 2004]). This may be an appropriate analogy for some aspects of economics but it is doubtful as a generic approach for all phenomena that economists are studying. Hence, analogies and metaphors, as a specific form of homomorphism, have strong limitations in their applications.

Qualitative Models

In terms of modelling metaphors or analogies will be considered part of so-called qualitative models, even though sometimes quantification is possible for them; whether models are qualitative or quantitative constitutes one of the two dimensions for a further categorisation of models (see Figure 3.6). Qualitative models describe reality with aspects and features (see Section 2.2) for systems, subsystems and elements, and can be divided into two types: classification and conceptual models; note that metaphors and analogies are principally conceptual models. A case in point is Newton's third law of motion, which states: 'for every action there is an equal and opposite reaction'. In essence this third law of motion is a conceptual model. Once qualitative models become enriched with parameters and their values, quantitative approaches become possible. Newton's law has been transferred later into formulae, which did allow calculations and further studying of related phenomena. Sometimes qualitative models suffice, especially for descriptions and explanations in social sciences; and sometimes we need quantitative models, for predicting possible future events, for example, estimating the time of arrival of flights. In most cases, qualitative models precede quantitative models.

As already discussed in Section 3.1 (taxonomic) classification, as one of the two types of qualitative models, is the act of placing an object, system, element or concept into a set or sets of categories (such as a subject index), based on its properties. This assumes that all objects, systems, elements or concepts in a specific set (or class) have similar properties from a certain perspective, at the aggregated level of the objects, systems, elements or concepts. This means that a person may classify the object or concept according to an ontology, which as a fundamental branch of metaphysics seeks to describe and categorise entities or posit basic categories within an overarching framework. Ontology has strong implications for the perceptions of reality and, therefore, may be linked to philosophical thinking. In this context, some philosophers, notably those of the Platonic school, argue that all nouns refer to entities. It also implies that ontology covers objects and events as well as constructs of the mind. However, other philosophers contend that some nouns do not name entities but provide a kind of shorthand way of referring to a collection (of either objects, systems and elements or constructs of the mind or events). This way of thinking has strong parallels with the distinction of aggregation strata. In this latter view, a society refers to a collection of persons with some shared characteristics (or properties in terms of system theories) and geometry refers to a collection of specific kinds of output resulting from intellectual activity. Any ontology must give an account of which wording refers to entities and which does not in addition to why and what categories result. When one applies this process to nouns such as electrons, energy, contract, happiness, time, truth, causality and god, ontology becomes fundamental not only to philosophy but also to many branches of science and activities of creation (such as design and engineering). Examples of taxonomic classifications include library classifications, scientific classifications of organisms, medical classifications, and security classifications. Thus, classification limits itself to describing properties as a (quasi-)static approach.

Where classification aims at describing (common) properties, conceptual models explain phenomena and, in that sense, emphasise studying relationships between elements or systems. This type of models might tell us what will happen next (whether it is based on empirical observations or heuristic laws). Derived from Engelbart [1962, pp. 128–129], developing conceptual models means specifying:

- The essential elements of subsystems of the system to be studied.
- The relationships of the elements that are recognised (i.e. aspects).
- The changes in the elements (i.e. content of system) or their relationships (i.e. structure) that affect the functioning of the system – and in what ways.
- The objectives and methods of research (or investigation).

Some will take it that conceptual models are broader and more fundamental than scientific theories in the sense that they set the preconditions for theory formulation. In fact, they might provide the conceptual and methodological tools for formulating hypotheses and theories, the process of inductive logic as pointed out by Popper [1966, pp. 52–55; 1999, p. 14], see Section 3.3. Once, a conceptual model has been proven through empirical studies, it becomes a theory, and if that happens under repeatable and 'objective' conditions, it is called a scientific theory. If the conceptual models are also seen to represent schools of thought, chronological continuity, or principles, beliefs and values of the research community, they become paradigms. A famous example of a paradigm is the Austrian School of Economics. Economists belonging to this school of economic thought advocate strict adherence to the principle that social-economic phenomena can only be accurately explained by showing how they result from the intentional states that motivate the individual actors and they emphasise the spontaneous organising power of the price mechanism. This paradigm then strongly influences research undertaken by those adhering to this school. Consequently, a conceptual model is always constructed in communication – it does not simply appear out of the blue sky.

For purposes of scientific research, a conceptual model, as a simplified system to study an other system, provides a working strategy, a scheme containing general, major concepts and their interrelations. Such models orient studies into phenomena towards specific sets of research questions aiming at formulating (substantial) theories. A conceptual model cannot be assessed directly empirically, because it forms the basis of formulating empirically testable questions and hypotheses. Ultimately, it can only be assessed in terms of its instrumental and heuristic value. If substantial theories prove to be useful in many circumstances, the underpinning conceptual model might be so too; however, the collection of sufficient empirical data for conceptual models or substantial theories to prove to their validity may take some time.

Even before embarking on some line of inquiry, whether scientific or practical, it may be important to argue about the merits of various conceptual models. The following are general (scientific) principles that can be used to judge the merits of a conceptual model:

- The scope of the conceptual model should suffice for the situations to study. Mostly, a conceptual model is more useful when it covers a wide range of situations as possible (this is called the principle of fecundity). Taking it further, when studying some phenomena, ideally they should be studied in all situations, and also under extreme conditions; for example, most economic behaviour is assumed to be rational but what about people that exhibit emotional behaviour. This implies that the boundaries of the application of the conceptual model should be sought; it also denotes that a broader scope is better because it subsumes narrower ones, other things being equal.
- The conceptual model should be limited in a meaningful way as a system (systematic power). For example, understanding information seeking by human actors, the proper system is not the provision of some service (such as a library and its customers) but rather an information actor immersed in his or her situation and information environment (for example, all information access systems). Hence, this principle explores whether the conceptual model is fit for purpose and has the ability to organise concepts, relationships and data in meaningful systematic ways.
- The conceptual model should be at a sufficient level of detail to study the phenomenon or provide an explanation (accuracy). This argument is akin aggregation strata. It means that a conceptual model should not be so abstract that it hardly describes the situation. For instance, modelling a car solely from how many people it can transport will hardly indicate its fuel consumption. However, if the model is extremely detailed, it might lead to the inclusion of irrelevant details. Using the example of the car

again, the colour scheme of the seats will not have any impact on the fuel consumption. It may well be that the development of a conceptual model will go through iterations before the right level of aggregation is established.

In addition to these three principles (scope, systematic power and accuracy), whether the conceptual model has been developed to study practical events or its aims at scientific discovery, when two competing conceptual models are compared the following criteria may be applied to judge their merits (for part derived from Wacker [1998]):

- Simplicity. This indicates that a conceptual model that is simpler tends to be better, assuming all other things being equal (this principle is commonly known as Occam's razor and also labelled the law of parsimony).

- Explanatory power. This criterion indicates the ability of a model to effectively explain the subject matter it pertains to. Particularly useful is to look at the details of what needs to be observed (but with greater detail, greater inaccuracy might be introduced). Sometimes this is called internal consistency or validity for the concepts within the model and external validity for the degree of homomorphism; generalisation would indicate to how many phenomena across systems and domains the model could be applied.

- Reliability. This measure points to the ability, within the range of the model, to provide valid representations across the full range of possible situations (and is strongly related to the scope of the model).

- Fecundity. This demonstrates the ability of the model to suggest problems for solving and hypotheses for testing; in other words, it represents the ability of a model to open new lines of inquiry.

To meet all these criteria, theoretical development or the construction of new conceptual models in any research area or investigation often requires conceptual and terminological development. Conceptual development may mean fulfilling, perhaps in a better way than before, the basic requirements for scientific concepts – precision, accuracy, simplicity, potential for generalisation, and suitability for expressing propositions. Moreover, effective conceptual models represent essential features (relationships) for the elements or subsystems of the domain that is investigated, related to behaviour induced by events.

Quantitative Models

Many times a model is required to provide answers in the form of hard numbers. Such quantitative models are mostly the result of the transformation of qualitative models into mathematical abstractions. Because of the required accuracy, these models often have a rather narrow scope of relationships and variables they might consider. In the realm of quantitative models four kinds are distinguished (three of these are derived from Ackoff [1962, p. 109]), see also Figure 3.7:

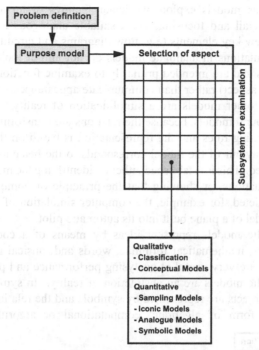

Figure 3.7: *Overview of qualitative and quantitative models. This expansion of Figure 3.6 shows the four basic quantitative models: sampling models, iconic models, analogue models and symbolic models.*

(1) Sampling models consist of a mere subset of mutually exclusive systems taken from a larger set of systems. The representation is based on the assurance that each relevant system within the universe had the same probability to be included in the sample. A sampling model resembles classification but differs in the sense that not all relevant properties have been identified, yet. However, sampling could also consider other criteria for selection. In that perspective, Flyvjberg [2006, p. 230] sets out that these case may include extreme or deviant instances, cases with maximum variation or critical instances (albeit this his writing is to be positioned in the context of research methodologies for cases; nevertheless, it seems applicable beyond the domain of case studies).

(2) Iconic models are linear transformations of a configuration of systems in the universe mostly based on a single aspect. The representation is based on the assurance that an iconic model retains the universe's topological characteristics for the specific aspect. Iconic models look like the real system but sometimes employ a change of scale or materials. They are used principally to communicate (design) ideas – for example, to the designer, to a customer (e.g. sketches, 3D prototypes) or to users. Iconic models represent real systems by, for example, scale models, photographs and graphical representations of networks.

(3) Analogue models explore particular features of an idea by stripping away detail and focusing, via a suitable analogous representation, on just a few key elements (e.g. flow diagrams and circuit diagrams). The representations in analogue models do not aim at looking like the real systems and are intended primarily to examine functions and behaviour (of one aspect) rather than communicate appearances. Analogue models like all other models are a simplification of reality. Some call these behavioural models, because the relations are transformations, equations or operating rules and the representation is based on the assurance that the behaviour of the model corresponds to the behaviour of the system modelled. This is established either by identifying the model's parameters and equations, or showing that the principle of homomorphism is not contradicted, for example, the computer simulation of an economy and the model of a plane built into its automatic pilot.

(4) Symbolic models represent ideas by means of a code (for instance, numbers, mathematical formulae, words and musical notation). These models are very useful at analysing performance and predicting events. Symbolic models are an abstraction of reality. In symbolic models the set of objects are represented by symbols and the relations are expressed in the form of algebraic, computational or algorithmic statements

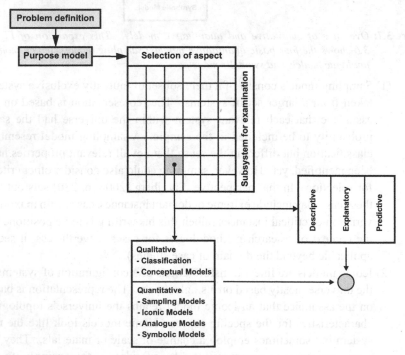

Figure 3.8: *Classification of models. The problem definition points to the aspects and subsystems to be considered. Models have two dimensions: (1) being descriptive, explanatory or prescriptive and (2) being qualitative or quantitative. On the second dimension, a further refinement leads to specific purposes of models.*

exhibiting no behaviour of their own. Symbolic models can be realised in or coupled with computing devices (or networks) in order to simulate with a behavioural model the actual behaviour of a system. For example, a formal statement about a social process must be translated into the algorithmic form of a programme acceptable to a computer, e.g. the contractual arrangements between buyer and suppliers in software used for managing business processes.

Even though these four types of quantitative representation differ quite substantially, all build on conceptual qualitative models (whether implicitly and explicitly) and have greater detail that may inevitably lead to loss of accuracy compared to the original qualitative model.

Overview of Models

The combined overview for the selection or development of qualitative and quantitative models is presented in Figure 3.8; in addition to the dimension of qualitative and quantitative models, models can be categorised as descriptive, explanatory or predictive. The search for a model starts with the problem definition, which should lead to the distinction of a subsystem and specific relationships of interest to the objectives of the study. A descriptive model seeks to support the process of thought, by providing consistent and appropriate terminology (it relates very closely to classification); it merely described observed properties, no matter whether there is ample evidence that they pertain to a problem definition or the description of a phenomenon. An explanatory model seeks to explain how on hindsight how a phenomenon of interest occurred; it identifies those events, state of systems and behaviour of systems that related to the phenomenon that is investigated. Predictive models allow researchers to predict the outcomes of a system's behaviour events, state of systems and behaviour of systems that related to the phenomenon. If designers propose a new system, predictive models indicate what the outcomes might or should be. Sometimes, a prescriptive theory gives directions or rules as to how something should work or be carried out. For example, a model may suggest how a menu of software should be laid out. Hence, the dimension of the explanatory power of a model – descriptive, explanatory or predictive – complements the dimension of qualitative and qualitative models.

Note that normally models move through successive stages of modification during the use for analysing and solving problems. With predictive models it becomes possible to forecast the future. These models are necessary to design solutions and can be either deterministic or stochastic models. An explanatory model precedes a predictive one. The understanding of why it happens leads to explanatory models. The underlying mechanisms become clear, most of these models have an analytical character. In turn, a descriptive theory comes before an explanatory one. A descriptive model simple describes what happens or states properties. The Linnaean taxonomy used by biologists can be considered as a descriptive model: it describes but has almost no impact on the understanding of why and how. The addition later of evolutionary mechanisms to the desriptive Linnaean taxonomy made it possible to develop explanatory models; the notion of natural selection combined with Mendel's [1865] experiments with wrinkled peas led to a range of explanatory models, up until Kaufmann's [1993] fitness landscapes. The explanatory models were followed by predictive models, such as adaptive dynamics [see Geritz et al., 1997; Meszéna et al., 2001], a game-theoretical approach to evolutionary processes. Hence, the development of models for a certain phenomenon should never be viewed from a static perspective but will be modified from refined from descriptive to explanatory to predictive models.

Given that models take centre-stage in many investigations, that also means that over the years quite a number of people have made statements about them (see the beginning of this section). Wiener's infamous statement of the model of a cat has already been mentioned. A few others have been added in Box 3.C and indicate both the necessity and the difficulty in developing any type of model (and theory). The statements underline that a model only fits with a certain problem or purpose. Figure 3.8 shows this notion undoubtedly: first a model limits itself to certain elements (from the universe) and aspects, and then on the two dimensions (qualitative and quantitative, and use). In all stages, higher accuracy comes along with a reduction in scope and fecundity. This dilemma of building and using models has been well-recognised but also indicates both the necessity of working with models and its inherent

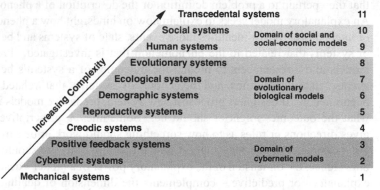

Figure 3.9: *The eleven levels of Boulding (1985). The domain of organisations and social systems moves at the ninth and tenth level, which indicates the importance of meaning, value systems, and symbolisation (levels also indicated by numbers). Models from evolutionary biology mostly dominate the fifth to the eight' level. The domain of systems theory and some other approaches in management science (e.g. information technology) find themselves at the second and third level. For example, that indicates that models from evolutionary biology might bridge the gap between some of the approaches management science based on cybernetics and the actual organisational domain. The fourth level in this figure moves between cybernetic and evolutionary models; for example, autocatalytic systems can be positioned here but have both teleological and cybernetic traits in their behaviour (for autocatalytic systems, see Chapter 8).*

weaknesses; the on-going development of insight and knowledge only adds to the temporal value of models.

3.5 Systems Hierarchy of Boulding

The systems hierarchy of Boulding presents another way of looking at modelling. Boulding [1956] has introduced a systems hierarchy to distinguish systems according to their complexity. He remarks that models at lower levels of his systems hierarchy are a prerequisite for models at higher levels but do not per se suffice for describing systems of a higher order; note the parallel with emergent properties and behaviour (concepts mentioned in Sections 2.2 and 2.6). Boulding [1956, pp. 202–205] discerns nine levels in his hierarchy of systems. However, later he reworked these nine levels and presented eleven levels in the systems hierarchy [Boulding, 1985], see Figure 3.9:

- Level of mechanical systems. Systems at this level are controlled by simple relationships and few parameters. A case in point is the Copernican revolution that introduced a new framework for the solar system (the sun being at the centre of the solar systems rather then the earth) and later permitted a simpler description of the planetary movements. In mathematical terms, the connections or relationships are seldom more complex than equations of the third degree. Examples are the laws of gravitation, Ohm's law (the relationship between voltage, current and resistance in electric circuits) and Boyle's law (the relationship between pressure and volume of a gas).
- Level of cybernetic systems. These systems are based on processes that maintain any given equilibrium and thus determine the behaviour of a system, within its limits. Most physical and chemical reactions and most social systems do in fact exhibit a tendency to equilibrium by using negative feedback (see Section 6.3 about feedback). The homeostatic system is an example of a cybernetic system and such systems exist throughout the empirical world of the biologist and the social scientist, according to Boulding [1956, p. 203].
- Level of positive feedback systems. At this level systems respond to perturbations by increasing their magnitude until a limitation is reached. After accelerating to that point of limitation, either breakdown or breakthrough happens. Examples are fires, which getter hotter as they burn and eventually extinguish themselves after all available resources are consumed, or learning, when the more one learns, the easier it gets and also novel insight may occur.
- Level of creodic systems. This fourth level in the systems hierarchy includes all systems that are capable of a structural change after being perturbated. An example is genetics, with the work of Dawkins [1989] showing the point. One could say that these systems are teleological in

BOX 3.F: HUMAN BODY AND SYSTEMS HIERARCHY OF BOULDING

Modelling and viewing the human body serves as an example for the systems hierarchy. Below are some descriptions for the different levels.

LEVEL 1: MECHANICAL SYSTEMS

The skeleton of the human body is an example of this level of simple, dynamic systems. Even though, some might consider the skeletal system complex, the bones can only move relatively to each other. Note that all muscular tissue and conjunctional tissue has been left out of the picture on the right.

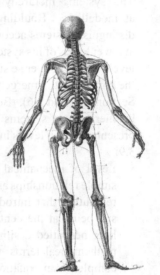

LEVEL 2: CYBERNETIC SYSTEMS

Maintaining balance and posture could be seen as a cybernetic system. Muscles and sensors act on tiny deviations to prevent humans from falling down. Another, more famous example is maintaining a constant body temperature.

LEVEL 4: POSITIVE FEEDBACK SYSTEMS

If the body moves to a new position, then it is not only homeostasis it tries to achieve (a stable position) but also a new point in the three dimensional space. Positive feedback features also in Chapter 5.

LEVEL 3: CREODIC SYSTEMS

For example, many of the basic reactions in our body concern autocatalytic sets. These are a collection of entities, each of which can be created by other entities within the set. Autocatalytic sets were originally and most concretely defined in terms of molecular entities. Chapter 7 will pay more attention to this phenomenon.

LEVEL 5–8: EVOLUTIONARY BIOLOGICAL SYSTEMS

At these three levels, we could consider the genetic evolution of human bodies: how they developed in the many different people and tribes with their own characteristics as well as the development of human kind (homo sapiens).

LEVEL 9–10: SOCIETAL SYSTEMS

An example at this level, how human beings interact in organisations. Not only do people learn from interaction but they also form collaborations to undertake ventures that would not have possible on their own.

nature and which could be called planned in a broad sense as they are guided by some kind of initial plan.

- Level of reproductive systems. At this fifth level genetic mechanisms guide both reproduction and growth, whether these are biological or social. Whereas in biological studies genes constitutes the basis for reproduction, Dawkins [1989, p. 192] has proposed memes, as unit for recombination in social systems. Memes constitute elements of a culture or system of behaviour that may be considered passing from one individual to another. Dawkins extends this concept to a wide variety of topics, such as ideas, artefacts, including people, products, books, behaviours, routines, knowledge, science, religion, art, rituals, institutions and politics. In organisational studies memes enjoy a high degree of popularity and it has become a popular term in social media.

- Level of demographic systems. Demographic systems, at the sixth level of the systems hierarchy, consist of populations of reproductive systems; a population is to be understood as a defined collection of comparable entities, not necessarily identical but similar enough to create a classification.

- Level of ecological systems. These systems at the seventh level are formed out of interacting populations of different species. Dynamic processes include not only population dynamics but also symbiosis (both beneficial and detrimental) and chain effects (for instance, the food chain). The Great Barrier Reef off the coast of north-east Australia serves as an example of an ecological system. But also management scientists have introduced the term ecosystem for pointing to interdependencies between firms, suppliers and customers.

- Level of evolutionary systems. Such systems can be both ecological, changing under the influence of mutation and selection, and artificial, obeying the same patterns but in the transferred sense of new ideas. These systems tend to evolve towards greater complexity.

- Level of human systems. According to Boulding [1985], the systems at the ninth level of his newer hierarchy, differ from other living systems because of the information processing capability of the brain; in that context he mentions that advance pattern recognition and communication abilities with speech, writing and use of sophisticated artefacts are distinctive marks.

- Level of social systems. These systems result from interaction between human beings and their artefacts; the social activity itself may be classified as belonging to economic, political, communicative and integrative systems. This interaction thrives on learning processes where evaluations and experiences are communicated throughout the system. At this tenth level, it concerns the content and meaning of messages, the nature and dimension of value systems, the transcriptions of images into a historical record, the subtle symbolisation of art, music, poetry and the complex gamut of human emotion.

- Level of transcendental systems. These philosophical systems are the ultimates and absolutes and the inescapable unknowables and they must also exhibit systematic structure and relationship, even though they contain a component of speculation.

From a biological evolutionary perspective and contemporary insight, the differentiation between the fifth to the tenth level may be blurred. Hence, Boulding's first hierarchy of systems [1956] and new ones should be interpreted for part in the Zeitgeist of their creation. In addition, if these levels underpin an investigation, they should be treated with care. Skyttner [2005, p. 110] gives an overview of other hierarchies of systems, such as the hierarchical levels of Miller. Each hierarchy has its own purpose and applications.

Nevertheless, the models based on lower levels and used for higher levels in any hierarchy may be have sufficient explanatory power to fuel understanding but they will also show deficiencies because of the differences between systems' levels. Again, models simplify the reality to serve a specific domain of research. Thus, observers and researchers will never be able to develop profound, comprehensive models that reflect reality; rather they have to make choices how to represent the problem domain in understandable pictures [Checkland, 1981, pp. 162–183]. The systems hierarchy serves a starting point for inspiring investigations at higher levels while at the same indicating their limitations.

So far, the concepts of Applied Systems Theory have covered all levels of the systems hierarchy of Boulding. Section 2.1 has generated definitions that are principally valid for all levels. From Section 2.2 on and throughout this chapter, the focus has been more on dynamics of systems (emergent properties should be considered as dynamic). In subsequent chapters, the focal point will move to the higher levels of this systems hierarchy.

3.6   Summary

Examining systems can start by looking at the whole (even before looking at the constituent elements) or by investigating the individual elements without considering the whole. Looking at the whole has the advantage that interrelationships are taken into account, which allows deductive reasoning (evaluating outcomes and performance of the whole to arrive at [root] causes situated at specific subsystems or elements and distinct aspects). Within this perspective, the blackbox approach is a particular way of looking at the system, by considering it as consisting of one element only. Nevertheless, deductive reasoning could turn into abductive reasoning when information is incomplete or incoherent, which leads to identifying the most plausible explanation. When investigating or hypothesising about the impact of subsystems or elements on the whole for selected aspects, i.e. inductive reasoning, one might forget about the interrelationships or the focus of the problem at hand. The choice for looking at the whole and deductive reasoning or examining parts and inductive reasoning is also influenced by the level of understanding of the behaviour of a system.

Whenever studying a system, the step of abstraction aims at avoiding unnecessary details by classification, aggregation or generalisation. Generalisation is a relation between class types, and classification is a relation between a class type and the objects belonging to this class. By using aggregation several connected elements are combined into one single element, this represents a typical feature of systems theories. When talking about aggregation for systems, the terms zooming in and zooming out are used. Abstraction requires observation and understanding of reality. All three modes for abstraction can be used in combination.

The distinction of a system within total reality is a step towards modelling and aims at understanding reality; hence, models simplify for the purpose of studying a system and, therefore, a model is never reality. A closer examination will always result in the choice of a subsystem and aspect for further investigation. Different types of models, along one dimension ranging from descriptive to predictive and along other dimension being either qualitative or quantitative, could be chosen or developed to fit with the system studied and the level of understanding available about the applications and limitations. Because models are always a simplification, the systems hierarchy of Boulding offers another way of looking at the validity of models from lower levels in this hierarchy for higher levels. Because models represent a lower level of complexity than the original object of study, the lower levels in the system hierarchy offer models for studying higher levels, if the investigator accounts for the differences and limitations (e.g. cybernetics to study management of social organisations). Hence, modelling reality comes along with a deliberate choice of subsystems and aspects, and with models that reflect the interpretations of reality.

References

Ackoff, R. L. (1962). Scientific Method: optimizing applied research decisions. New York: Wiley.

Arthur, W. B. (1994). Inductive Reasoning and Bounded Rationality. The American Economic Review, 84(2), 406–411.

Bakan, J. (2004). The Corporation: The Pathological Pursuit of Profit and Power. London: Constable.

Baker, F. (1973). Introduction: Organizations as Open Systems. In F. Baker (Ed.), Organizational Systems, General Systems Approaches to Complex Organizations (pp. 1–25). Homewood, IL: Richard D. Irwin.

Beer, S. (1959). Cybernetics and Management. New York: Wiley.

Bertalanffy, L. v. (1973). General System Theory. New York: George Braziller.

Boulding, K. E. (1956). General Systems Theory. Management Science, 2(3), 197–208.

Boulding, K. E. (1985). The World as a Total System. Beverly Hills: Sage.

Box, G. E. P., & Draper, N. R. (1987). Empirical Model-Building and Response Surfaces. Oxford: John Wiley & Sons.

Cannon, W. B. (1932). The Wisdom of the Body.

Checkland, P. (1981). Systems Thinking, Systems Practice. Chichester: John Wiley & Sons.

Darwin, C., & Wallace, A. R. (1858). On the Tendency of Species to form Varieties; and on the Perpetuation of Varieties and Species by Natural Means of Selection. Journal of the Proceedings of the Linnean Society of London. Zoology, 3, 46–50.

Dawkins, R. (1989). The Selfish Gene. Oxford: Oxford University Press.

de Rosnay, J. (1975). Le Macroscope: Vers une vision globale. Paris: Éditions du Seuil.

Eigen, M. (1973). The Origin of Biological Information. In J. Mehra (Ed.), The Physicists's Conception of Nature (pp. 594–632). Dordrecht: Reidel.

Eldredge, N., & Gould, S. J. (1972). Punctuated Equilibrium: An Alternative to Phyletic Gradualism. In T. J. M. Schopf (Ed.), Models in Paleobiology (pp. 82–115). San Francisco: Cooper & Co.

Engelbart, D. C. (1962). Augmenting Human Intellect: A Conceptual Framework (AFOSR-3223). Menlo Park.

Feiring, C., & Lewis, M. (1987). Equifinality and Multifinality: Diversity in Development from Infancy to Childhood. Paper presented at the Biennial Meeting of the Society for Research in Child Development, Baltimore, MD.

Flyvbjerg, B. (2006). Five Misunderstandings About Case-Study Research. Qualitative Inquiry, 12(2), 219–245.

Fourastié, J. (1949). Le Grand Espoir du XXe siècle. Progrès technique, progrès économique, progrès social. Paris: Presses Universitaires de France.

Frigg, R., & Hartmann, S. (2006, 27th February, 2006). Models in Science. Stanford Encyclopedia of Philosophy. Retrieved from http://stanford.library.usyd.edu.au/entries/models-science/

Geritz, S. A. H., Metz, J. A. J., Kisdi, E., & Meszéna, G. (1997). Dynamics of Adaptation and Evolutionary Branching. Physical Review Letters, 78(10), 2024–2027. doi: 10.1103/PhysRevLett.78.2024

Gettier, E. L. (1963). Is Justified True Belief Knowledge? Analysis, 23(6), 121–123.

Hagen, E. E. (1973). Analytical Models in the Study of Social Systems. In F. Baker (Ed.), Organizational Systems, General Systems Approaches to Complex Organizations (pp. 73–84). Homewood, IL: Richard D. Irwin.

Hodgson, G. M. (1993). Economics and Evolution - Bringing Life Back into Economics. Cambridge: Polity Press.

Hofstadter, D. R. (2000). Gödel, Escher, Bach: an Eternal Golden Braid (20th anniversary ed.). London: Penguin Books.

Kast, F. E., & Rosenzweig, J. E. (1974). Organization and Management, a Systems Approach. Tokyo: McGraw-Hill.

Kauffman, S. A. (1993). The Origins of Order: Self-Organization and Selection in Evolution. New York: Oxford University Press.

Kickert, W. J. M. (1993). Autopoiesis and the Science of (Public) Administration: Essence, Sense and Nonsense. Organization Studies, 14(2), 261–278. doi: 10.1177/017084069301400205

Mendel, G. (1865). Versuche über Pflanzenhybriden. Paper presented at the Verhandlungen des naturforschenden Vereines, Brno.

Mesarović, M. D., Macko, D., & Takahara, Y. (1970). Theory of Hierarchical, Multilevel, Systems. New York: Academic Press.

Meszéna, G., Kisdi, É., Dieckmann, U., Geritz, S. A. H., & Metz, J. A. J. (2001). Evolutionary Optimisation Models and Matrix Games in the Unified Perspective of Adaptive Dynamics. Selection, 2(1-2), 193–210.

Morgan, G. (1997). Images of organization. Thousand Oaks: Sage Publications.

Nola, R., & Sankey, H. (2001). After Popper, Kuhn and Feyerabend: Recent Issues in Theories of Scientific Method. Dordrecht: Kluwer Academic Publishers.

Peirce, C. S. (1901). On the Logic of Drawing History from Ancient Documents Especially from Testimonies.

Popper, K. (1999). All Life is Problem Solving. London: Routledge.

Popper, K. R. (1966). Logik der Forschung. Tübingen: J.C.B. Mohr.

Rosenblueth, A., & Wiener, N. (1945). The Role of Models in Science. Philosophy of Science, 12(4), 316–321. doi: 10.1086/286874

Selye, H. (1973). Homeosatis and Heterostasis. Perspectives in Biology and Medicine, 16(3), 441–445.

Selz, O. (1913). Über die Gesetze des geordneten Denkverlaufs, erster Teil. Stuttgart: Spemann.

Silvert, W. (2001). Modelling as a Discipline. International Journal of General Systems, 30(3), 261–282.

Simon, H. (1959). Theories of Decision-Making in Economics and Behavioral Science. The American Economic Review, 49(3), 253–283.

Simon, H. A. (1947). Administrative behavior. A study of decision-making processes in administrative organization.

Skyttner, L. (2005). General Systems Theory: Problems, Perspectives, Practice. Danvers, MA: World Scientific Publishing.

Timpf, S. (1999). Abstraction, levels of details, and hierarchies in map series. In C. Freksa & D. M. Mark (Eds.), Spatial Information Theory - cognitive and computational foundations of geographic information science (Vol. 1661, pp. 125–140). London: Springer.

Wacker, J. G. (1998). A definition of theory: research guidelines for different theory-building research methods in operations management. Journal of Operations Management, 16(4), 361–385. doi: 10.1016/S0272-6963(98)00019-9

West Churchman, C. (1979). The Systems Approach: Revised and Updated. New York: Dell.

Wilson, E. O. (1998). Consilience: the unity of knowledge. New York: Alfred A. Knopf.

4 Generic Approaches to Problem Analysis and Solving

The previous two chapters have laid the foundation of systems theories and already mentioned that these concepts are used for analysis, solving problems and decision making; to this end, this chapter will go into more detail about how to analyse problems, how to find solutions and how to make decisions. In doing so, it takes a rational approach. This contrast for part with reality when people often feel pressured, some associate decisions with emotions and past experiences, some procrastinate and others do not know how to put all information together in sense-making overview; hence, they tend to respond with a decision that seemed to work before, even for relatively new situations, or with a decision that may relate poorly to facts and information available. This dilemma has been investigated on from different perspectives early on when decision-making was researched; one of the most notable is the work by Herbert Simon [e.g. 1959]. As much as possible relevant concepts and methods from these original thoughts about how to solve problems and how to make decisions have been incorporated in this chapter. Thus, the approach to problem analysis and solving presented in this chapter supports structured problem solving; however, it does not aim at providing an exhaustive list of all theories, methods and tools.

The steps of this structured approached to problem solving together with some thoughts on applications and limitations are presented in this chapter. The first section explores what types of decision making can be distinguished; the next sections will pay attention to one of these three types: non-programmed decisions, i.e. those decisions that occur as one-off events. Section 4.2 starts with the stage of problem analysis for non-programmed decisions. Some other descriptions of problem solving skip this phase; cases in point are the method of Saaty [2000] and the problem solving skills mentioned by Despo and Epaminondas [2016]. Based on the finding of the root cause during the analysis, Section 4.3 goes into detail about the generation and weighing of alternatives. This leads to the actual decision making as described in Section 4.4; this section also considers group decision making. How the selected alternative is then further detailed and implemented is the topic of Section 4.5. The evaluation of the solution appears in Section 4.6. Based on the preceding sections, Section 4.7 provides the overview of non-programmed decision making. Finally, Section 4.8 makes some further comments about this process of problem solving and decision making. In this chapter, the steps for non-programmed decision making are related to the basic concepts of systems theories and the systems approaches.

© Springer International Publishing AG 2017
R. Dekkers, *Applied Systems Theory*,
DOI 10.1007/978-3-319-57526-1_4

4.1   Types of Decision Making

In general, three types of decisions can be distinguished. The first type is
the so-called programmed decision, for which solutions or interventions are
available in a structured manner; in such a case the decision maker selects
the best alternative from a range of solutions or interventions that have been
defined before the situation to be resolved emerges. The second type is non-
programmed decisions [March and Simon, 1958, pp. 169–182] or problem
solving; for these types of decisions the problem is often ill-defined and
not necessarily appropriate solutions are at hand at the start of the problem-
solving cycle. The third type of decisions is the one during emergencies and
crises; in such cases, the decision-making is constrained by a time horizon
and information is incomplete, possibly inaccurate and not structured. All
three types of decision making are described in more detail now.

Programmed Decisions

In the case of programmed decisions there are solutions readily available
from past experiences or derived from extant knowledge to solve structured
problems; in the context of decision making, the class of structured problems
refers to those problems for which the information for the decision making
is available. Programmed decisions are also called routine or repetitive
decisions and they can be handled by established rules or procedures. An
example is the turning of a vehicle in traffic; after the driver opts to turn into
a side road, the decision to stop for other traffic heading on is determined
by rules and procedures (though they may vary from country to country).
These types of decisions are often called for at certain points in a standardised
process and are decided based on identifiable parameters (see Figure 2.2 and
Section 2.2), relevant to the decision. An example is the length of a queue
and the number of servers (points at which the service is provided); when the
queue exceeds a certain length, the decision may be taken to open additional
servers to keep the waiting line within a set standard for its length. Typically,
programmed decisions do not require much consideration or discussion, and
can generally be automated to ensure consistency and reduce time for decision
making. Such happens with queues in microprocessors and information
systems; they measure the queue length to intervene by prioritising or adding
additional parallel streams of capacity. But even in retail environments or
service providers, sensors or indicators at appropriate points could measure
the length of a queue to determine the number of windows that should be
opened. Thus, programmed decisions typically concern situations for which
appropriate alternatives are available and information about the impact of
these alternatives is at hand.

Conversely, this also means that structured problems are familiar to those
making the decisions and are clear with respect to information needs. For
these reasons, programmed decision making can be best applied to routine
problems that can be anticipated. The creation of these routines results in

the formulation of rules, procedures and policies. However, programmed decisions are not necessarily confined to simple issues, such as vacation policies or waiting times in queues. They are also used to deal with more complex issues, for instance the types of tests that a doctor needs to conduct before performing a major surgery on a patient with diabetes. This also implies that those solving structured problems have awareness and experience with the type of information that is required and which alternatives are available or decisions needs to be made. The awareness can be informed by knowledge, training and experience; training often concerns the rules, procedures and policies that inform the decisions and solutions in addition to providing 'simulations' for this type of decisions. For example, fire drills will aid stewards and emergency services in applying the rules and procedures to a simulated event; in addition, these drills will allow evaluating the effectiveness of the rules and procedures in the case of specific events. This means that those involved in the decision and problem solving have acquired knowledge, training and experience about which content of information leads to which decision in the case of programmed decisions.

In the case of applying this type of decision making to organisations, it is often noted that programmed decisions are taken at the lowest level of the organisational hierarchy. Whereas it might be true that they are found more often at those lower levels of management, higher levels in the hierarchy are also involved in taking programmed decisions. An example is the decision to increase or decrease capacity of a plant; sales and marketing information is conjoined with information about supply chains and capabilities of the manufacturing units to inform these decisions. Thus, programmed decisions are found at all levels of management and work in an organisation, though the scope and impact of these decisions may very.

Non-Programmed Decisions

Whereas programmed decisions rely on the known effects and impact of defined solutions based on pre-determined information and sources of information, non-programmed decisions aim at developing (novel) solutions that meet the demands of a unique situation presented as an unstructured problem with unstructured information sources. Examples of these non-programmed decisions are: the solving of vibrations in a bridge caused by winds, the reduction in delays in a rail network and the improvement of the performance of organisations. These unstructured problems are also called ill-defined problems or in specific cases wicked problems (for a pivotal writing on the latter type of problems, see Rittel and Webber [1973]). Wicked problems are those problems that are difficult or impossible to solve because of incomplete, contradictory, and changing requirements that are often difficult to recognise and, moreover, may involve stakeholders with differing views on the problem and its resolution. A case in point for wicked problems is global warming. Despite that the origins of global warning are not fully known, it is clear that mankind has been contributing to this phenomenon.

Because there is a broad range of factors, including growth of populations, deforestation, increased reliance on carbon fuels and excessive consumption, it is difficult to establish how to counter the looming effects. Moreover, there are many actors involved having their own stakes; among them oil and gas companies, governments of economies (particularly, ones reliant on traditional fossil fuels), advocates of alternative sources for energy and citizens. These actors have their own perspective on how global warming should be addressed resulting in a plethora of information, reports and solutions (see Section 11.4 on the concept of 'boundary critique', which aims at reconciling these different perspectives). The opacity about what the problem actually constitutes, which information is relevant and what stakeholders' influences direct the resolution makes that non-programmed decision making is often riddled with ambiguities and information deficiencies.

Because of these unique features, problems as non-programmed decisions can be resolved in different ways. Some may say that these often ill-structured, one-shot decisions can be handled by techniques such as instantaneous judgment, intuition and creativity. However, these problems can also be solved by heuristic approaches in which logic, common sense and a systematic approach to trial-and-error are used. Such an approach for non-programmed decision making related to systems thinking is presented in Section 4.2 ff. Characteristic for approaches based on heuristics is that they separate the analysis of the context and the problem situation from the selection and implementation of the solution. This is because just relying on the manifestation of the problem could result in firefighting, which in turn may lead to side effects that again trigger a cycle of firefighting. Consider a manufacturing plant; if its output decreases with demand across periods being relatively constant, one could easily suggest to introduce overtime. However, overtime is costly and may thus reduce the productivity of the plant, while perhaps the underlying problem is poor quality control (such phenomena actually happened when Japanese manufacturing facilities became more competitive during the 1970s and 1980s). Thus, figuring out what constitutes the core of the problem that needs to be resolved is an important feature of non-programmed decision making.

Within organisations, these non-programming decisions occur when dealing with the implementation of strategies, the analysis of processes in organisations, the design of organisational structures and the design of products and services. That means that even though this type of decisions are commonly faced by management levels higher in the hierarchy, many within organisations may be confronted with the detailing of solutions, their implementation, and their effects on the organisation and the interaction with the environment; even when they were not involved in the analysis and the selection of the solution. Training focused on decision-making should assist those to think through problems using a logical, non-programmed approach. In this way decision makers and actors in the decision-making process can acquire the skills how to deal with extraordinary, unexpected, and unique problems as complement to instantaneous judgment, intuition and creativity.

Decisions in Crises

An other type of decision making, so-called decision making in crises, may be necessary when unexpected problems occur that can lead to unwanted situations or disasters if not resolved quickly and appropriately. This leads to three features of such problem solving and decision making. The first is that these decisions are made based on incomplete and inaccurate information, hence decision making under the type of uncertainty implies that the effects of the solutions are not fully known; this can be associated with abductive reasoning (Section 3.3) in terms of the most likely cause to be identified based on the information at hand. It may also mean that insight will change over time with more relevant information and information with a higher degree or accuracy becoming available, sometimes even after the decision is taken. Second, the lack of analysis is for part caused by the incomplete and inaccurate information and for part caused by insufficient time for a thorough investigation at the time that a decision has to be taken. The constraint of time, often calling for an immediate solution, is the third feature. This may lead to solutions being chosen to avert the crisis, but these are not necessarily sustainable solutions on the long run. In addition, the unknown impact of solutions (due to the state of available information and the lack of analysis) may also trigger a trial-and-error strategy. In such instances a solution is tried and when it does not work an other solution is implemented. However, the successive trialling out of solutions may also increase the urgency, for which the solutions were aimed at reducing it. Thus, decision making in situations of crises will lead to search strategies for the most effective solution by implementing them rather than knowing on beforehand that the crisis situation will be reduced or averted; this is caused by insufficient time for acquisition of all relevant information and for analysis of the situation.

Decision making in crises differs from the cases that an emergency falls under the category of programmed decisions or that a wicked problem presents itself in terms of non-programmed decisions. For the first category of programmed decisions sometimes what we denote as emergencies are factually programmed decisions. The dispatching of an ambulance following a distress call is an example; if it was indicated that the person called for is having a heart attack a standard protocol is in place to ensure treatment in a swift and effective manner. Although the individual event may present itself as a crisis, the decision making and the solutions are programmed. For the second category of decision making, non-programmed decisions, the difference with decision making in crises is the lack of adequate information and analysis in addition to a tight constraint in time. The difference between a wicked problem and decision making in crisis is that in the case of problems being wicked the requirements are often incomplete, contradictory and changing, and that stakeholders who are involved have differing, and possibly diverting, views on the problem and its resolution. Because of the vagueness and contradictions surrounding them, wicked problems may go by unresolved and, eventually, turn into a crisis. A case in point is the state of the British rail transport system. After the privatisation in 1993–1994 of British Rail, the investments in improvements and trains fell behind and was incoherent; in 2016 and 2017 it was noted that the performance of rail transport providers did not match expectations, that the rail network was insufficiently for the demand and that the age of the trains was one of the oldest in Europe. Hence, a solution, which was advocated from the perspective of particular stakeholders, eventually caused crises that had to be addressed, but also led to unwanted effects; in the case of British Rail the decision to privatise resulted later in failures, delays, inadequate investments and even continuous efforts to restructure the governance. Thus, wicked problems, as one form of non-programmed decision making, can turn into crisis if not resolved and in such cases, sometimes, a (continuous) state of firefighting if the routines and procedures of programmed decision making are insufficient.

4.2   Problem Analysis

In the case of non-programmed decision making the key to successful resolving is analysis of the problem. Such analysis should lead to defining root causes or which matters need to be resolved. First, the search for root causes may be necessary because it can be that programmed decisions are becoming less effective. For example, for a specific customer complaint a defined routine how to act is available but the organisation has been increasingly experiencing these complaints. Another reason to start unprogrammed decision making could be that programmed decision making combined with other information leads to reconsidering the current situation. For example, a tachogenerator for speed control of the drum in washing machines might cause recurrent problems, while an improved version or an alternative technology has become available. Another type of unstructured problem solving is when a future state of a system has to be defined. A case in point is the impact of information and communication technology on the provision of information to travellers. For all these three scenarios, the analysis as a first step of non-programmed decision making starts with defining the problem to find out what needs to be resolved (in many cases, the root cause or causes).

Problem Definition

Based on the concepts of Applied Systems Theory, the first element of a problem definition is the delineation of the system(s) or subsystem(s) that is (are) investigated and the determination which aspect (or aspects) need to be looked at. An example of this demarcation about which subsystem and aspectsystems to be considered is the distribution of parcels from a company to its customers. This part of the particular problem definition concentrates on the resources as system for the aspect of logistics. However, such delineation also implies the aggregation stratum that is being considered. In this case, the aggregation stratum is set at the level of the resources for logistics and not at the company as a set of aggregate resources. Thus, a problem definition should state which system or subsystem and which aspects are considered; the demarcation of the system or subsystem implies also an aggregation stratum.

Furthermore, a comprehensive problem definition contains at least one criterion (or requirement). Using the same example from the previous paragraph about a logistics system what to look for depends on the criterion being used in the first instance; it could be about the number of complaints about the wrong parcel being delivered or about the reliability of delivery (or even a combination of both). In addition, a criterion also implies a target state for a specific aspect of a system. In the case of the distribution of parcels from a company to its customers, the performance objective for number of complaints might state that less than 1% of complaints about delivery per total orders should be achieved and that the reliability of delivery is set to be at least 97%. If the actual performance is 2.5% complaints per total orders and the actual reliability of delivery is 92%, then such justifies an investigation. Often there are signals of weakness; these signals are indications that the performance of a system is weakening gradually but individually these indicators are not a direct cause for concern. In the example it could have been that previously the performance objectives were met. At a certain moment, it should have been noted that the number of complaints are increasing and that the reliability of delivery has been slipping but not to the extent that an enquiry should be made; in such a case, both indicators are to be seen as signals of weakness for an investigation. An additional criterion could be a timeline; this could include the time by which a solution has to be implemented. The timeline could be imposed externally, for instance, those set by regulators in specific industries, or derived from internal requirements, for example by management. This means that a problem definition not only refers to which subsystem and aspectsystem are being considered, but also states at least one criterion that is not met or signals of weakness, and it could include a timeline for resolution

The requirements can be evaluated by looking at a higher aggregation stratum and how subsystems are interconnected; this is symbolically depicted in Figure 4.1. Since a subsystem is part of a whole, this higher level imposes requirements on the subsystem under consideration. In the case of the logistic system for the delivery of parcels to customers, the performance objectives for may have been derived from a competitive strategy. In the case of a technical system, the requirements for subsystems are likely derived from the performance requirements and functions of the total system. To arrive at a more complete picture, also connections between subsystems should be considered. For the case of the logistic system, that could be the production system and the order processing system. Thus, requirements for a subsystem

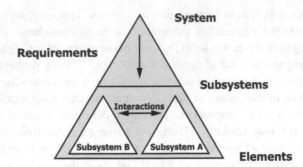

Figure 4.1 *Symbolic representation of the use of aggregation strata for defining problems (derived from Figure 3.3). The higher level of aggregation sets requirements for subsystems A and B. When subsystem A is being investigated, also its interrelationships with subsystem B determine how it should function.*

can be derived from both higher levels of aggregation and interrelationships with other subsystems relevant to the aspect being considered.

An alternative way of defining a problem is found in the approach called Kepner-Tregoe problem solving and decision making [Kepner and Tregoe, 1981]. For example, an initial problem definition states: 'the server crashed.' In terms of this method by Kepner and Tregoe, a more adequate problem definition should include more information on the object, purpose, focus, environment and viewpoint. Such should result in an unambiguous and easily understood statement. In this case, a revised problem definition might be: 'the e-mail system crashed after the support engineer applied hot-fix ABC to Exchange Server 321.' Related to this method by Kepner and Tregoe is the so-called '5 whys technique'. In this technique five questions are asked to define the problem:

- Who is experiencing the problem?
- Why is this important, why is this being done?
- What are the effects, symptoms, errors, defects or something that was not expected to happen?
- When does the problem occur or when did it start happening?
- Where does the problem occur?

This technique with these five questions is used until to the point where there is no explanation for the problem; using it in combination with Kepner-Tregoe problem solving and decision making accelerates the process for defining the problem.

However, it can be that a problem is not always a problem. Take for example an organisation functioning well and new leadership comes in. The new management sets out a new strategy for the company, not necessarily because there was a problem. Hence, a problem situation is created without there being any divergence of a target state. Because such may happen especially when there are different stakeholders and changes in personal views and objectives, a complementary stakeholder analysis (see Sections

8.6 and 10.4) will be helpful to identify if the problem is caused by actual facts, perceptions of the problem or changed perspectives.

Analysing Problems

After defining the problem as first step of solving problems the analysis as second step can start; see Figure 4.2 for the cycle of analysis. With the problem definition established, modelling of the current situation or the future situation comes first in the cycle of analysis. The model used for the analysis can be covering a single theory, methodology, conceptualisation and method or can be an amalgamation of theories, methodologies, conceptualisations and methods; for generic deliberations on modelling, see Section 3.4. Notably, for the purpose of the analysis, the modelling sets out which data needs to be collected. Consequently, the analysis is based on the evaluation of the collected data against the criterion (or requirements). This should lead to conjectures. These conjectures are set off against the problem definition and the cycle of analysis starts again based on a redefinition of the problem.

This cycle of analysis halts when a root cause has been found. Allegedly, the term 'root cause' appeared for the first time in an article in *The Lancet* [1905, p. 1507] about the payment structure and related working condition of colliery surgeons. Commonly, the root cause describes the depth in the causal chain of causes and effects where an intervention could reasonably be implemented to improve performance or prevent an undesirable outcome. In other words, causes are sought until a particular cause will be reaffirmed by further evidence and, therefore, further analysis will not add more insight to what actually causes the problem stated in the problem definition; hence, searching for other causes will not make sense anymore.

In addition to systems theories, there are other methods that are frequently used. An example is the fishbone diagram, aka cause-and-effect diagram and Ishikawa diagram, see Figure 4.3. Fishbone diagrams are typically worked

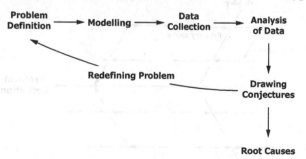

Figure 4.2 *Cycle for problem analysis. The cycle of analysis starts with the modelling after the problem definition. Consequently, which data needs to be collected is determined by the content of the model. The analysis itself evaluates the data against the criteria set out by the problem definition. Based on the conjectures it can be determined whether root causes have been found; if not, the problem needs to be redefined and the cycle of analysis is followed again.*

right to left, with each large 'bone' of the fish branching out to include smaller bones containing more detail. The left side of the diagram is where the causes are listed. The causes are broken out into major cause categories. The causes identified will be placed in the appropriate cause categories as the diagram is built. The right side of the diagram lists the effect. The effect is written as the problem definition for which a solution is sought. It should be noted that this method may result in listing many causes to which then efforts are devoted to eliminate the enumerated causes to find the root cause; this allocation of resources can be avoided by following a model-driven approach, see previous paragraphs. Another well-known method is the Pareto analysis. It is named after Vilfredo Pareto (1848–1923), an Italian economist-sociologist, who observed that 80% of Italy's wealth belonged to only 20% of the population. This law, now known under a variety of labels such as Pareto's principle, the 80-20 rule and the principle of imbalance, states that the majority of effects is related to a minority of causes. This notion is now commonly used to identify the top portion of causes that need to be addressed to resolve the majority of problems; see Figure 4.4. However, it can be limited by its exclusion of possibly important problems that may be small initially, but which will grow in significance with time. A third method is fault tree analysis, a top-down, deductive failure analysis in which an undesired state of a system is analysed using Boolean logic to combine a series of lower-level events. Fault tree analysis focuses on identifying root causes for failures of systems, subsystems and elements; see Section 3.3 for deductive reasoning and systems. This method is used in many disciplines, but mainly in the domains of safety engineering and reliability engineering to understand how systems can fail, to identify the best ways to reduce risk and to determine event rates of a safety accident or a particular system level (functional) failure. It is used in the aerospace, nuclear power, chemical and process, pharmaceutical, petrochemical and other high-hazard industries;

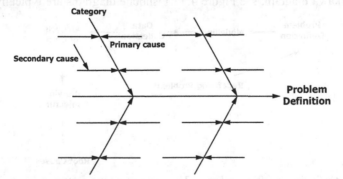

Figure 4.3 *Fishbone diagram. For finding root causes, categories are devised to which potential causes are attributed. These causes are divided into primary causes, secondary causes, etc. By obtaining information about the relevance of each of the causes it is possible to pinpoint relationships between the causes and to find underpinning root causes.*

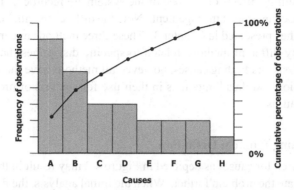

Figure 4.4 *Generic example of Pareto analysis. Causes are found on the horizontal axis and presented as a histogram in order of frequency of observations. Typically, the frequency of observations is listed on the left vertical axis and the cumulative percentage on the right vertical axis. In this case, causes A and B are the majority of observations.*

but it can also be applied in other domains, such as social service systems and project management, for the identification of risk factors. The fault-tree analysis is also used in software engineering for debugging and is closely related to cause-elimination techniques that are used to detect software bugs. Thus, fault-tree analysis is a principle method for design and engineering of products and services and for other applications of risk management, too, but it can also be applied to analyse performance gaps or failures of systems. The fishbone diagrams, Pareto analysis and fault-tree analysis are three examples of generic methods that can be applied in combination with systems theories, on their own or in any combination.

In addition to three generic methods for the analysis of problems, there are methods for specific domains. For example, in strategic management a structured planning method for identifying strengths, weaknesses, opportunities and threats is for projects or business ventures; this type of analysis, mostly known under its acronym SWOT, can be carried out for a company, product or industry. It involves specifying the objective of the business venture or project and discerning the internal and external factors that are favourable and unfavourable to achieve that objective. In addition to this method, in strategic management often a framework is used to analyse and monitor the macro-environmental factors: political, economic, social, technological, environmental and legal; it is mostly known under its acronym PESTEL. Very differently, for design and engineering of products and services failure mode and effects analysis (more commonly known by its abbreviation FMEA) is a frequently used method. It was developed by reliability engineers in the late 1950s to study problems that might arise from malfunctions of systems. It involves reviewing as many components, assemblies and subsystems as possible to identify failure modes, and the causes and effects of these modes. For each element the failure modes and their resulting effects on the rest of the system are recorded. This technique is also used in project management. Note the difference with holistic systems approaches presented in Chapter 3. These three methods underline that there are many different methods related to specific disciplines that can be used for analysis and finding causes; however, all methods have their own specific application and thus limitations in their use for analysis of problems to find (root) causes.

Redefining Problem Definition

This search for causes as depicted in Figure 4.2 may result in the necessity to investigate the problem further. When the initial analysis, the first cycle, does not pinpoint a root cause, the cycle of analysis has to be performed again. This is only possible by redefining the problem definition. This redefinition may concern the subsystem to be investigated, the aspects that are being looked at, the aggregation stratum for which models are developed and data collected, the criteria (requirements) or sometimes the signals of weakness. Such redefinition of the problem is the starting point of a new analysis, which may end up with looking at other theories, conceptualisations, methods and tools to pinpoint causes. The redefining of the problem is not necessary when the root cause has been found and thus the cycles of analysis will have to be repeated until a root cause is found.

When engaging in successive cycles of analysis it is important to avoid drifting and to avoid narrowing down to trivial problems. Drifting means that successive cycles of analysis tend to consider different aspects, different subsystems and different criteria than set out by the original problem definition; it can be that this drifting has little to do with the problem being solved. When the analysis is narrowing down it means that it is concentrating on marginal causes; these marginal causes only contribute little to the original problem definition, resulting in that nothing really is resolved. Drifting and narrowing down can be avoided by focusing on the original problem definition during the consecutive cycles of analysis and by reviewing the chain of evidence created by the analysis.

4.3   Finding and Weighing Alternatives

Taking the root cause as starting point, the next steps of problem analysis and solving are finding and selecting alternatives and principle solutions that solve or mitigate the root cause. There are different techniques and methods for doing so and some of these cover both the generation and selection of alternatives; when discussing a particular method, it has been attributed to the step of problem solving where it is most likely used or where it has most weight.

Generating Alternatives and Principle Solutions

The first approach to generating alternatives and principle solution is analytical in nature and based on extant knowledge. Sometimes the alternatives logically arrive through the root cause and then it is not necessary to probe for a range of alternatives related to this cause. In other cases it may be that looking into extant knowledge is helpful for finding alternatives, which solutions, methods and tools have been put forward already and what is known about their effectiveness in relation to the root cause. An other source of alternatives might be existing solutions that already have been implemented or considered before. In the first instance, when solving problems, one would look for principle solutions; these are conceptual solutions that substantially differ from each other. The example of going from one place to another by a person may clarify the generation of principle solutions. For a person going somewhere there are many principle solutions: walking, biking, motorised transport (mopeds, scooters, motorbikes, cars, vehicles), motorised transport provided by others (taxis, bus, trams, subways, trains), maritime transport (rowing boats, surfing, ferries, ships) and aeronautic transport (balloons, sailplanes, helicopters, planes, rockets); the more detailed solutions in parentheses are the manifestations of the principle solutions (see also value engineering and the controlled convergence method in this section and Section 4.5 on detailing principle solutions and alternatives). This non-exhaustive list of principle solutions shows how very different they are. However, this list can also be divided into sources of energy for transport (note that transport is a mechanical motion in the first place): human energy, gravity, electricity, fossil energy, wave energy and solar energy. The appropriate categorisation of the principle solutions depends on the problem definition. In this case of transport for a person, the question is whether the problem definition was directed at the transport of one person or at reducing the environmental impact; both problem definitions have a different starting point for generating and evaluating principle solutions. This approach of finding principle solutions and generating alternatives based on extant knowledge is also called 'vertical thinking'; this path to generating alternatives for decision making later relies on the logic of the root cause, and depends on the knowledge base of those involved and accessibility of information about the proposed solutions.

A second path to developing principle solutions is the theory of inventive problem solving, mostly known by its acronym TRIZ; it is a tool developed by Genrich Altshuller that aims at solving problems in engineering, particularly those with seemingly contradicting requirements. He has written about this in a number of books, including some that are written in novel-style [e.g. Altshuller, 1996]. The tools are based on solutions that have already been used successfully before. At the heart of the methods are 40 inventive principles for solving contradictions rather than seeking a compromise or seeking a trade-off; to this purpose a contradictions matrix has been developed. This matrix has been derived from known and patented solutions; it lists 39 factors that could impact negatively on each other and for each impact there are a number of inventive principles, usually three or four out of the 40, that can be used to resolve it. Thus, the tool is based on analogous solutions for a problem. For more complex problems, a tool called ARIZ (algorithm of inventive problem solving) has been developed, consisting out of 85 step-by-step procedures to do so. Some companies, among them Samsung, have adopted this tool and use it throughout the organisation to support solving technological and organisational problems. The use of TRIZ, and some its complementary tools, such as ARIZ, leads to principle and, sometimes, innovative solutions.

A third path to developing alternatives is brainstorming. Although this method might refer to activities of an individual, originally it is a group creativity technique by which efforts are made to find a conclusion for a specific problem by gathering a list of ideas spontaneously contributed by its members. Whereas there are many variations for this technique, allegedly, its roots go back to advertising executive Alex F. Osborn who began developing methods for creative problem-solving in 1939; he outlined the method in a book chapter [Osborn, 1948, Ch. 33]. At present there are many variations of the method and a few are mentioned here:

- Free association. This is the most obvious brainstorming technique. The key to this method is thinking freely and without judging thoughts, while maintaining an association to the problem. An effective way to do this is to think quickly. A popular method is to make flash cards. Participants create a quick thumbnail sketch or write down a word on a card, next they turn the card over so they are not distracted by it, then move on to the next card and repeat the process. This 'out-of-sight-out-of-mind' approach to brainstorming allows continuously wiping the slate of thoughts clean to generate ideas.

- Team idea mapping method. This method of brainstorming works by the method of free association. It may improve collaboration and increase the quantity of ideas, and is designed so that all attendees participate and no ideas are rejected. The process begins with a well-defined topic. Each participant brainstorms individually, then all the ideas are merged onto one large idea map. During this consolidation phase, participants may discover a common understanding as they share the meanings behind their ideas. During this sharing, new ideas may arise by association, and they are added to the map as well. Once all ideas are captured, the group can prioritise and take action, where appropriate.

- Lateral thinking. The term 'lateral thinking' was introduced by de Bono [1967] and has become one of the well-known methods; his method for creativity is based on using six hats, each representing a perspective on the problem and solutions:
  - White hat: neutral information. This perspective focuses on collecting facts and information needed.

- Red hat: emotions and hunches. This hat entails the perspective of uncovering emotions and feelings and sharing fears, likes, dislikes, loves and hates.
- Black hat: judging and evaluating. This angle focuses on being the devil's advocate or why something may not work. It spots the difficulties, challenges and failures.
- Yellow hat: optimism and positive views. This viewpoint explores the positives, and probes for value and benefit.
- Green hat: ideas and creativity. This role provides possibilities, alternatives and new ideas as an opportunity to express new concepts and new perceptions.
- Blue hat: big picture and control. This position manages the thinking process, the generation of ideas and their evaluation.

The hats are used in different stages and by different participants. There are also other variations for lateral thinking, such as imagining that the problem for which a solution is sought behaves like an 'animal'.

- Nominal group technique (originators: Delbecq and Van de Ven [1971]). In this method, the participants are asked to write their ideas anonymously. As a next step the facilitator collects the ideas and the group votes on each idea; this process is called distillation. After distillation, the top ranked ideas may be sent back to the group or to subgroups for further brainstorming. For example, one group may work on the colour required in a product. Another group may work on the size, and so forth. Each group will come back to the whole group for ranking the listed ideas. Sometimes ideas that were previously dropped may be brought forward again once the group has re-evaluated the ideas.
- Group passing technique (aka brainwriting [Geschka et al., 1976, pp. 49–50]). Each person in a circular group writes down one idea and then passes the piece of paper to the next person, who adds some thoughts. This continues until everybody gets his or her original piece of paper back. By this time, it is likely that the group will have extensively elaborated on each idea. The group may also create an 'idea book' and post a distribution list or routing slip to the front of the book. On the first page is a description of the problem. The first person to receive the book lists his or her ideas and then routes the book to the next person on the distribution list. The second person can log new ideas or add to the ideas of the previous person. This continues until the distribution list is exhausted. A follow-up or 'read out' meeting is then held to discuss the ideas logged in the book. This technique takes longer, but it allows individuals time to think deeply about the problem.

Often brainstorming is set off against vertical thinking, taken as arriving at logic and analytical solutions, because of finding novel and innovative alternatives through more creative processes.

Weighing Alternatives

Already captured by some of the techniques for generating alternatives and principle solutions, weighting or prioritising them is the next step. Note should be taken of the following. Whereas analysis is aiming on finding the root cause, or exceptionally root causes, and for that reason focuses on specific aspects (and perhaps subsystems), the weighing of generated alternatives takes normally integral aspects and criteria into account. This means that very different from the stage of analysis, more criteria, requirements and aspects have to be taken into account. For example, take an ecosystem in which a species is introduced as an intervention to increase biodiversity. This new species may use food resources that were at the disposal of incumbent species and thus compete for these resources. Ultimately, this may cause incumbent species to extinction. Therefore, if the intervention does not take into account the availability of food resources for all species, including the newly introduced one, it may have an adverse effect; if the cause was lack of biodiversity, then the intervention may seem effective but not when evaluated against multiple criteria. This example of an ecosystem shows that the phase of weighing alternatives is different from the stage of analysis by the number of criteria, requirements and aspects considered.

In this spirit, value engineering is a systematic method to improve the value of goods and services by considering its functions and the use of resources for its functions, usually expressed in costs. Value engineering dates back to the World War II, when Harry Ehrlicher, Jerry Leftow and Lawrence Miles developed its methods, due to shortages of skilled labour, raw materials and components. Thus, alternatives could lead to reduced costs and improved products. To this purpose, value engineering identifies the function(s) and evaluates alternatives for this function; see Figure 4.5. First, the function or functions of the product or service need to be identified, which is the input for generating principle solutions (see Section 5.4 for a more detailed explanation of the term function). This set of solutions is compared against constraints. Take for example, a student on shoe-string budget. The first function of transport it to get as close as possible to the destination; a second

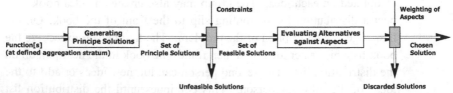

Figure 4.5 *Process for value engineering. After generating principle solutions, there are measured against constraints; those solutions that are not meeting the constraints are considered unfeasible. After this first step of selection, the remaining set of feasible solutions are then weighed on different aspects, criteria or requirements before one alternative is chosen; normally this is the aspect that ranks highest. Typically for value engineering is to evaluate the function (or functions) against utilisation of resources.*

function will be the protection against weather conditions. In this case, the constraints are cost and time. A student constrained by budget will not able to afford a helicopter flight to go to the university a few kilometres away, even though it is one of the possible principle solutions for transportation from the dormitory to the university. The example shows that set of solutions is evaluated on relevant aspects with regard to feasibility. The final step is weighting of functions and prioritising the alternative, ideally resulting in chosen solution; this weighting is typically done by setting of the value of the function against the use of resources. In the case of the student's transport to the university, this is the effectiveness of the functions transport and protection versus the cost. For example, using a bike might get a student closer to the teaching rooms but might be less comfortable under certain weather conditions. And the expense of a bike is less than the cost of a bus journey; this also contingent on the period and intensity of use. Depending on the weighting of the function protection (e.g. against weather conditions), the bus might be the preferred option or the bike. Hence, value engineering allows maximising the value for the customer (or user) in terms of fulfilment of function(s) against the resources used (most often taking the form of costs).

Another method that does not reach a decision right away is dialectic decision making, aka the Socrates method. Perhaps more applicable in the social sciences, it aims at generating two or more competing proposals. For each of the proposals, the underlying assumptions are identified, and the advantages and the disadvantages are determined. Consequently, the decision can be that one of the two alternatives is chosen or a new proposal is generated or a compromise is forged. For the latter, Dekkers et al. [2013, p. 330] observe that in practice that 'managerialism' seems to override the opinion of experts; especially in situations where management methodologies or governance, such as staged decision making, are introduced, that appears to be more likely to happen. This particularly the case for infrastructure, product and service development, in which technological considerations often leave less space for compromising. But possibly, this extends to all kinds of decisions that transcend disciplinary boundaries; the weighing of aspects, which relates to the subjectivity of the observers, see also Section 2.4 for this discussion. Thus, the approach of dialectic decision making does not necessarily lead to an optimal decision, but may be helpful to work with stakeholders.

[Pugh's] Controlled Convergence Method

Selecting solutions also appears in the controlled convergence method, originated by Pugh [1981]; later it was popularised by others, such as Ward et al. [1995], under the label set-based concurrent engineering. Pugh's controlled convergence method is based on the subsequent narrowing down of alternatives to a selected solution while at the same time detailing it, see Figure 4.6 for a symbolic overview. At each stage progress of concepts and design are set-off against criteria and requirements; with progressive

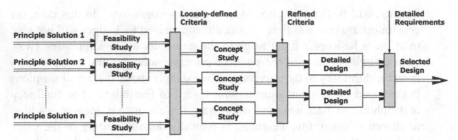

Figure 4.6 *Principles of the controlled convergence method. First, principle solutions are generated. Then these principle solutions are evaluated against loosely-defined criteria. During next stages selection of sets of feasible solutions are further developed, while the criteria are becoming more refined and detailed. This process is continued until a final design of a solution emerges.*

insight these criteria become more detailed, too. The advantage of this method is that not an early selection of a specific solution or concept leads to a lock-in that will cause problems downstream for the implementation. However, the point-based method is most commonly used for new product and service development. This is because the disadvantage of the controlled convergence method is that during early stages of this method 'projects' run in parallel, drawing on resources, whereas in the point-based method there is only one project. Concentrating on essential challenges for each concept considered in parallel rather than trying to do everything for every 'parallel project' can partly circumvent this disadvantage. However, it is claimed that the controlled convergence method for developing solutions improves the integrity of the solution, improves the implementation and reduces the lead-time from the development of principle solutions and concepts to the detailed solution being in use.

4.4   Decision Making

After weighing the criteria the next step in problem solving is making the decision which alternative or principle solution to take forward for implementation. The premise for this section is that more than one alternative is available; in the case that only one alternative is at hand, then it still needs to be checked whether it meets all criteria. The method of value engineering and the controlled convergence method, discussed in the previous section, already encapsulate the stage of decision making. Thus, in this section, the additional common methods and aspects of decision making are discussed: multiple-criteria decision making, decision trees, satisficing, case-based reasoning, decision making in groups and the Abilene paradox.

Multiple-criteria Decision Making

One of the most common approaches for decision making is multiple-criteria decision making, sometimes called multiple-criteria decision analysis and multi-attribute analysis. This technique is used for comparing options and alternatives for achieving objectives (there is some overlap with the previous section about weighing alternatives and principle solutions). To this purpose the first step is developing objectives that need to be achieved, followed by the generation of alternatives, options and solutions (in that sense, it has similarities to value engineering, see Section 3.3). The alternatives are compared and weighted on criteria derived from the objectives. Subsequently, the option that scores best on all criteria or that achieves the highest overall score is selected. An example is the purchase of a car; such as purchase could be evaluated against purchase price, operational costs (for example, fuel consumption and maintenance costs), comfort, image, transport capacity, etc. It should be noted that in making the decision, there might not only very complex issues involving multiple criteria, but there are also multiple parties who are deeply affected by the consequences and may weigh the criteria differently. Even for the purchase of the car that may apply, when family members have different views on the purpose of a car and the weighting of the criteria. It should be noted that the appraisal of alternatives can be done either quantitatively or qualitatively (or even combined). Hence, multiple-criteria decision making aims at evaluating the options against criteria derived from objectives and allows subjective weighting of each to select the most appropriate solution.

Table 4.1 *Sensitivity analysis of site for potential repositories of nuclear waste based on multiple-criteria decision making. The top rows indicate how each of the scenarios is weighted on the four criteria. The bottom eight rows show the score of each site (i.e. solution). In this case, the scenarios represent different perspectives of stakeholders. (Source: Department for Communities and Local Government [2009, p. 98])*

Weights for different perspectives	Base	Equal	Local	National Environmental	Economic
Cost	100	100	0	0	200
Robustness	20	100	10	20	40
Safety	10	100	50	20	0
Environment	10	100	50	20	10
Overall results					
Dounreay	81	76	60	74	82
Site 2	82	76	56	74	83
Site 3	82	72	57	68	85
Site 6	85	80	68	77	86
Site 7	85	77	66	73	86
Sellafield 8	87	77	71	72	88
Offshore West Shallow	64	60	75	58	63
Offshore West Deep	55	64	83	68	50
Offshore East	16	29	58	36	15

An example of the methods for multiple-criteria decision making and the subjective weighting is found in a manual of the Department for Communities and Local Government [2009, pp. 90–101]. It is about the case of appraising potential sites that could serve as repository for radio-active waste at the end of 1980s. Without going into much detail about the project, Table 4.1 shows the outcome of the evaluation for each site. The top rows show the criteria that inform the appraisal of the potential nine sites; from an originally longer list of sites, only these were considered feasible. The base case represents the weighting of the group directly involved in the preparation of the decision making after being informed by the various actors involved and the collection of data. The equal case was constructed to show the influence of weighting. The three right-hand columns display the perspectives of the local communities, the environmental perspective and the economic perspective. From the table it can be derived that a number of options for sites are very closely positioned to each other, across and within perspectives. Ultimately, that means that such a decision might be based on other criteria than listed or subject to interpretations of data.

Principally, structuring complex problems well and considering multiple criteria explicitly leads to more informed and better decisions. There have been important advances in this field since the rise of the multiple-criteria decision-making in the early 1960s. A variety of approaches and methods, many supported by specialised decision-making software, have been developed for an array of disciplines, ranging from politics and business to the environment and energy. Over the course of time, some other approaches have been added, such the analytic hierarchy process and the use of fuzzy sets. The analytic hierarchy process [Saaty, 1990, 2000] converts subjective assessments of relative importance to a set of overall scores or weights; to that purpose it asks actors involved in the decision making how important one criterion is to another. Note that some serious concerns have been raised about the method and that extensions have been proposed and alternative methods propagated; however, it is beyond the scope of this book to go into more detail. Also, fuzzy sets as algorithm for use in multiple-criteria decision making has received criticism. Because fuzzy sets are based on the membership of a set not being crisp, probability values are assigned. Whereas many methods and their extensions have been proposed, it is not clear whether the application of this mathematical approach will lead to better decision making. Therefore, multiple-criteria decision making is partially subjective, through the evaluation of alternatives on each criteria and the relative weighting of criteria, and it is also sensitive to how it is performed, without or with specialised decision making software.

Decision Trees

Another way for supporting decision making is called the decision tree. This technique uses a flowchart-like structure for depicting alternatives; see Figure 4.7. However, it is mostly used by calculating the impact of branches

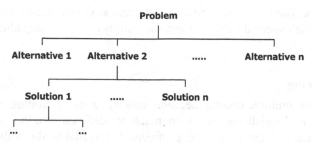

Figure 4.7 *Generic overview of decision trees for depicting alternatives. Principle solutions or alternatives are found at the first level. At the second and subsequent levels these become more detailed.*

and comparing the alternatives in these branches on computed results. Usually the branches are also compared taking into account their chances of appearance, in which each internal node represents an appraisal on an attribute; for example, whether a coin flip comes up heads or tails. This also implies that each branch represents the outcome of a chance event. Figure 4.8 shows an example taken from operations management in which alternatives for the investment in a factory are compared with the status quo (keeping things the way they are is often also an alternative). The example also shows that decision trees tend to focus on a single attribute or measure; in terms of systems theories that means a specific aspect. In practice this means often concentrating on financial aspects. In the example, other criteria, such as flexibility and future capabilities for expansion, are not captured, whereas in practice these other criteria may also determine the decision to be taken.

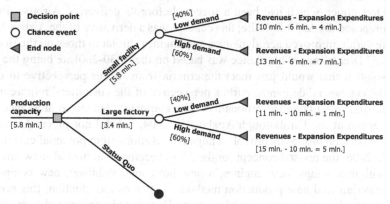

Figure 4.8 *Example of a decision tree. In this case a company has to make the decision to whether to build an additional facility to its current facility for a new product line. For this purpose, it has two options: a large factory and a small facility. It could also decide to keep the status quo, implying not manufacturing the new product line. There is also a chance attached to the development of the market for the new product line: 60% for high demand and 40% for low demand. As can be derived from the decision tree, the option for a small facility is the most advantageous one. (Adapted from Jacobs and Chase [2014, p. 125])*

Thus, decision trees are used as a visual and analytical decision support tool, where the expected values (or expected utility) of competing alternatives are calculated.

Satisficing

Whereas multiple-criteria decision making aims at finding the optimal alternative, satisficing is an approach to decision making or cognitive heuristics that entails searching through the available alternatives until an acceptability threshold is met. The term satisficing, a combination of satisfy and suffice, was introduced by Cyert et al. [1956], although the concept appeared first posited in Simon's [1947] book *Administrative Behavior*. Simon used the concept of satisficing to explain the behaviour of decision makers under circumstances in which an optimal solution cannot be determined; the finding of the optimal solution is called maximising. He pointed out that human beings lack the cognitive capabilities to maximise: rarely all outcomes can be evaluated with sufficient precision, usually the relevant probabilities of outcomes are not known, and humans possess only a limited memory and limited cognition. Simon denoted this take on decision making based on these limitations as bounded rationality. That means that bounded rationality as limitation to decision making may only exacerbate the choice for a non-optimal solution through satisficing; note that satisficing and bounded rationality are related but separate concepts.

An example of satisficing and bounded rationality is the start of the development of the Airbus A350 airplane in the beginning of the 2000s. When airlines pushed Airbus to provide a competitive airplane to the Boeing 787 Dreamliner, which had been a success before its deliveries, Airbus initially proposed the A330-200Lite; this concept was a derivative of the Airbus A330 featuring improved aerodynamics and engines similar to those on the Boeing 787 Dreamliner. This choice was based on the A330-200Lite being the first solution that would just meet the criteria from Airbus perspective to meet the customers' demands; Airbus perspective of the customers' requirements proved to be bounded rationality. The company planned to announce this version at the Farnborough Airshow in 2004, but did not proceed, and the next design proposals provoked negative reactions from potential customers. In 2006, the revised concept for the A350 became an almost all-new aircraft, with new wings, new engines, a new horizontal stabiliser, new composite materials and new production methods. After its introduction, this revised concept became a commercial success. Hence, Airbus managed to avoid the trap of satisficing and bounded rationality, though it can be reasoned that the intervention of customers was necessary to establish this.

Case-Based Reasoning

In making decisions also lessons learned from the past and experience can play a role; in this sense, case-based reasoning is the process of solving new problems based on the solutions for similar past problems. This principle of re-using existing solutions to new problems is apparent in daily life. For example, a car mechanic who repairs an engine by recalling another car that exhibited similar symptoms with its engine is using case-based reasoning. Also, a lawyer who advocates a particular outcome in a trial based on legal precedents or a judge who creates case law is using case-based reasoning. Essential to these examples is that applying case-based reasoning builds on a degree of similarity between the case under review and previous cases, so that the solution is valid for the new problem, too. This means that the concepts of isomorphism and homomorphism (see Section 3.4) and the principles of generalisation (see Section 3.1) should be applied in advance of declaring an existing solution from an old problem (or old problems) valid for a new situation. Hence, this validity of this similarity should be investigated before applying the existing solution to the new problem. To a certain extent, the concept of TRIZ (see Section 4.3) is an example of case-based reasoning. So, too, anyone who copies working elements of nature as a database of solutions to problems; this is also called practicing biomimicry. Thus, case-based reasoning might also be sometimes considered as a kind of drawing analogies to solve problems; see Section 3.4 for analogies. Although case-based reasoning is common practice, those applying it should be well aware of its limitations.

Decision Making in Groups

In addition to structuring decision making in terms of the process followed, it also takes place in groups rather than by individuals. Group decision making (aka collaborative decision making) happens when individuals collectively make a choice from the alternatives before them. The decision is then no longer attributed to any single individual who is a member of the group. This is because all the individuals and social group processes, such as social influence, contribute to the outcome. The decisions made by groups are often different from those made by individuals. Group polarisation is one clear example: groups tend to make decisions that are more extreme than those of its individual members, in the direction of the individual inclinations. Therefore, decisions made in groups may differ from those made by individuals.

There is much debate as to whether this difference results in decisions that are better or worse. According to the idea of synergy, decisions made collectively tend to be more effective than decisions made by a single individual. Groups can better evaluate decisions from different perspectives than individuals. Moreover, decision making in groups allows considering different aspects and disciplines; the use of multi-disciplinary teams in project management are an example of this. However, there are also examples where the decisions made by a group are flawed, such as the infamous decision to invade the Bay of Pigs (Cuba) by the American government in 1961, the incident on which the groupthink model of group decision making is based. Groupthink occurs when a group makes faulty decisions because group pressures lead to a deterioration of mental efficiency, reality testing and moral judgment [Janis, 1972, p. 9]. Groups affected by groupthink tend to ignore alternatives and to take irrational actions that dehumanise other groups. A group is especially vulnerable to groupthink when its members are similar in background, when the group is insulated from outside opinions, and when there are no clear rules for decision making. Thus, group decision making has its merits in terms of perspectives and aspects that may be considered, but could also lead to flaws caused by behaviour in groups.

This means that factors that impact other social group behaviours also affect group decisions. For example, groups high in cohesion, in combination with other antecedent conditions (e.g. ideological homogeneity and insulation from dissenting opinions) have been noted to have a negative effect on group decision-making and hence on group effectiveness. Moreover, when individuals make decisions when part of a group, there is a tendency to exhibit a bias towards discussing shared information (i.e. shared information bias), as opposed to unshared information. Hence, group decision making can lead to limiting views of which information is relevant, particularly when groups tend to display a high degree of cohesion and are insulated from other opinions.

Abilene Paradox

Such bias may lead to what is called the Abilene paradox [Harvey, 1974]. A group of people may find themselves in this paradox when they collectively decide on a course of action that is counter to the preferences of many (or all) of the individuals in the group. It normally involves a common breakdown of group communication in which each member mistakenly believes that their own preferences are counter to the group's and, therefore, each member does not raise any objections. A common phrase relating to the Abilene Paradox is a desire not to 'rock the boat.' This differs from decision making in groups in the sense that the Abilene paradox is characterised by an inability to manage agreement.

4.5   Implementation of Solutions

After an alternative or principle solution has been chosen, the next step is the implementation of the solution; this can be divided into the detailing of the solution and the effectuation of the solution. Note that the controlled convergence method (Section 4.3) extends to the phase of detailing solutions.

Detailing of Solution

The chosen alternative or principle solution needs to be worked out in greater detail before it can be implemented. It is hardly the case that an alternative (or principle solution) is ready for implementation. That means that in terms of aggregation strata zooming in for details, and, at the same time, that the interrelationships with other subsystems and aspectsystems needs to established (see also Figure 4.1). An example of zooming in is the design of the principle solution to use pneumatics instead of hydraulics for the landing gear of the Fokker F-27 aircraft. Because the installation of a pneumatic landing gear differs from a hydraulic one, different design choices had to be made. A case in point are relief valves used in pneumatic systems to prevent damage to pipes and components; they act as pressure limiting units and prevent excessive pressures from bursting lines and blowing out seals but in hydraulic systems have less functions. An example of detailing interrelationships with other subsystems and aspectsystems is the design of interfaces in computer systems. The introduction of a new module for calculating parameters may also mean that the user interface needs to change. Because of visual representation this could even lead to the calculation of additional parameters to be included in the design of the original module that is revised. Consequently, detailing leads to consideration of specifications for both elements of a system (or subsystems) and defining how the solution interacts with other subsystems on a broad range of aspects.

When detailing solutions trade-offs in criteria may be necessary. It could be that the feasibility results in reconsideration to what extent the criteria can be met. An example of trade-off is the size of our brain. Whereas larger brains may be more advantageous, they are also consuming more resources, such as blood flows and energy, make childbirth difficult and are easier damaged. Thus, for a given structure of the body, the size is a compromise. Such trade-offs not only happen in biological systems, but may also happen when designing technological systems, organisational systems and societal systems; although in the latter cases trade-offs are made by the designers of these systems or stakeholders that are involved in the detailing of the solution.

Effectuation of Solution

For a detailed solution to be effective an impact analysis on the total system may be helpful. This is particularly the case for interventions in biological systems, technological systems, organisational systems and societal systems. From a systems perspective, interventions, i.e. the implementation of solutions, results in changes in relationships between subsystems and elements and possibly alterations in the content of the system. The purpose of an impact analysis is comparing the existing state of the system and the future state of the system, i.e. the detailed solution, on a range of aspects. A case in point is the building of a new power plant; not only will a business case be made and the technological feasibility scrutinised, but also the environmental impact of the construction and operation of such a plant will be considered under normal circumstances. The concerns raised will then be continuously monitored, possibly even after its commissioning. Thus, an impact analysis normally highlights concerns for the actual implementation of a solution, moving from one state of a system to the next state.

Also, the impact analysis of the gap between the current and future state of a system implies that possibly humans need to be trained in the case of implementation of solutions; this may be the case for ecological, organisational, societal and technical systems. The skills and the capabilities required for the new state as an intervention may differ from those in a current state. Thus, training prepares human as elements of a system or as the environment of a system for dealing with new relationships in a system and possibly new contents.

The introduction of new relationships and possibly new contents may also mean that so-called 'teething problems' occur. These teething problems are deficiencies that were not foreseen when choosing and detailing the solution. However, these need to be resolved in order for a system to work properly and meet performance criteria. These teething problems may appear in different manifestations and can be resolved in one of the two ways as outlined in Section 4.1: non-programmed decision making and decision making in crises, depending on the impact and the time constraints; however, it should be avoided that trial-and-error strategies lead to a continuous state of firefighting.

4.6 Evaluation of Solutions

After the implementation of a chosen solution, monitoring should take place whether it actually solves the root cause, the gap in performance noted during the problem definition and the signals of weakness for the system. Although much attention can be given to careful analysis, sound decision making and diligent detailing of the solution, the only proof will be 'eating the pudding'. An example where it went wrong all the way is the introduction of a new business model for a range of supermarket chains under the corporation called Laurus [de Hoo, forthcoming]. The new concept was to introduce one brand name instead of the many supermarket chains it owned and one distribution organisation instead one for each supermarket chain; the transformation should have resulted in structure of the company identical to its main rival. Separate from a flawed analysis on which the decision took place, signals about weak performance of the new concept were ignored. Eventually, this lead to bankruptcy, supermarket chains were sold off and the corporation became defunct. Therefore, during and after the implementation of a solution in a system, monitoring should address to what extent the original problem definition (or altered in case of and incorrect starting point for problem analysis) has been resolved and to what extent the implemented solution is mitigating effects noted.

In addition to recording whether the original problem definition and the root cause are resolved, the implemented alternative should be monitored on integral performance criteria. Because typically, an alternative is generated through addressing a root cause, the performance at other criteria is only assessed when making the decision. But as much as it is not sure whether the implemented solution addresses sufficiently the problem definition for the system, it is also not certain what effects an intervention has on all relevant aspects and criteria. Such a dilemma can be observed when addressing some solutions to environmental pollution; the capturing of solar energy can only be achieved by devices for which the production itself may cause environmental impact. Thus, ideally, the effects of solutions should be monitored on a range of aspects and criteria.

Further points to monitor during the implementation of an intervention arise from the impact analysis, mentioned in Section 4.5. The impact analysis highlights concerns that arise moving from one state to another for a system, particularly with regard to the resources as a system. An example is the training of operators for an information system. In this case, the number of operators that have received training, whether this training is effective and how much effort is need for the training are points to be tracked. Thus, an impact analysis not only sets out how the implementation of a solution or intervention needs to be prepared, it also indicates what points should be monitored during the implementation in addition to aspects and performance criteria.

4.7 Overview of Process for Problem Solving and Decision Making

Figure 4.9 presents an overview of all the steps of problem analysis and solving as a structured approach to non-programmed decision making. The graphical summary shows nine steps:

- Defining a problem in terms of subsystem, aspects, aggregation stratum, criteria, performance gap or signals of weakness (Section 4.2).
- Analysing a problem using models (Section 3.4) and supported by techniques, such as Pareto analysis and fishbone diagrams (Section 4.2).
- Defining a root cause: the initiating cause of either a condition or a causal chain that leads to the original problem definition (Section 4.2).
- Generating alternatives or principle solutions that eliminate or mitigate the root cause (Section 4.3).
- Weighing alternatives or principle solutions with the purpose of comparing them on a set of aspects and criteria (Section 4.3).
- Selecting an alternative or principle solution that is the most suitable, as a decision making process (Section 4.4).
- Detailing the chosen alternative or principle solution to fit it with the contingencies of its implementation and use (Section 4.5).
- Implementing the detailed alternative or solution so that it can be used (Section 4.5).
- Evaluating the implemented solution to find out whether the problem has been resolved actually and if further improvements are necessary (Section 4.6).

Note that Figure 4.9 contains three iterations. The first iteration that can happen is when the analysis does not yield any root causes; in such a case a redefinition of the problem is necessary. If the lack of performance or the signals of weakness still persists, this should be addressed by either zooming out or considering different aspects. Second, if no solution for a root cause can be found, then again this should normally be addressed by either zooming out or considering different aspects. Third, if the implemented solution is not bridging the lack of performance or the signals of weakness do not disappear or unwanted side-effects emerge, then either a different solution needs to be chosen or the problem needs to be redefined. These three iterations, particularly the last one, suggest that problem analysis and solving may be a continuous process rather than a one-time off for non-programmed decisions.

4.8   Some Further Notes

This chapter presented a rational view on decision making for non-programmed decisions; this view might be even valid for decision making in crises to a certain extent. To some it will be an extension of the well-known problem solving technique of Dewey [1910] called reflective thinking. It consists of six steps, similar to the ones introduced in this chapter. There also some opposing notions on decision making. The first one is intuitive decision making. This type of decision-making is based on implicit knowledge relayed to the conscious mind at the point of decision through affect, emotion or unconscious cognition. It could be that intuitive decision-making relies more on the mind's parallel processing functions, whereas deliberative decision-making relies more on sequential processing. A second notion is trial-and-error, which is characterised by repeated, varied attempts which are continued until success is achieved or until the agent stops trying. This approach is far more successful with simple problems and in games, and is often resorted to when no apparent rule applies. This does not mean that the approach need be careless, for an individual can be methodical in manipulating the variables in an attempt to sort through possibilities that may result in success. Nevertheless, this method is often used by people who have limited knowledge in the problem area. This means that there are also differing views on decision making, such as intuitive decision making and trial-and-error.

Separate from other methods for decision making, the steps in Figure 4.9 are not complete on three points. First, it is described as a sequential process with some limited iterations; in practice it may possibly be that these stages are overlapping and, hence, that there are more iterations taking place. In that case the approach in this chapter should be seen as a nominal model for decision making rather than an absolute one. Second, for organisations the link to strategy is missing. To that purpose, the breakthrough model (Section 10.3) makes the link with processes of foresight and strategy, though the approach to problem solving is implicitly present in this model. Third, social influence and political decision making have not been considered, albeit they are mentioned here and there in this chapter. Such influences

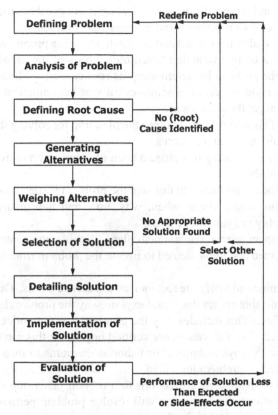

Figure 4.9 *Overview of steps for problem analysis and solving. The first step is defining the problem, after which analysis takes place. The analysis should end in defining the root cause; if not the problem needs to be redefined. After the root cause has been found, the fourth step is generating alternatives and principle solutions. This is followed by weighing of the alternatives. If the weighing of the alternatives and the decision making does not yield a feasible alternative, the problem needs to be redefined. The selected alternative or principle solution is detailed before it is implemented in the eighth step. Finally, after implementation, the performance of the solution and its possible side effects are monitored; if these are unsatisfactory another solution may need to be chosen or the problem redefined for a new cycle of analysis and problem solving.*

are also determined by acceptability of the solution to stakeholders; this has been discarded. Therefore, the model for decision making is incomplete due to the lack of overlapping of stages and the limited description of iterations, the application for specific problems lacks relevant links, such as strategy for organisations, and it does not fully account for the position of stakeholders.

Also, other rational approaches exist that aim at problem analysis and solving. Most notably, the approach of Kepner and Tregoe (Section 4.2) is one of them. An other approach is 'eight disciplines problem solving'. This method was developed at Ford Motor Company to approach and to resolve problems, and is typically employed by engineers or other professionals. The method focuses on product and process improvement, for which its main purpose is to identify, correct and eliminate recurring problems. It establishes a permanent corrective action based on analysis of the problem and on the origin of the problem by determining its root causes. Although it originally comprises eight stages, or 'disciplines', it was later augmented by an initial planning stage; these stages are:

- Plan. This initial stage involves the planning for solving the problem and determining the prerequisites.
- Use a team. During this phase a team of people with relevant knowledge is established.
- Describe the problem. In this step the problem is specified by identifying the who, what, where, when, why, how and how many; this is done preferably in quantifiable terms.
- Develop an interim containment plan. In this stage containment actions are defined and implemented to isolate the problem from anyone affected by it.
- Determine and verify the root causes and escape points. During this phase all applicable causes that could explain why the problem has occurred are identified. This includes why the problem was not noticed at the time it occurred. Then all causes are verified or proved; this can be done by use five the '5 whys technique' or fishbone diagrams to map causes against the effect or problem identified.
- Verify the permanent corrections for problem. This step verifies whether the solutions and correction will resolve problem permanently. It may involve testing before implementation.
- Define and implement the corrective actions. This stage involves decision making about the best course of action and consequently implementing the selected corrective actions.
- Prevent the recurrence and further system problems. During this phase the management systems, operation systems, practices and procedures are modified to prevent recurrence of this and similar problems.
- Congratulate the team. The final step is the recognition of the collaborative efforts of the team involved in the solving of the problem; this may include a formal thanks by the organisation.

This method 'eight disciplines problem solving' has become a standard in particular industries, such as the automotive and assembly industries, that require a structured problem solving process using a team approach. A third approach is the PDCA (plan–do–check–act or plan–do–check–adjust) cycle, which is an iterative four-step management method used in business for the control and continual improvement of processes and products. It is known under many names, such as the Deming circle, Deming cycle, Deming wheel, Shewhart cycle, Shewhart control and plan–do–study–act (PDSA). It is often related to the use of specific tools for analysis that have been mentioned before, such as the fishbone diagram and Pareto analysis (Section 4.3), and other techniques, for instance statistical analysis. These three examples of other approaches differ from the more structured and formal way of solving problems presented in this chapter; particularly they make less use of systems approaches presented in Chapter 3.

4.9   Summary

The generic approach to problem analysis and solving presented in this chapter is based on a rational approach, using basic concepts of systems theories. The approach is directed at non-programmed decisions. Particular notions of the presented approach to problem analysis and solving are:

- The defining of a problem in terms of subsystem(s), aspects, aggregation strata, criteria, performance gaps or signals of weakness.
- The emphasis on analysis that leads to identifying root causes. The view on the analysis is that it should be led by modelling, rooted in concepts of applied systems theory. Among other methods that can support the analysis are Pareto analysis, fishbone diagrams and statistical analysis.
- The separated phases from generating alternatives, weighing alternatives and decision making. The generation of alternatives and principle solutions can be done by vertical thinking, TRIZ and lateral thinking (e.g. brainstorming). However, some methods for weighing alternatives, such as the controlled convergence method and value engineering, transgress these individual steps by the inclusion of decision making.
- The methods and interpretations of decision making. The methods that have been discussed are: multiple-criteria decision making, satisficing and case-based reasoning. Often decisions are made in groups, for instance, multi-disciplinary teams to consider and weigh different aspects of the decision as much as possible. However, there are potential side-effects of decision making in groups, such as group polarisation, groupthink and the Abilene paradox.
- The stage-wise development of solutions. The controlled convergence method also encapsulates this development of a solution in steps, reducing potential solutions along the way. Detailing of the solution leads to possible trade-offs between the requirements for the system and the solution and to defining interrelationships with other subsystems.
- The monitoring of the implementation. The original problem definition, the identified root cause, integral performance requirements and an impact analysis serve as base for monitoring.

The approach to problem analysis and solving presented in this chapter is also characterised by iterations that may lead to redefining the problem, successive cycles of problem analysis to find the root cause, reconsideration of solutions discarded in earlier stages of problem solving and triggering a new cycle of non-programmed decision making when the solution falls behind on expectations or has unexpected side-effects after implementation.

References

Altshuller, G. (1996). And Suddenly the Inventor Appeared. Worcester, MA: Technical Innovation Center.

Anonymous. (1905). The Present State of Medical Practice in the Rhondda Valley. The Lancet, 166(4290), 1506–1507. doi: 10.1016/S0140-6736(00)68498-2

Cyert, R. M., Simon, H. A., & Trow, D. B. (1956). Observation of a Business Decision. The Journal of Business, 29(4), 237–248.

de Bono, E. (1967). New Think: The Use of Lateral thinking in the Generation of New Ideas. New York: Basic Books.

de Hoo, F. (2017). Zwarte Plafonds, Rode Cijfers – de teloorgang van Konmar: van olé naar ojee. Eindhoven: De Hoven.

Dekkers, R., Chang, C. M., & Kreutzfeldt, J. (2013). The interface between "Product Design and Engineering" and manufacturing: A review of the literature and empirical evidence. International Journal of Production Economics, 144(1), 316–333. doi: 10.1016/j.ijpe.2013.02.020

Delbecq, A. L., & Van de Ven, A. H. (1971). A Group Process Model for Problem Identification and Program Planning. The Journal of Applied Behavioral Science, 7(4), 466–492. doi: 10.1177/002188863710070004

Department for Communities and Local Government. (2009). Multi-criteria analysis: a manual. London: Department for Communities and Local Government.

Despo, K., & Epaminondas, E. (2016). Teaching a 'Managing Innovation and Technology' Course: Ideas on How to Provide Students the Knowledge, Skills, and Motivation to Encourage Entrepreneurial Success. International Journal of E-Entrepreneurship and Innovation, 6(1), 38–55. doi: 10.4018/IJEEI.2016010103

Dewey, J. (1910). How We Think. Boston, MA: D. C. Heath and Company.

Geschka, H., Schaude, G. R., & Schlicksupp, H. (1976). Modern Techniques for Solving Problems. International Studies of Management & Organization, 6(4), 45–63. doi: 10.1080/00208825.1976.11656211

Harvey, J. B. (1974). The Abilene paradox: The management of agreement. Organizational Dynamics, 3(1), 63–80. doi: 10.1016/0090-2616(88)90028-9

Jacobs, F. R., & Chase, R. B. (2014). Operations and Supply Chain Management - Global Edition (14 ed.). New York: McGraw-Hill/Irwin.

Janis, I. L. (1972). Victims of groupthink: A psychological study of foreign policy decisions and fiascoes. Oxford: Houghton Mifflin.

Kepner, C. H., & Tregoe, B. B. (1981). The New Rational Manager. Princeton, NJ: Princeton Research Press.

March, J. G., & Simon, H. A. (1958). Organizations. New York: John Wiley & Sons.

Osborn, A. F. (1948). Your Creative Power. New York: Scribner.

Pugh, S. (1981, 9–13 March). Concept selection: a method that works. Paper presented at the International Conference on Engineering Design, Rome.

Rittel, H. W. J., & Webber, M. M. (1973). Dilemmas in a general theory of planning. Policy Sciences, 4(2), 155–169. doi: 10.1007/bf01405730

Saaty, T. L. (1990). How to make a decision: The analytic hierarchy process. European Journal of Operational Research, 48(1), 9–26. doi: 10.1016/0377-2217(90)90057-I

Saaty, T. L. (2000). Fundamentals of Decision Making and Priority Theory With the Analytic Hierarchy Process. Pittsburgh, PA: RWS Publications.

Schein, E. H. (1996). Three Cultures of Management: The Key to Organizational Learning. Sloan Management Review, 38(1), 9–20.

Simon, H. (1959). Theories of Decision-Making in Economics and Behavioral Science. The American Economic Review, 49(3), 253–283.

Simon, H. A. (1947). Administrative behavior. A study of decision-making processes in administrative organization.

Soelberg, P. O. (1967). Unprogrammed Decision Making. Industrial Management Review, 8(2), 19–29.

Ward, A., Liker, J. K., Cristiano, J. J., & Sobek II, D. K. (1995). The Second Toyota Paradox: How Delaying Decisions Can Make Better Cars Faster. Sloan Management Review, 36(3), 43–61.

5 Processes

This chapter explores further the mechanisms of change for systems, based on the approaches to describe and to examine systems in Chapters 2 and 3. The change of states constitutes the realm of processes. The growth of trees, engines burning fuel in order to power cars or printing documents are examples of trees. But also transformative processes, such as falling in love or studying this book and learning from it, are examples. Whether it concerns recurrent processes or those for establishing new structures, conceptualising and modelling adequately these processes serves as base for resolving many problems (see Chapter 4 for the generic approach for problem analysis and solving). The topics of this chapter and also later chapters means moving away from the inclusion of static, descriptive concepts found in Chapter 2 and 3 to dynamic notions.

Many processes are teleological in nature. Thus, they serve a purpose: they contribute to related elements in the environment of a system or they aim at preserving the state of a system through interaction with the environment. Especially, engineers and designers or humans, creating social-economic systems or any system for living, usually want to optimise or improve processes to achieve objectives whether that concerns the output of a process (for example, durable consumer goods) or how processes are conducted (think about environmental pollution). Therefore, the creators of systems need to be aware of the systems' objectives and behaviour in order to purposefully optimise or improve. Even though this chapter described processes in the widest sense possible, the focus is on recurrent processes.

To this purpose, the chapter extends the basic concepts of systems to processes and relates these to the behaviour and the analysis of systems. First, this chapter will explore the conceptualisation of processes as interaction between flowing elements and resources. Section 4.2 will address the differences between homeostatic processes and adaptive processes. The topic of Section 4.3 is the distinction between primary and secondary processes. Then the chapter will continue with discussing the concept of function and the relationship of function with processes. Section 4.5 is dedicated to systems of resources. Section 4.6 will elaborate on the behaviour of processes. Next, Section 4.7 will link the blackbox approach to the analysis of processes. The final section of this chapter contains an overview of alternative methods for mapping processes, esp. for mapping business processes.

5.1 Processes as Interaction

The underlying principle of processes is that systems change over time (static systems hardly exist although that also depends on the period of observation). This may happen without human intervention, such as the growth of trees or

R. Dekkers, *Applied Systems Theory*,
DOI 10.1007/978-3-319-57526-1_5

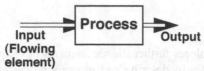

Figure 5.1 *Process as change of state. The initial state of the flowing element (input) is transformed to the final state (output). The difference between the input and the output defines the transformation (process).*

the corrosion of steel structures; but the building of a steel structure from raw material is definitely a human activity (this is perhaps why Checkland [1981, p. 110–111 ff.] refers to human activity system as synonymous to processes). No matter the source or cause of changes, the transformation of a system from one state into another is called a process (see Figure 5.1). That means that the original state of elements and their relationships have converted into a new state of elements and relationships. For example, stapling sheets of paper is a process; before the execution of the process, there were loose sheets, after stapling there is a pack, which means that the spatial orientation of the sheets has changed and they have become inseparable. Thus, in generic terms, the input, consisting out of flowing elements, has state A with certain properties and the output state has B with any of these properties changed. The difference between state A and state B of the flowing elements defines the process, or alternatively, the effect of the process on the flowing elements.

This also means that if no change of state for the flowing elements has taken place, then no process was executed. The following example might demonstrate that this has a powerful meaning in what is examined. Consider the checking of the quality of a product at a workstation in a production line (assuming that the product is not transported to conduct the quality check). If the problem definition covers the analysis of recurring quality problems then the quality check should be considered a process; the state has changed that some properties of the product comply with requirements, whereas before this check the compliance was not assured yet. Another problem definition focusing purely on logistics, transport and handling of goods in this factory, would consider that no change has taken place; hence, the quality check is not a process in the sense of the logistic problems to be solved. This example indicates that what is considered as a process is also defined by the problem's perspective.

To achieve a target state of its output, the execution of any process requires resources to bring about the changes in state of the flowing elements. The stapling of sheets of paper can only be done with a kind of stapling device. Hence, in its most basic form a process is modelled as a flowing element interacting with a resource (Figure 5.2). For example, if the flowing element is a tree trunk, changing into logs for the fireplace, the resources consist of a saw, an axe and human labour. Note that the box depicting the process is not a system but merely an abstraction of a place and time for the interaction between flowing element and resource. In some cases, when the resources are trivial to the problem, then they may be left out of the depiction. When

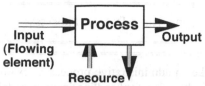

Figure 5.2 *Process as interaction between flowing element and resource. The transformation of input into output requires the presence of resources. The changes of the state of the flowing element correspond the changes of state of the resource.*

modelling the reduction rate of a gearbox, resources such as lubricants would be omitted (but not when modelling the heat transfer processes of that very same system). The change of the state of the flowing element relates to the change of state of the resource; thus processes represent resources as a system and flowing elements as a system having an effect on each other.

Also the question which system should be regarded as the flowing element and which system as the resource depends entirely on the problem definition and the chosen aspect. If the process consists of drilling a hole in a metal part, assembly workers might view the process as the transformation of a metal part without a hole into one with a hole, whereas the tooling department might look at the metal parts merely as resources that wear out the drill bits. That implies also that what is an 'end' to one process, might be a 'means' for another process; this is commonly known as the means-end hierarchy (see Figure 5.3). Applications of this hierarchy can be found in consumer research, for example the chain of consumables necessary for the use of durable products, but it also links to processes and the purpose of systems. Particularly, by looking at what the purpose of the output is will derive meaning for the process (Section 5.4 discusses this more extensively). Thus, the definition of systems guided by a problem definition determines how to look at processes.

5.2 Types of Processes

In addition to the examples in Section 5.1 being mostly recurrent processes two more categories of processes deserve attention: adaptive processes and homeostatic processes. The recurrent processes exert every time the same

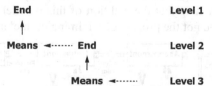

Figure 5.3 *Symbolic representation of the means-end hierarchy at two levels. The end at the first level has a means of achieving it. However, this means for the end at the first level becomes itself an end at the second level. Subsequently, the end at the second level needs a means, which is positioned at the third level. This sequence of ends and means can have more layers than depicted here.*

Box 5.A: Example of the Three Types of Processes

Suppose that we take an administrative process as example for these three processes: a simplified process for handling insurance claims with respect to car insurance.

RECURRENT PROCESSES

The recurrent process consists of a claim registration followed by a claim assessment, settlement and notification (see figure below). Each of these subprocesses necessary for handling the claim related to a car accident call on different resources. Only the second and fourth steps require a customer advisor (not automatically the same advisor, though; that depends on the practices and work allocation in a specific insurance company).

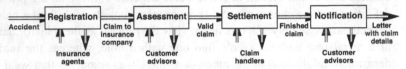

HOMEOSTATIC PROCESSES

A homeostatic process maintains equilibrium with the environment. This could be achieved by managing the available capacity of customer advisors in response to the influx of 'accidents' (not claims themselves, since they only appear as output from the first process), see figure below.

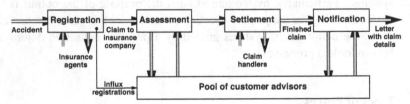

ADAPTIVE PROCESSES

For the insurance claim process an adaptation would be allowing the customers to fill out their own claims on-line (see figure below); for this step no resources in terms of advisors or agents are needed from the perspective of the company as an entity. Please note that the addition of this parallel subprocess is the adaptive process and not the process of 'claiming online' itself.

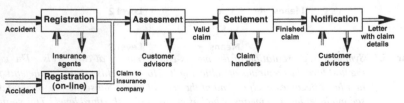

interaction between the flowing elements and resources, ideally resulting in recurrent output. However, it might also be that through the execution of a process the state of flowing elements and resources changes more permanently, for which there are two basic ways to respond:

(a) through the changes in parameters that affect the capability of a system to interact. This leads to interventions within the current structure to maintain a specific state;

(b) through the alterations in the structure (which introduces sometimes new elements, removes existing ones and affects also the properties and parameters of relationships).

The first ones are called homeostatic processes and the second ones adaptive processes (see Box 5.A).

For systems at the lower levels of the systems hierarchy of Boulding (see Section 3.5), the second option only becomes possible through external intervention: the designer of the system intervenes. Systems at higher levels of the systems hierarchy have the capability to exert an intervention themselves and change their structure, although the mechanisms to establish these interventions might differ substantially at the different levels. But already at the second level, the design of the system allows an interaction with the environment, which can be viewed as regular interactions – with regard to the processes – and as perturbation – with regard to the conditions under which systems operate.

Homeostatic Processes

Characteristic for the processes of systems at the second and higher levels is that they are homeostatic processes that deal with perturbations. Homeostatic processes occur when a system has to regain equilibrium after exposure to effects caused by stimuli or perturbations from the environment of the system (see Section 3.3 on homeostasis). In this sense, homeostatic processes point to dynamic self-regulation or, in other words, to the condition of a system when it is able to maintain its essential variables within limits acceptable to its own structure in the face of unexpected disturbances. Examples of homeostatic processes are the cruise control of a car, the human body keeping its body temperature constant, the generation of human offspring as part of the demographic system to maintain an equal divide between man and woman and the assurance of the quality of products and services by companies. Maintaining a homeostatic state not only requires returning to a defined state but it also determines the interaction of a system with its environment. For any of the previous examples, an external influence exists that might lead to perturbations of the system, on which it has to act. Hence, homeostatic processes for maintaining an internal, constant state are triggered by stimuli external to the system.

Moreover, the influence of these stimuli has to be mitigated if they move the system away from its equilibrium; this points to the various concepts of control that will be elaborated in Chapter 6. These corrective processes are internally oriented, even though the stimuli are external to the system. A well-known case in point is that homeostasis occurs when the temperature of the human body rises in case of exposure to the sun. One of the processes, which bring the body temperature back to its normal value, is perspiring. An important resource enabling this process consists of the liquids in the human body. Consequently, homeostatic processes aim at maintaining a given state of a system within a defined range.

Often that stable state or that stable behaviour is essential to the structural stability of a system; this is also known as morphostasis: the process in complex system-environment exchanges that tends to serve or maintain a system's structure or state. For example, the size of the pupil of the human eye is negatively correlated with the intensity of light entering the retina thus keeping the amount of light within the limits of optimal processing of visual information. Too much light will destroy the light sensitive cones of the retina. Similarly, the blood sugar content and many other chemical quantities are balanced within the human body. Homeostatic processes never exceed striving for a current state (including the related structure to maintain the balance). The implication of morphostasis might also be that systems tend to 'repair' the perturbations in structure rather than neutralising the external influence. Examples are tissues in the human body, family structures and technological developments in society. But also, the 'balance of power' idea in the international political system denotes a homeostatic mechanism whose outcome presumably neither country desires by itself. Also in families, homeostasis may become pathological when family members no longer prefer that state yet cannot escape it as a consequence of the way they interact with one another. All these indicate that interventions take place in the context of a structure to maintain the structure. Homeostasis, in these cases of morphostasis, depends on the structure between elements as a continuous process, which make it impossible to move away from that given 'point of equilibrium'.

Adaptive Processes

Whereas homeostasis concerns states or behaviours of systems and processes, adaptive processes relate to structure and organisation of these systems. One of the biological adaptive processes is morphogenesis, the shaping of an organism by embryological processes of differentiation of cells, tissues and organs, and the development of organ systems according to the genetic 'blueprint' of the potential organism and environmental conditions. In evolutionary biology, it is believed that these processes, sometimes happening in the early stages of growth of an individual are responsible for the variation in species. A famous example of these processes is the 'placement' of limbs, etc. that are guided by so-called Hox genes; these Hox genes can cause human beings to have five or six toes, even though the latter is rare. Adaptation in technical systems, the lower levels of the systems' hierarchy, becomes only possible through external interventions; hence, it makes more sense to speak about interventions than about adaptive processes. For instance, by adding applications in a computer or tablet, its capabilities for providing or processing information changes, but this can only be done if a human being signals the installation of such an application. Whether as a result of external interventions or as an outcome of internal processes, generically speaking, adaptive processes aim at changing the structure of the systems as internal processes.

Adaptive processes occur in case that adaptation to the need from the environment of the system is necessary, and, therefore, they are externally oriented. Simplifying this statement: the structure of the system follows the requirements set by the environment. This is most common in new product and service development, engineering and some approaches in management sciences. For example, the design of organisational structures and information systems follow this pattern. To a certain extent, biological processes follow this pattern, too; ultimately, species adapt to the environment. In this way of viewing, the analysis opens up for deterministic approaches, which allows comparing one state of the system with another or one system with another. One of the underlying thoughts of this way of thinking is that it becomes possible to achieve equilibrium with the environment; that can be considered as balancing the internal structure and external structure of a system.

In short, to distinguish between these types of processes we state that adaptive processes in a system occur when the values from relevant aspects from the system environment are being considered as normative but at the same time are outside the remit of the boundaries for the homeostatic processes. For adaptive processes to occur that means mostly that these values from relevant aspects change beyond control, so to say. However, in biological processes there is a phenomenon called neutral or random genetic drift; it denotes the idea that some random mutations come at a quite regular, almost constant rate over time, and that the type of mutation is random and remain untouched by selection [Kimura, 1983]. In other words, these mutations do not serve any function or do not change any function of the already existing genes in a population. As a result they do not disappear through selection. Being the opposite of adaptive processes, homeostatic processes arise when values of relevant aspects within a system are being regarded as normative but within reach of the boundaries of these homeostatic processes. The internal processes of the system aim at maintaining this steady state and in a way resist changing. Adapting occurs at the thin borderline between adaptive and homeostatic processes, but that will be discussed in Chapter 9.

Depicting Processes

Not only did this section introduce two specific instances of processes, as seen in Box 5.A, it also introduced different flowing elements. The three types of flowing elements that will be used from now on are shown in Box 5.1. The depiction of the primary process does not have to be materialistic (products). It can also be either flowing elements in the form of essential information or energy flowing throughout a system; since energy cannot be captured directly as flowing element, it has a different symbol but is still used for the primary process. In that sense, the primary process serves as an abstraction of reality. When the process comprises the transformation of information about the primary process, it has a different symbol again. This is the case for control processes; information is then not directly the primary process but about a characteristic of the flowing elements or energy. The use of three different symbols makes it easier to distinguish the different purposes of processes.

5.3   Primary and Secondary Processes

Since processes aim at changing the state of the flowing elements, viewed as system, this chapter so far has focused on the transformation of input into output by using resources to achieve this change. This transformation, when we talk about the primary process, means changing the state of the flowing elements from an initial condition (input) to a final condition (output) observed at a given aggregation stratum. Thus, resources are necessary to sustain this primary process; these resources as a system undergo a change, too, when used for the primary process. Therefore, secondary processes are those that recuperate resources that have been deployed for primary processes. Hence, primary and secondary process are linked to each other through the use of resources as a system.

Box 5.1:   Symbols for Processes (Applied Systems Theory)

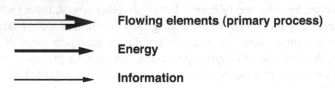

Flowing elements (primary process)

Energy

Information

In Applied Systems Theory, three distinct symbols are used for processes. The symbol for flowing elements is different from the other two; note that the symbol can be used also for indicating information as the primary process when it concerns the 'core' process of the entity (Box 5.1 demonstrates this for the claims). The energy flow is distinguished from a regular primary process due to its different characteristics. Finally, the symbol for 'information' differs from the other two, because it is related to information about the primary process, mostly used for control of primary processes (flowing elements and energy); the use of this symbol appears mostly in Chapters 6 and 7.

Primary Process

The primary process is the conversion of flowing element from one state into another for which is needs resources. The primary process is often linked to the purpose or objective of a system. Let us consider a jet engine of an aircraft in which we define this engine as a system. The primary process of the system is providing the relevant part of its environment (the aircraft) with thrust and aims at creating kinetic energy through burning fuel in compressed air resulting in high speeds of gas leaving the engine. As another example, the primary process of a petrochemical plant is the conversion of crude oil into intermediaries or consumables, such as car fuel. But none of these processes exists without a resource inducing this transformation, the engine or the plant, for the examples given. That makes it sometimes more difficult to distinguish what the system is when trying to identify it. In the case of examining an organisation, is it necessary to look at the organisation as a system of resources that produces output or to look at the primary process and the resources of the organisation as ensuring the output. Though perhaps by some considered pencil-licking, these two perspectives represent very different views on modelling of the same organisation. Therefore, the specific view on the primary process as interaction between flowing elements and resources has a profound impact on analysing and resolving problems.

The principle of processes as interaction between flowing elements and resources requires that in order to change a property or parameter of the flowing elements a property or parameter of the resources should change too. For instance, by following a lecture, a student will acquire knowledge but that demands a lecturer to prepare a presentation; for this presentation the lecturer had to gain knowledge to know what should be taught in which way and in which order. For both examples in the previous paragraphs, the execution of the primary process causes parts of the jet engine and the equipment of the petrochemical plant to wear off; thus maintenance and overhaul is necessary. That means that the interaction between resources and flowing elements

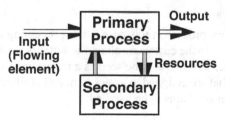

Figure 5.4 *Position of the secondary process in relation to the primary process. Similarly to the primary process, the secondary process transforms its flowing elements, the resources, needed for the primary process, from one state to another. When deployed by the primary process the initial state of the resource should potentially allow the interaction with the flowing element. Please note that the secondary process itself is also a process and requires in turn resources for its execution; for reasons of simplification that has been omitted in this figure.*

results in changes properties of the resources, which are also considered a system.

Secondary Processes

Therefore, secondary processes act on resources to recondition the state of resources so that they can be used for the primary process again (see Figure 5.4). For example, maintenance processes aim at restoring the state of the jet engine and the petrochemical plant so that the continuation of the primary process is ensured. As another illustration, when we would discuss the process of pumping fuel to the jet engine during flight, this would be a secondary process. After all, the process of pumping fuel delivers the resource to ensure the continuation of the primary process, and, therefore, enables it to deliver the thrust for making an airplane move forward. Box 5.B illustrates the application of secondary processes to the example of Box

---

Box 5.B: Processes for Insurance Claims

The example of the simplified process for handling insurance claims continues. From Box 5.1 the homeostatic process is taken.

PRIMARY PROCESSES

The primary process in this case is the processing of 'accidents' into notifications to the customers about the settlement of claims. The pool of customer advisors is to be seen as a reservoir of available customer advisors that are assigned to assessment and notification depending on the number of claims to be handled.

SECONDARY PROCESSES

In this case, one secondary process is the training of customer advisors (as resources) to keep them up-to-date with regulations and procedures. Also the induction of newly-hired advisors constitutes a secondary process.

5.A; in this case the secondary process consists of the training of advisors, which allows these to process claims appropriately. Hence, the primary and secondary process are linked through the flows of the resources.

A secondary process can become a primary process on another aggregation stratum; this is the means-end hierarchy (see Section 5.2 and Figure 5.3). In another case in which we would be interested in the process of pumping fuel from one of the fuel tanks to the combustion chamber we would characterise this fuel pumping process as the primary process. A secondary process, and, therefore, the supplier of the resource, would be for example driving this pump by a hydraulic motor. This way, each secondary process enables the primary process it is related to fulfilling its function.

5.4   Process and Function

But what exactly constitutes a function that a process fulfils? The function defines the purpose of the output of a process (note that this concerns flowing elements or a system of flowing elements). This means that output of a process (whether created by a designer or naturally evolved) satisfies a need or contributes something towards a greater entity (usually its environment); the abstraction of that contribution is called the function. West Churchman [1979, p. 13] demonstrates this notion for the automobile:

If you begin by thinking about the function of the automobile, that is, what it is for, then you won't describe the automobile by talking about the four wheels, its engine, size and so on. You will begin by thinking that an automobile is a mechanical means of transporting a few people from one place to another, ...

Note that in this example of a car, it is seen as system. However, the process of transporting results in people or passengers being at another place, a change of state for the human beings; thus, the function of transport is a spatial relocation. An other example is that if somewhere in a process a reduction of the rotational speed is needed, the function of this part of the process would be described as reducing rotational speed (note that in most cases a function can be described with a verb and a noun); this is the function of a mechanical gearbox; however, the same function can be achieved by using magnetics. Thus, in terms of the generic processes depicted in Figure 5.4, a function is realised by a system of resources; note the link of the function to ontology as mentioned in Section 3.4. This also implies that functions may be realised by different systems of resources, akin the step of finding principle solutions in Section 4.3. The examples of the mechanical gearbox and the magnetic one for the function of reducing rotational speed demonstrate this. From this perspective, the conceptualisation of function plays a central role in system thinking by relating systems of resources to processes and to problem analysis and solving. Thus, generally speaking, a function is always a rather abstract description of the output of a process; by not no talking about the system of resources, such as specific hardware for technical systems, specific organisms in biological systems or specific departments of an organisation system, the options how that function can be fulfilled remain open and allow further decision-making or evolutionary pathways.

Eventually, an output has to be produced through the execution of the process, that is the transformation of the input into the output. This is a specialisation of the function (Figure 5.5), the translation of the function into an activity (making the output available to the environment constitutes an event, see Sections 2.5–2.7; in the case that the output is made available internal to the system it will be called an activity). Returning to one of the two examples in the previous paragraph, the function of reducing rotational speed can be translated into employing a gearbox. It is important to notice that this is a choice out of (probably) many possibilities, based on more or less well-defined criteria. Instead of a mechanical or magnetic gearbox, a designer could have chosen a belt drive or a chain drive to realise the same function. As soon as the process is formulated, the necessary resources become apparent, which are different for the mechanic gearbox, the magnetic gearbox, the belt drive and the chain drive. A similar example could be given for the administration of a company; it is irrelevant to a certain extent whether financial statements are generated manually or automatically, albeit that the resources differ substantially in both instances. In that sense, it is the process that relates the function to the resources. Thus, the function describes merely the effect of the process, not the way the output is actually manufactured as a process.

Usually, for design of solutions (see Sections 4.3–4.5) the need of the environment is known (or created) through interaction with the environment. This means that the freedom of the designer lies in defining a suitable function and deciding on the most efficient way to implement it. Analysing a process involves looking at the existing process and the output. Generating an abstracted description of the intended function of the process is one of the hardest skills to master. But having formulated the function, most of the time, it becomes possible to formulate alternative processes or to identify alternative systems of resources for the same function. By considering

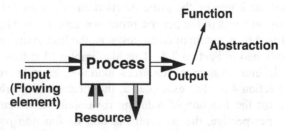

Figure 5.5 *Function as an abstraction of the output of a process. The function comprises of the contribution of the system to its environment, in particular for processes, the meaning of the output produced by the resources by the transformation of flowing elements.*

different processes and different resources the behaviour of the process can be influenced, thus enabling optimisation.

This purpose of optimising the behaviour of processes and thus of systems can also be found in the thinking about teleology. Teleology concerns the supposition that there is design and purpose – a directive principle –, or finality in the works and processes of nature, and the philosophical study of that purpose. In this perspective, teleology depends on the concept of a final cause or purpose inherent in all systems (though termed 'beings' in

---

Box 5.C: FUNCTION AND TELEOLOGY FOR INSURANCE CLAIMS

The example of the simplified process for handling insurance claims continues; see Box 5.B for the primary and secondary processes.

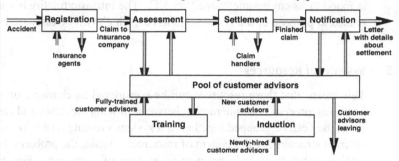

FUNCTION

The primary process delivers settlement as outcomes. For customers, these financial settlements have the purpose of reducing the risk of peaks in expenditures due to 'unexpected' events.

EXTRINSIC FINALITY

In this case, the extrinsic finality is found in the settlement being a 'means to an end'. The settlement is used by the customers either to cover unexpected expenditures related to car accidents or to smoothen expenditures over a longer period of time; different customers may have different perspectives on this. Insurance may also be necessary for legal reasons, for the purpose of simplification not dealt with here.

INTRINSIC FINALITY

But also for this example, the continued existence of the entity 'insurance company' provides employment and thus livelihood for the employees. Training of employees, being better in handling insurance claims, contributes to the sustainability of the systems of resources (employees).

teleology). There are two types of such causes, intrinsic finality and extrinsic finality:

- Extrinsic finality consists of a being realising its purpose outside the being for the utility and welfare of other systems. For instance, minerals are 'designed' to be used by plants, which are in turn 'designed' to be consumed by animals.
- Intrinsic finality consists of a system realising its purpose by means of a natural tendency directed toward the perfection of its own nature. In essence, it is what is 'good for' a being. For example, physical masses obey universal gravitational tendencies that did not evolve, but are simply a cosmic 'given'. Similarly, life is intended to behave in certain ways so as to preserve itself from death, disease and pain.

The concept of extrinsic finality from teleology resembles that of function, the purpose of the realisation of a process (or in abstracted form function) is found in its environment; see Box 5.C. The intrinsic finality is somewhat related to purpose of the systems resources, which will be discussed next.

5.5   Systems of Resources

This means that these resources can be considered as elements or systems, because a process is an interaction between flowing elements and resources. Thus, they can be grouped together as a system, creating an environment and external structure for the system of resources. Again, the problem definition sets out what to include or exclude as part of a system. For example, take suppliers delivering to a manufacturer. When considering the supply chain from raw materials suppliers to products to customers, the suppliers constitute part of the resources and are thus part of the system of resources under review. However, when examining the control of deliveries to the company, than the problem's perspective leads to the exclusion of suppliers as internal elements and they become part of the environment of the system of resources (i.e. the resources as constituting the manufacturer). Reasoning in this way demonstrates that systems of resources follow the definition of systems and, therefore, their contents depend on the problem definition which process is considered.

The interaction with flowing elements as a system indicates that resources provide, or facilitate, the activities through which inputs are converted into outputs [Miller and Rice, 1967, pp. 28–29]. The resources required for any execution of processes are physical entities, in the case of technical and biological systems, or are artificial constructs of the human mind, such as for organisational systems. The extent to which resources exist or do not exist constitutes the major internal constraint on process definitions and performance. Considering suppliers as part of the resources for the entire supply chain or as part of the environment has a profound effect on how problem with regard to the performance of supply will be resolved. Hence, the performance of a process links univocally to the system of resources deployed and will not exceed its capability.

Therefore, constraints arising either from the environment or from within the system of resources itself need to be reviewed to determine whether they are in fact inviolable. Whether that concerns technical, biological or organisational processes. A relaxation of constraints could lead to new processes or better performance of old ones; but there is no corresponding defined process for the evaluation and also for criteria to judge the performance to know what was standard has become sub standard over the course of time. The resources necessary for a process facilitate the transformation of input into output, whereas at the same time they inhibit the conversion through their constraints. Hence, to change the conversion as a process, a designer can either alter the span of the resources (i.e. include more resources or exclude resources) or replace the resources.

In systems of resources with more than one process and no adequate determination of priorities, the performance of one process acts as a constraint on the performance of another. Large systems are differentiated into constituent systems, each of which has its discrete primary process. Furthermore, the environment of any constituent system is comprised of other constituent systems and the whole, and, therefore, the constraints on definition and performance in constituent systems include those imposed by other constituent systems. The greater the differentiation of a large, complex system, the more numerous the constraints that are imposed.

Events at the boundaries of systems of resources cause activities within the system, sometimes exceeding the constraints. This can be the input of the flowing elements or the output that does not fulfil its function any more. For example, orders cause activities within an organisational system to fulfil demands by customers. Or the events might be changes in the environment of the system, the external resources and the way they interact with internal resources; for example, customers asking for different products and services. In any case, events activate internal processes of maintaining homeostasis or they lead to adaptive processes, which require changes in the structure of the system, to cope with these events.

5.6 Behaviour and Processes

Hence, processes display behaviour taken as reacting to events akin the behaviour of systems (see Section 2.6). When talking about behaviour it is quite common to use it in the context of the behaviour of persons and technological systems. It is common to use a phrase, such as: 'he behaved rather annoying last night'; expressing discontent in this case, indefinitely it refers to activities and effects of these. The concept of behaviour applies also the other objects, for example, the behaviour of a car subject to certain road conditions, such as slippery surfaces. Apart from these examples, why take interest in behaviour of systems and the processes they are subjected to? The fact that we experience and recognise behaviour apparently holds that we are able to observe a change in at least one of the aspects of a system. The change of this state is called behaviour of the system. As stated before, a process may also display behaviour. Whereas changing the state of the flowing elements itself is already behaviour related to the systems of resources, processes may also display temporal conduct; think about the wear and tear of furniture, domestic appliances and cars. These examples show that the behaviour of a process as conceptualisation is nothing different than the change of the behaviour of the system, for example, originating from disruptions of resources.

Furthermore, behaviour can only occur for dynamic systems. This is simply because only dynamic systems have interactions with the environment (see Sections 2.5 and 2.6). This is automatically the case for processes: the system of flowing elements interacts with the system of resources; the flowing elements are input from the environment and the output to the environment is flowing elements with a changed state. Having said so, it is possible to make a distinction between static and dynamic behaviour for processes. Static behaviour of a process occurs when the output of the process depends on the value of the relevant aspect of the input. Of course, the value of the output at a certain point in time has to be matched with the corresponding input and the duration of the process to make a correct comparison possible. In case of static behaviour of processes there should be a linear relation between the value of the relevant aspect and the temporal dimension of variations in the input and the output. Different from static behaviour, the dynamic behaviour of a process not only depends on the point of time and the input value of the considered aspect; the output also depends on, for example, buffering actions inside the process ('memory') or perturbations. When we assume that the output of the process has to be of constant quality, dynamic behaviour might jeopardise that goal. If still the goal stands to reach a constant quality of output, this raises the issue to what extent processes are capable to maintain that quality of the output; to this purpose will Chapter 6 deal with mechanisms for control of processes. Whereas processes are by definition dynamic with regard to the system of flowing elements and the system of resources, the behaviour of the process may be either static or dynamic.

Whereas naturally a process may be classified as having either static or dynamic behaviour, it could also be that behaviour caused by one or more processes within a dynamic system of resources might be static for one aspect

Figure 5.6 *Process as blackbox. Inputs are transferred into output without knowing any further details about the internal process. The investigator aims to relate variations in the input of flowing elements to the variations of the properties of the output.*

but dynamic for another aspect. Therefore, in order to prevent deviations to arise in the actual problem resolution, the problem definitions should clarify what aspect(s) to focus on for the system of resources. The level of aggregation may also influence the characterisation of behaviour, in terms of being static or dynamic. For example, the observation that the human body is in a stable condition is underpinned by the dynamics of the internal processes whether it is constant temperature or is stable motion. Henceforth, behaviour depends on the aspects chosen for the investigation of a system and on which aggregation stratum the analysis takes place.

5.7   Processes and Blackbox Approach

Similarly as in Section 3.2, we can apply the blackbox approach to processes, too, as an effective way to understand its behaviour. Whereas the blackbox approach examines the external structure of a system without identifying any of the internal elements, in the case of processes the input and output are examined. This approach supports a study by not looking at the activities within the process and their relationships, thus creating space to focus on the behaviour of the aggregate process. Hence, at one level of aggregation we may consider a process as a blackbox, too, for which its inputs and outputs serve as the external structure (see Figure 5.6).

By using the blackbox approach for analysis a study will aim at relating changes in one or more inputs of a process to alterations in one or more outputs. For example, practising physicians deploy this method by deducting from the behaviour of the human being as a process, e.g. temperature, pain and coughs, what internal causes bear relevance to well-being. Through purposeful dosing of medicine and assessing their efficacy, doctors draw conclusions about their earlier inferences. The reaction to medicines generates indications about the internal processes of the system of resources, the human body. There is a strong relationship between the exertion of stimuli in the form of input and the behaviour of a process as observed in its output, though the identification of these links requires observations.

Similarly, to the use of the blackbox approach for systems, the higher the number of inputs and outputs for a process, the harder its gets to arrive at meaningful inferences; even to the extent, that it might become impossible to determine with certainty how the process responds to these changes in external relationships. Although capturing the behaviour of a process becomes more difficult in case of stochastic changes in relationships, tuning of attributes and relationships of the system of resources belongs to the possibilities to alter the behaviour. In terms of processes, a study has to relate the variances in the input to the variances in the output. Through inductive and deductive reasoning, akin inductive logic (see Section 3.3), the behaviour of the process might be revealed as reaction to changes in external stimuli (events).

When during a later stage of a study, the behaviour of a process has revealed itself, the necessity might arise to open the blackbox; by doing so, the

---

Box 5.D: Blackbox Approach, Behaviour for Insurance Claims

For the example of the simplified process for handling insurance claims the primary and secondary process are taken from Box 5.B. The original processes are depicted in the blackbox; normally, like in Figure 5.5, the blackbox does not reveal the more detailed 'internal' processes.

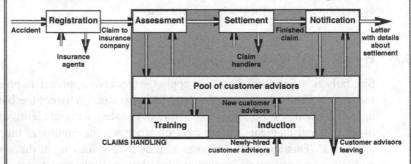

Claims Handling as Blackbox

Without going into detail, the blackbox allows examining the process at an aggregate level. In this case, the input of the blackbox is a claim to the insurance company and the output a letter with a settlement. Similarly, newly-hired insurance advisors enter the blackbox and some customer advisors are leaving.

Studying Behaviour

By looking at the input and output, the observer will learn whether claims are processed within a certain lead-time. Additionally, that can be linked to customer advisors entering and leaving the blackbox. That could lead to inferences about the available capacity of customer advisors for handling insurance claims. However, for example, if productivity improves, which is an adaptive process, the throughput of handling insurance claims improves, in which case there are more customer advisors leaving the blackbox than entering. Conversely, the observation of input, output and resources might lead to preliminary indications about the performance of internal processes.

---

activities at the lower aggregation stratum might serve again as blackboxes. This zooming in allows first defining more precisely what parts to investigate based on the actual behaviour of the process; note that this applies for the related system of resources, too. When the behaviour corresponds with the constraints imposed by the function and requirements of the output, then no further investigation is necessary, except for curiosity. When optimisation is required or possibly an intervention in the structure of the activities and the system of resources, then zooming in and considering activities as blackboxes or zooming in at the subsystems and elements of the system of resources as blackboxes again, might reduce the overload on information. Hence, stepwise process mapping, assisted by the blackbox approach and zooming in, provides a powerful tool for analysing the performance of processes.

5.8 Business Process Mapping

This section pays attention to process mapping, particularly as used in organisations and business environments to support the analysis of performance of processes and the evaluate the effectiveness of improvements. Some authors draw attention to the importance of process mapping. For example, Biazzo [2000, pp. 103–104, 111] classifies alternative approaches to business process analysis along two dimensions, strategy and focus. When the analysis looks at the behaviour of actors, the approach for a pragmatic construction is action analysis and for a rational construction strategy coordination analysis. And when the analysis concerns systems, the approach is either social grammar analysis (pragmatic) or process mapping (rational). Biazzo sees process mapping exclusively as a rational approach focused on systems of resources, comprising of defining boundaries, describing inputs and outputs of processes, representing workflows, conducting interviews with those responsible for the various activities, studying available documentation, creating a model and step-by-step revising this model for the purpose of analysis. He stresses the importance of selection the proper approach and remarks that practitioners pay insufficient attention to the social context of work. With a different take, Bond [1999] reasons that business process modelling should precede the design of an information system, that way aligning the information system with the organisational requirements. In this respect, Lee and Dale [1998, p. 215] indicate that business process management intends to align the business processes with strategic objectives and customers' needs but requires a change in a company's emphasis from functional to process orientation. Preiss [1999, pp. 42–45] pays explicit attention to the role of process improvement in the context of extended enterprises by modelling. He concludes that sufficient tools are available for analysis. The statements of the authors underline the importance of business process modelling; process maps are intended to represent a process in such a way that is easy to read and understand for interconnecting activities as part of processes, linking systems of resources, performing analysis and evaluating interventions.

Given that modelling of process is important, there are a number of different methods of process mapping in addition to the modelling approach of Applied Systems Theory. Five proven methods that reside mostly in development of information and communication technology are described in the rest of this section, although many other formal and ad hoc methods exist.

Structured Systems Analysis and Design Methodology

The first one of those five methods, Structured Systems Analysis and Design Methodology, is used for the analysis and design of information systems. It is an open standard, which means it is freely available for use by organisations; many companies offer support, training and case tools for this design method. Organisations use this method because they expect that the use of a disciplined approach will eventually improve the quality and the reliability of the systems they produce. Because of the this expectation, many organisations have been willing to incur the considerable expense and effort coming along with the implementation of the Structured Systems Analysis and Design Methodology, for example, training of staff.

The Structured Systems Analysis and Design Methodology combines project management and modelling of information processes. For the purpose of managing the development of a project for information systems, it provides a framework for allocating modules, stages and steps. For the modelling of information processes it has three key techniques, which are described by Davis [1992] as being:

- Logical Data Modelling. This technique is used for identifying, modelling and documenting the data requirements of a business information system. This technique should identify a Logical Data Structure and should generate the associated documentation. The Logical Data Structure represents the entities of an information system (objects in their widest sense about which a business needs to record information) and their relationships (necessary associations between entities). Please note that in terms of Applied Systems Theory, this Logical Data Structure covers both the contents and the structure of a system (see Section 2.2).
- Data Flow Modelling. This method is used for identifying, modelling and documenting how data are captured, transported and stored in a business information system. This step results in Data Flow Diagrams supported by appropriate documentation. Data Flow Diagrams represent processes, data storage, external entities (which send data into a system or receive data from a system) and data flows (routes by which data flow in, through and out of the system); this carries similarities to the modelling of processes and activities in this chapter.

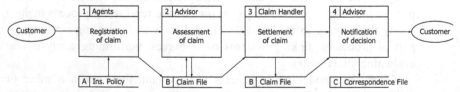

Figure 5.7 *Data Flow Modelling used for mapping process of insurance claims, based on the generic process of Box 5.A. Note that the example has been greatly simplified to demonstrate its basic principles. This data flow model should be used in conjunction with Logical Data Modelling and Entity Event Modelling.*

- Entity Event Modelling. This is the procedure for identifying, modelling and documenting the business events, which affect each entity and the sequence in which these events occur. An Entity Event Model consists of a set of Entity Life Histories (one for each entity) and appropriate supporting documentation; this model can be seen as documentating the states and changes in flowing elements and systems of resources.

The widespread acceptance of the Structured Systems Analysis and Design Methodology in the domain of information systems may lie in the fact that it does not rely on a single technique. Each of the three system models provides a different viewpoint of the same system, each of which are required to form a complete model of the system (see Figure 5.7 for an example of Data Flow Modelling). Within the method each of the three techniques are cross-referenced against each other to ensure the completeness and accuracy of the model. This creates a number of advantages according to its advocates:

- Its structured analysis provides a clear picture of requirements that those that are involved will understand.
- It can be used by both experienced and inexperienced staff because of depicting sequences of activities.
- It supports the planning and control of projects for information systems.
- Its structured approaches leads to comprehensive specifications that results in higher quality of the information systems being built.

However, the Structured Systems Analysis and Design Methodology also has some disadvantages, for example:

- It can be considered 'generic', because it describes the details that need to be considered in physical design, without describing in detail what needs to be produced.
- It can also be 'theoretical', because it describes an ideal approach that may not be relevant for most developments.

The Structured Systems Analysis and Design Methodology is a widely used method for the development of computer applications and has been adopted across private and public organisations since its origins in the 1980s.

International DEFinition Method

The second method that this section looks at is the International DEFinition Method. This method was developed during the 1970s by the US Department of Defence, particularly for the US Air Force, before the Structured Systems Analysis and Design Methodology. It is mostly known by its abbreviation: IDEF. The method is designed to model the decisions, actions and activities of processes in an organisation or system of resources. Although developed over thirty years before, it was not until 1993 that the Computer Systems Laboratory of the National Institute of Standards and Technology released IDEF0 as a standard for function modelling. Peppard and Rowland [1995, pp. 173] describe as it having started life as a software development tool, although it is now an accepted tool for process mapping within manufacturing and service organisations.

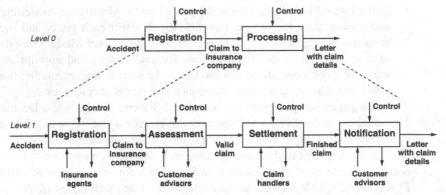

Figure 5.8 *Hierarchical view of IDEF0 for process of insurance claims, based on the generic process of Box 5.A. Processes are decomposed in activities at lower levels, akin the aggregation strata in Applied Systems Theory. The diagrams provide an overview of input and output for processes, their resources and their control (control mechanisms will appear in more detail in Chapter 6).*

The IDEF0 diagrams are used for process mapping and visualisation of information in the form of inputs, controls, outputs and resources; note the parallel with primary processes in this chapter and control processes in Chapter 6. Even though the method supports the hierarchical decomposition of activities (see Figure 5.8 for application to the example), it can only limitedly express process execution sequence, iteration runs, selection of paths, parallel execution and conditional process flows; these are all necessary for designing computer and information systems. The use of hierarchical decomposition has similarities with the distinction of aggregation strata (Section 3.1). The IDEF3 diagrams, as a further development of IDEF0, overcome the above mentioned weaknesses by capturing all temporal information, such as precedence and causality relationships associated with the processes in enterprises, and, thus, providing a basis for constructing analytical design models.

As seen from Figure 5.8, mapping using this standard generally involves two or more levels. The first level, the high level map, identifies the major processes – read primary process – by which the company operates [Peppard and Rowland, 1995]; for example, controlling operations, quoting customers, delivery to the customer, supporting customers and support services for an organisation. The second level map breaks each of these processes into a sequence of steps, and then breaks those steps down again until the appropriate level is reached for analysis and design. This distinction of levels aims at keeping an overview while each diagram does not present overwhelming detail.

There are a number of strengths and weaknesses associated with IDEF0. The main strength is that it is an effective method detailing the system activities, using aggregation strata. The descriptive activities of a system can be easily refined into increasing detail until the model is as descriptive as necessary for the analysis and design of business processes. However, IDEF0 diagrams are perceived to be concise so that they are only understandable if the user is an expert on the process being mapped. Furthermore, the models also tend to be interpreted as a sequence of activities. The models may limitedly represent embedded sequencing, a necessity for the design of information systems, however if this is not originally intended, hence, the user may need to interpret this. These weaknesses make that the method should be used with care and also might need sufficient explanation so that all involved can understand it.

ASME Mapping Standard

The third method, the mapping standard of the American Society of Mechanical Engineers (ASME), dates even further back than IDEF0. It roots in the work of Frank Gilbreth (1868–1924), an early advocate of scientific management and pioneer of motion study, who presented to members of the ASME in 1921 on 'Process Charts – First Steps in Finding the One Best Way'. Later, in 1947, the organisation adopted a symbol set as the ASME Standard for Process Charts derived from Gilbreth's original work.

Nowadays, this mapping standard is widely used in manufacturing and increasingly popular in office and service environments; particularly, the version known as value stream mapping has gained popularity in the context

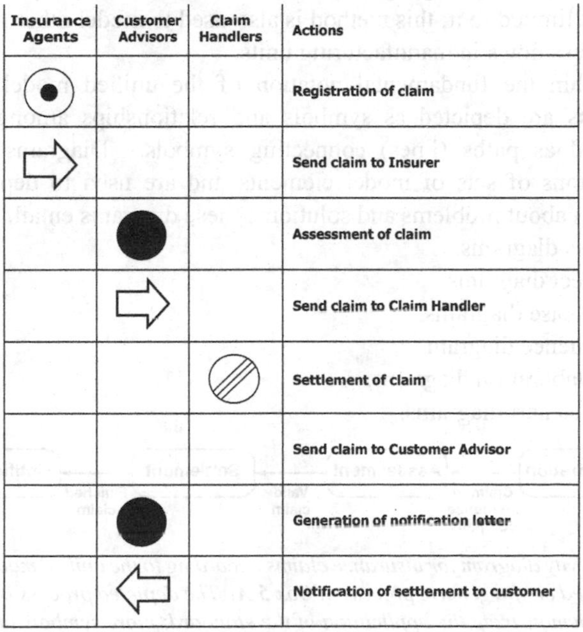

Insurance Agents	Customer Advisors	Claim Handlers	Actions
⊙			Registration of claim
⇨			Send claim to Insurer
	●		Assessment of claim
	⇨		Send claim to Claim Handler
		⊘	Settlement of claim
			Send claim to Customer Advisor
	●		Generation of notification letter
	⇦		Notification of settlement to customer

Figure 5.9 *Mapping of process for insurance claims according to ASME Mapping Standard, based on the generic process of Box 5.A. The depicted process is a simplification to demonstrate the application of the standard.*

of the five principles of lean thinking [Womack and Jones, 1996, pp. 16–26]. The original, official ASME mapping standard has eleven different symbols in process diagrams; for example, it has symbols for 'do' operations, handling operations, inspection and storage. An example of its use is shown in Figure 5.9. It is suited for detailed level mapping and has the distinct advantage that 'inherent in its use is an evaluation of whether a step is value adding. Only one of the columns contains value adding steps and thus the areas of waste or non-value-adding activity are clear' [Peppard and Rowland, 1995]. Thus, the ASME mapping standard and its contemporary adaptation – value stream mapping – aim at reducing waste for optimising business processes.

Unified Modelling Language

The unified modelling language, the fourth method in this section, is a standardised, general-purpose modelling language in the domain of software engineering, which includes modelling of business processes. It combines data modelling (which are entity relationship diagrams), modelling of workflow (business modelling), object modelling (objects are variables, functions and data structures in programming) and component modelling for the purpose of developing and building computer applications. Those developing software see the unified modelling language as a language and not a methodology, because its use is independent from programming language. Although the unified modelling language is generally used to model computer applications, it is not limited to it; this method is also used to model other processes, such as process flows in manufacturing units.

Within the fundamental notation of the unified modelling language, concepts are depicted as symbols and relationships among concepts are depicted as paths (lines) connecting symbols. Diagrams are graphical projections of sets of model elements and are used to depict knowledge (syntax) about problems and solutions; these diagrams entail:
- Class diagrams.
- Object diagrams.
- Use case diagrams.
- Sequence diagrams.
- Collaboration diagrams.
- State chart diagrams.

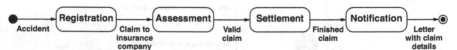

Figure 5.10 *Activity diagram for insurance claims according to the unified modelling language, based on the generic process of Box 5.A. The depicted process is a simplification to demonstrate the application of the standard; more symbols are used to create activity diagrams normally, such as symbols for decisions. In addition, the unified modelling language uses eight more diagrams to depict processes and other related state changes in a computer application.*

- Activity diagrams.
- Component diagrams.
- Deployment diagrams.

All these diagrams serve different purposes and depict different facets of a computer application; an example of an activity diagram is found in Figure 5.10. At the same time, they need to be collated to ensure that all diagrams are consistent in relation to each other; that is often done by using tools and developers' software that support the development of computer applications.

Whereas the unified modelling language has turned into an industry standard for computer applications, it carries also a number of disadvantages:

- The number of diagrams increases the complexity to keep an appropriate overview and the interrelationships between the diagrams are difficult to interpret.
- The standard provides less guidance for the actual systems development (for example, the coding for programming is not included). Thus, the automatic generation of software systems is not possible.
- The method encourages an object connection architecture rather than interface connection architecture.

Notwithstanding these disadvantages, the unified modelling language should be viewed as a software modelling language with an emphasis on graphics and dynamics to capture state changes of elements and subsystems that allows interacting with stakeholders.

Soft Systems Methodology

Although not primarily designed for business process mapping, or particularly, information technology, the soft systems methodology as the fifth method in this section has found widespread use for improving business processes of organisations. In that respect, Checkland [1981] claims that this particular systems theory and its practice apply to many problem areas. According to him, the approach lends itself particularly well to dealing with complex situations, where those involved lack a common agreement on what constitutes the problem and finding common ground needs to be addressed. To this purpose, stakeholders engage with each other guided by an analyst or facilitator in this approach; in that sense, this participatory stance has many parallels with action learning [for example, Raelin and Coghlan, 2006]. The soft systems methodology and its related approaches have found widespread recognition and are taught at many academic institutions.

The approach of soft systems methodology has some similarities with Applied Systems Theory. It also looks at systems as being part of total reality; to that purpose, the so-called root definition describes the relationship of a system with its environment. And it recognises the function of a system (see Section 5.4), or better process, through what is called CATWOE:

- Clients (Who are the beneficiaries or stakeholders of this particular system?).
- Actors (Who are responsible for implementing this system?).

- Transformation (What are the inputs and what transformation do they go through to become the outputs?).
- Weltanschauung (What particular worldview justifies the existence of this system?).
- Owner (Who has the authority to abolish this system or change its measures of performance?).
- Environmental constraints (Which external constraints does this system take as a given?).

Many will see this mnemonic as a checklist for goal definition. However, it might be necessary to use the 'checklist' as appropriate. According to Checkland [1981], systems thinking is a way of modelling rather then a technique, see Figure 5.11 for an example, and it applies as a methodology to handle complex situations.

Furthermore, the most distinct difference with Applied Systems Theory is that Soft Systems Methodology uses seven stages for solving a problem (Chapter 4 presents problem analysis and solving as consisting of nine steps):

- Entering the problem situation.
- Expressing the problem situation.
- Formulating root definitions of relevant systems (using CATWOE).
- Building conceptual models of Human Activity Systems (defined as an assembly of people and other resources organised into a whole in order to accomplish a purpose).
- Comparing the models with the real world.
- Defining changes that are desirable and feasible.
- Taking action to improve the real world situation.

Because of these seven stages and its participatory approach, soft systems methodology serves well as a process for process evaluation and a modelling approach at the same time.

5.9   Summary

One of the most common purposes of modelling by using Applied Systems Theory is about changes in the state of systems, also called processes. In that perspective, a process can be described as the transformation of the input(s) into the output(s), or as the interaction between the flowing element(s) as a system and the system of resource(s). That means that a resource acts on flowing elements to achieve outputs in which at least one property of the flowing element has changed. To this purpose, generically speaking, a

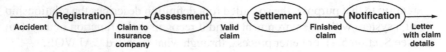

Figure 5.11 *Human Activity System for insurance claims according to soft systems methodology, based on the generic process of Box 5.A. The depicted process is a simplification to demonstrate the application of the approach.*

system of flowing elements and a system of resources is needed to describe processes.

In this sense, on its own aggregation stratum, the resource can become the flowing element; this is the recursive scheme of primary and secondary processes. Secondary processes aim at maintaining the state of resources for primary processes. When these secondary processes become the focus of attention for problem analysis and solving, this is captured by the means-end hierarchy; this denotes that what is a means at a specific aggregation stratum becomes an end at the next lower level of aggregation.

Moreover, the function of a process, as an abstraction of the output, defines its contribution to the environment. Moreover, there might different principle solutions to achieving a specific function. Then, a designer of a system performs a specialisation of the function to define a process with its system of resources; that means that analysing an existing process includes generalising it into a function. Choosing and constructing a process to fulfil a certain function requires an understanding of the process' behaviour; one systems approach for this purpose is the use of blackboxes for analysing processes.

References

Biazzo, S. (2000). Approaches to business process analysis: a review. Business Process Management Journal, 6(2), 99–112.

Bond, T. C. (1999). Systems analysis and business process mapping: a symbiosis. Business Process Management Journal, 5(2), 164–177.

Checkland, P. (1981). Systems Thinking, Systems Practice. Chichester: John Wiley & Sons.

Davis, W. S. (1992). Tools and Techniques for Structured Systems Analysis and Design. Reading, MA: Addison-Wesley.

Johnson, R. A., Kast, F. E., & Rosenzweig, J. E. (1964). Systems Theory and Management. Management Science, 10(2), 367–384.

Kimura, M. (1983). The Neutral Theory of Molecular Evolution. New York: Cambridge University Press.

Lee, R. G., & Dale, B. G. (1998). Business process management: a review and evaluation. Business Process Management Journal, 4(3), 214–225.

Miller, E. J., & Rice, A. K. (1967). Systems of Organisation: Task and Sentient Systems and their Boundary Control. London: Tavistock Institute.

Peppard, J., & Rowland, P. (1995). The Essence of Business Process Re-engineering. Hemel Hempstead: Prentice Hall Europe.

Preiss, K. (1999). Modelling of knowledge flows and their impact. Journal of Knowledge Management, 3(1), 36–46.

Raelin, J. A., & Coghlan, D. (2006). Developing Managers as Learners and Researchers: Using Action Learning and Action Research. Journal of Management Education, 30(5), 670–689.

West Churchman, C. (1979). The Systems Approach: Revised and Updated. New York: Dell.

Womack, J. P., & Jones, D. T. (1996). Lean Thinking. New York: Simon & Schuster.

6   Control of Processes

It may well be that the output of the primary process, one of the topics in the previous chapter, does not match with pre-defined outcomes or expectations; that deviation evokes the need for control, the subject of this chapter. Consider the following example of driving a car, where the output is driving a car at a given speed. Would it suffice to push the gas pedal of a car into a certain position, await it to accelerate to a speed that the driver had in mind and cruise constantly at the same speed? Most likely not, because all kinds of deviations may occur during travelling, such as other cars on the road, the state of the road, unexpected traffic jams, etc. The occurrence of these types of disturbances is why this chapter will focus on control processes. These control processes connect directly to the changing state of processes, flowing elements and systems of resources; see the previous chapter. In this perspective, processes deliver an output, which fulfils a function, and this out should not only be aligned with the function but also meet requirements and constraints. For example, when preparing and eating food we know that cooking does not always deliver the same result, even though we adhere to strict processes; some fast-food restaurants have battled this by submitting detailed instructions to workers how to prepare dishes they serve and meticulous control of ingredients used. But if there is variation in the ingredients, those controls might have limited reach and those that cook have to adapt recipes and instructions; that is also control. Thus, this chapter discusses basic concepts of control for processes.

These control processes, as known today, are rooted in technological developments dating back to antiquity. An example of one of those early applications is how vertical windmills – as in the design of some of the famous Dutch windmills – are kept in the direction of the wind. However, a more formal analysis of the field started with an analysis of the dynamics of the flyball governor (James Watts' final step in the development of the steam engine), conducted by the famous physicist Maxwell in 1868 entitled 'On Governors'. This described and analysed the phenomenon of 'hunting' in which lags in the response by the system can lead to overcompensation and unstable behaviour. This caused a flurry of interest in the topic. Another notable application of dynamic control was manned flight. The Wright brothers made their first successful test flights in 17th December, 1903; by 1904 they succeeded in controlling flights for substantial periods with their Flyer III (more so than the ability to produce lift from an aerofoil, which was known). Control of the airplane was necessary for its safe and economically successful use. By World War II, control theory was an important part of fire control, guidance and cybernetics as military applications. Also the so-called space race to the moon depended on accurate control of the spacecraft. For this reason, many technological advances have relied on the development

© Springer International Publishing AG 2017
R. Dekkers, *Applied Systems Theory*,
DOI 10.1007/978-3-319-57526-1_6

of adequate control theory. However, control theory is not only useful for technological applications, but also for fields like biology, economics, organisations and sociology. In the context of these other applications and systems theories, it appears as regulator in Bogdanow's [1928, p. 102 ff.] work and the writings of Ştefan Odobleja between 1929 and 1937 (according to Iancu [2009]). A more extensive background to the incorporation of control mechanisms in systems theories appears in Dekkers [2015]. This chapter builds on the latter publication to discuss the various concepts of control of processes for a broad range of applications, such as biology, economics, engineering and management.

To clarify the concept of control, Section 5.1 presents a few examples of control processes. Through these examples the chapter will arrive at a definition of control processes and outline the conditions for effective control. Subsequently, Sections 5.2–5.5 describe the distinct four main types of control processes. And Section 5.6 focuses on when to apply which type of control. Section 5.7 will discuss echelons of control processes. Section 5.8 on Ashby's famous law of requisite variety concludes this chapter.

6.1   Generic Concept of Control

Continuing with what control constitutes, examples are found all throughout daily life. For example, the temperature of a room is measured and compared with the temperature as set by using a thermostat. The heating system will be activated when the actual temperature differs from the set temperature. Another application is a cruise control system in a car. The driver activates the cruise control system as soon as the desired speed has been reached so that the speed will be constant to this set value. In case of a slope in the road or a sudden head wind, the engine will adjust its power in order to annul the difference between the speed as set by the driver and the actual speed. Although more complex, this is also the basic concept of an autopilot in an aircraft controlling vertical and horizontal speed, altitude and heading. One of the differences in the latter context is that the pilot often sets the desired value of the parameter in question prior to this value has been reached. Another example is a manager telling an employee what to do right away to make the deadline for a specific order. But also manually adding a touch of paint on a product after a robotic painting process took place, in order to accomplish the desired finish of the painted surface, is a matter of control. The result of the robotic painting process is being measured, compared with a desired result, and if necessary adjusted. These examples show that many applications of control exist.

Moreover, these examples show that an intervention exerted by a control process should always be accompanied by a primary other than the transformation of data necessary for the control process. That means that control processes intervene in on-going primary processes; for the examples in the previous paragraph, the primary processes are heating a room, driving

---

Box 6.1:   Defining Control Processes

A control process is a process, whose purpose is solely to, if necessary, intervene in a transformation process in terms of adjusting the values of the relevant aspects to the desired values. It might consist of measuring, comparing, assigning (allocation) and intervening. Intervention takes place to adhere to standards.

---

a car, flying an aircraft and processing orders. Thus, the primary process and the control process are always interrelated, regardless of the fact that we can treat the processes for control separately from the primary process of the system. Or more precisely, the objective of a control process is controlling a primary process by intervening either in the execution of the primary process or in the flowing elements or in the system of resources.

Essentially, control processes also constitute processes as defined in Section 5.1. Control processes have their own resources and their own flowing elements; they cause a change of state of their flowing element (data) and therefore they, at least partly, determine the behaviour of the flowing element in terms of changes of values of the relevant aspects related to the primary process under consideration. This change can be induced in two ways. The first case is when the system itself generates the target state for the primary process, as happens in organisations. Or the environment determines that target state; technological systems, such as cars and planes, do not set their own objectives, for example, reaching a destination, but human beings external to these systems do. Once the target state has been set, the control processes generate interventions to reach this defined target state; hence, control processes are processes transforming data about the state of flowing elements, the process and the system of resources into interventions into the state of the flowing elements, the process and the resources.

After denoting the purpose of control processes and their characteristics, it becomes possible to formulate a definition for a control process as written down in Box 6.1. This definition also entails that control processes rely on three conditions:
- a normative state for the properties of eiter the flowing elements, the process and the system of resources should be given;
- the possibilities to measure this normative state should be present;
- the possibilities to influence the behaviour of the process to achieve this state should be present.

Only when these three conditions are met, it becomes possible to exert control (see Box 6.A for an example of conditions for control). The normative state defines the standards for the control processes and the aspects or properties that will be measured. Effective interventions should allow reaching this normative state. An example of an ineffective intervention appeared once

BOX 6.A: EXAMPLE OF HYDRO-ELECTRIC POWER GENERATION

One of the ways to generate electric power is by converting the energy of (fast) moving water into electricity. There are many famous examples of dams built for this purpose, the largest one being the Three Gorges Dam in China, the Itaipu Dam across the border between Brazil and Paraguay (see picture; source: Wikipedia [2013]) and
the Guri Dam in Venezuela. Those and other dams generate electricity to the distribution net (see figure); however, electricity itself is difficultly stored.

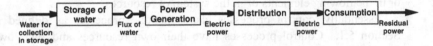

NORMATIVE STATE

Therefore, the power generation should not be less than the actual consumption but also not more. The normative state is that the hydro-electric power generation by the dams should match the actual power consumption.

MEASUREMENT OF NORMATIVE STATE

To be in balance, this means that both the generation and the consumption of electricity should be measured. Electric power generation and power consumption are relatively easy to measure; it only requires the measurement of the voltage and the current.

CAPABILITIES FOR INTERVENTION

There are a number of interventions possible to match the power generation with the energy consumption:
- Switching on and off turbines in the dam (symbol in flowing elements between 'Storage of water' and 'Power Generation').
- Regulating the flux of water to each of the turbines (same symbol).
- Reducing the consumption of electricity (for example, by asking air-conditioning to be switched off at hot days when there are low levels of water in the basin behind the dam).

In this example of hydro-electric power generation all three conditions for control are met. This means it is possible to control the amount of power generated and to a certain extent also the power consumed; these possibilities for control will be discussed in Boxes 5.B–5.E.

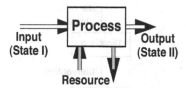

Figure 6.1 *Target states for control. Control aims at keeping the output - the properties of the flowing elements after the execution of the primary process (State II) - within predefined limits. Variations in the input (as reflected in State I) and the resources may cause aberrations of the target state.*

when quality problems were experienced with the motor for the tape drive of video recorders: the electrical connections made out of cupper were internally broken. To reduce the quality problems extra tests were introduced. However, these additional tests did not change the occurrence of quality deficiencies. What seemed an effective intervention turned out to have no effect at all: the wiring was once more bended resulting in new faulty connections; this resulted from know properties of cupper wire. Whereas the additional quality check eliminated faulty connections, at the same time it introduced faulty connections. Later on, this could be corrected by changing the design of the connection and wiring system for the motors. This example illustrates that effective control depends on the capability for measuring properties and the effect of interventions that potentially will result in achieving the normative state.

In this respect of reaching a normative state, Beer [1959, p. 22] remarks that the big feature of natural, and especially, biological, control mechanisms is that they are simply homeostats; see Section 3.3 on homeostasis. A homeostat is a control device for holding some variable between certain limits. The classical biological example is the homeostasis of blood temperature, which varies little, although the human body might pass from getting supplies out of refrigerated storeroom into a fully heated function room. A homeostat holds a critical variable at a desirable level by a self-regulatory mechanism. This means to say the value is always at its intended level to a know standard of approximation, and that there is a compensatory mechanism which edges it back towards that this standard whenever it begins to wander away. But homeostats assume that control is always a kind of self-regulatory mechanism, which is not necessary the case as shown in the next sections when control mechanisms are explored in more detail.

6.2 Control and Directing

Basically, four main mechanisms of control exist: directing, feedforward control, feedback and completing deficiencies. These control mechanisms act on the state of the flowing elements or the state of the process; the process itself transforms flowing elements in State I to elements having State II (see Figure 6.1), which will be omitted for the purpose of simplification from the

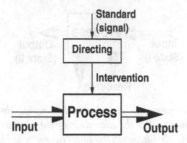

Figure 6.2 *Directing. A control signal, the standard, is converted into interventions for the process (or input). Observe that no measurement takes place, the control process relies on the adequate translation of the standard into an one-off intervention (or directives).*

later figures that depict control mechanisms. Bookbinding is such a process: single sheets, leafs as they are called, are sewn together into sections, which are bounded together and completed with a book cover. In this example, State I is the loose leafs and the book cover and State II the completed book that can be shelved. However, the objective of control is to maintain the State II, the state of the flowing elements, as output within predefined boundaries. This means that the purpose of control is that after an intervention has taken place the desired state of the output will be achieved.

Directing represents the simplest form of control. A (one-time) signal is passed to the process as a directive and one might assume that the desired state will be reached (see Figure 6.2). It corresponds to pushing the gas pedal of a car to a pre-set position corresponding to the speed that needs to be reached. The actual speed that the vehicle reaches may depend on other conditions, such as the type of road surface, the inclination of the road and the wind direction. However, the controller does not intervene in the case of directing, since the desired state has been translated into a one-off directive that governs the primary process.

Hence, this type of control process, directing, refers to an intervention in terms of giving a directive to the primary process. It is especially called a directive because of its one-off characteristic of being generated. This means that if the state of the output needs alteration the only event required is a new signal or standard; see Box 6.B for the example throughout the chapter. Using the case of driving a car again, the gas pedal is set for a specific speed no matter the actual conditions and the speed of the car will only change if a new normative speed is 'entered' in the control system. Therefore, directing means generating a one-time intervention by the control process for the primary process. The intervention may occur in terms of timing, location, aspects and intensity, according to a certain standard, which is the primary input for the directing process. In case one solely applies directing as the only control process for the process under observation, there is no monitoring of relevant aspects after the execution of the intervention. Thus, there is no measuring process, which verifies whether any exceeding value of any

---

BOX 6.B: DIRECTING – CASE HYDRO-ELECTRIC POWER GENERATION

In the case of hydro-electric power generation, as described briefly in Box 6.A, the intervention could be the switching on and off of water turbines coupled to generators that convert the energy of the rushing water into electricity. How many of these water turbines coupled to generators are operating at the same time determines the output of electric power. That allows the operators to match the power generation to the demand for electric power.

If this matching is based on demand patterns for hours during the day, weekly variation and seasonal influences, it will be called directing; in essence, these demand patterns are based on predictions rather than actual measurements of power consumption. In such a case, the switching off and on of water turbines is independent of actual demand. In the figure below this intervention is depicted by the use of a valve. Any deviation of the pattern of consumption will not result in a correction, meaning that there is too much power or too less power distributed.

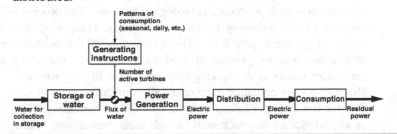

parameter belonging to any aspect occurs. The lack of a measuring process is a relevant characteristic of directing.

The generation of the intervention as an activity for directing might consist of setting the value of a parameter based on the structure of the flowing elements, on the structure of the process or on the structure of the systems of resources (or any combination thereof). After this intervention, the primary process should produce the desired output without further interference, e.g. setting the temperature of a house; in general, directing generates norms for processes seen as blackboxes no matter their internal structure. Another possibility is changing the structure of the process – often found in organisations. Whatever primary process and which type of intervention, after setting the signal or standard no correction will take place. The controller must know exactly which signal produces which results, i.e. the setting of the value of the standard or introducing a new structure for the process; this implies that the controller possesses a causal model that relates the directives to the desired output of the process. The control of most processes does not comply with this prerequisite of stability due to disturbances in the input of flowing elements, the deployment of resources and the variations in the throughput (process variation).

Thus, principally, directing cannot classify as a control mechanism for maintaining homeostasis. The principle of directing assumes that a one-time signal (or interventions) will suffice for reaching a desired state of the output of the process; however, that requires a model that univocally links the directives to the target state of the output. For example, a homeostat aims at keeping a variable with certain pre-set limits. Hence, it reacts on deviations from the variable through which the regulatory mechanism intervenes in the input, the resources, the process or the output. Therefore, if it is necessary to maintain homeostasis, then directing as control mechanism is not adequate and other control mechanisms need to be used.

6.3   Feedback as Control Mechanism

As one of these other control mechanisms that can be deployed to maintain homeostasis, feedback measures the output of a process and intervenes in the input, the resources or the conversion process itself. Feedback is observed or used in (complex) systems, such as engineering systems, architectural systems, economic systems and biological systems. The process of feedback consists of the following activities: measuring, comparing, actuating and intervening (see Figure 6.3). These activities of the control mechanism enable the primary process to produce the required output by adjusting parameter values of the flowing elements or modifying parameter values of the resources or adjusting parameter values of the primary process (see Box 6.C for the example throughout the chapter), if necessary. The following is an instance of feedback used in web-based workflows. Feedback loops are established by Internet service providers for unwanted messages. When subscribers click the 'This is Spam' button in their web mail clients, the feedback loop sends a message back to the Internet service provider letting

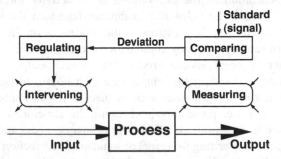

Figure 6.3 *Feedback. Deviations in the measurement of parameters of the output's state lead to interventions in parameters of either the input or the process or the resources (the latter are not depicted). The comparison might include calculations to make it possible to compare the standard with the measurement. The intervention depends on a model to convert deviations into interventions.*

Box 6.C: Feedback – Case Hydro-Electric Power Generation

In addition to directing for the case of hydro-electric power generation, see Box 6.A, a feedback loop could also match the power generation with the actual demand for electric power. This means that the actual power consumption should be measured and compared with seasonal patterns; that leads to switching on and off of turbines. However, this feedback loop tunes the electric power consumption to the predicted patterns set by directing. Note in the figure how the feedback is used in conjunction with directing (see Box 6.B).

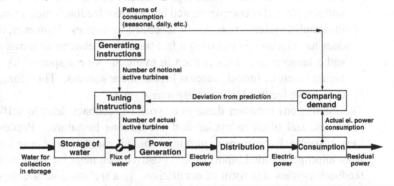

---

them know to filter these specific messages next time. The control of a system by feedback requires getting information from the output back to the input of a system or to the process parameters. Alternatively, the feedback might measure the state of the process and intervene in the input or resources. In both cases, this involves the replacement of the open, linear chain of cause and effect familiar in most science by a circular causality, a closed loop that implies the merging of causes and effects. Hence, feedback represents the measuring of output and then undertaking corrective actions upstream.

Although using the same principles in cybernetics, in control theory feedback is a control mechanism whereby some proportion of the output of a primary processor, or in general the function, is passed back (feedback) to the input or sometimes to the conversion process. Often this is done intentionally, in order to control the dynamic behaviour of the system. The difference between the more generic principle of cybernetics and this technological approach of control theory is that the comparison and actuation are more or less integrated and that the intervention is directed mostly at the input of the primary process.

Feedback as a control mechanism, measuring the state of the output (flowing elements) or the state of the process and then taking corrective action upstream, comes in two forms. Feedback may be negative, which tends to neutralise the impact of disruptions on the output, or positive, which tends to increase the effect of disturbances in the output. Both manifestations of feedback have their own applications:

- The one generally used is negative feedback, and this acts to counter the gap between the actual value and a reference value. For example, if the external temperature rises (the disturbance), the internal temperature of a refrigerator will move upwards (the actual value) and the control process will activate a cooling system to maintain the refrigerator temperature (the reference value). The ultimate aim of negative feedback is to maintain equilibrium.

- The less-known form is positive feedback, which responds so as to increase the magnitude of any particular perturbation, resulting in amplification of the original signal instead of stabilisation. Often this is undesirable and is complemented by negative feedback measures, leading in complex systems to a mix of feedback influences. However, this form does have its uses in ensuring a fast transition between an unwanted state and a target state. This is seen in evolution where species try to reach higher levels of fitness, success breeds more success. Therefore, the aim of positive feedback is reaching a new state.

The interactions between these two types of feedback lead to self-limiting processes, and often to cycles and oscillations in nature. Processes that include feedback are prone to hunting, which is oscillation of output resulting from improperly tuned inputs of first positive then negative feedback. Audio feedback typifies this form of oscillation. In a technical sense, these might also be caused by the intervention and signal being in phase; that means that an amplification of the input as an intervention occurs at the moment of that input being at its peak, resulting in the amplification. Without going into too much detail, it requires feedback systems to be designed in such a way that interventions do not lead to instability, even for social organisations. Therefore, feedback applied in a specific situation requires an appropriate understanding of the relationship between the characteristics of the primary process, the impact of negative and positive feedback, and the effects, timing and the position of interventions.

Traditionally, feedback has found many applications in *electronic engineering*. The processing and control of feedback is designed into many electronic devices and may also be embedded in other similar technologies. The most common general-purpose control system is the so-called proportional-integral-derivative controller. Each term of the proportional-integral-derivative controller copes with the aspect time for the behaviour of a process in a different way. Without going into much technical and mathematical detail, the proportional term handles the present state of the primary process, the integral term handles its past state, and the derivative or slope term tries to predict its future state. An everyday example of a proportional-integral-derivative controller is setting the temperature of a shower head; you sense the difference in temperature (comparing it with the temperature in mind), adjust the hot water flow according the response to earlier changes while predicting how much adjustment is need to achieve a certain (agreeable) temperature for taking the shower. If the deviation is inverted on its way round the control loop, the system is said to have negative feedback; otherwise, the feedback is said to be positive. The feedback process might consist of specific terms from the proportional-integral-derivative controller as well as having specific settings for each of the terms to meet the specific characteristics of a process.

Feedback has been applied in *mechanical engineering* as well. Going back to ancient times, float valves were used to regulate the speed of Greek and Roman water clocks. Another example is the fantail of the windmill; in 1745 blacksmith Edmund Lee invented the 'self-regulating wind machine', a fantail and a set of gears, to keep the face of the windmill pointing into the wind. Later, other self-regulatory mechanisms were contrived for windmills to control speed and load. Self-regulatory mechanisms appeared in steam engines as well; the centrifugal governor by James Watt in 1788 to control the speed of his steam engine was one factor in its development that made this type of power one of the symbols of the later Industrial Revolution. Steam engines also use float valves and pressure release valves as mechanical regulation devices, see the mathematical analysis of the flyball governor by Maxwell [1868], mentioned in the introduction of this chapter. In addition to using regulating the speed of steam engines themselves, steam was initially used for controlling the rudders of ships. For example, *The Great Eastern* was one of the largest steamships of its time and employed a steam-powered rudder with feedback mechanism designed in 1866 by J. McFarlane Gray (according to White [1900, p. 669]). Later, Farcot [1873] coined the word servo to describe steam powered steering systems. After that hydraulic servos came into use for positioning guns. A next notable development was the first autopilot designed by Elmer Ambrose Sperry, Sr. in 1912. It was Nicolas Minorsky [1922] who published a theoretical analysis of automatic ship steering and described the proportional-integral-derivative controller. The utilisation of mechanical feedback continued, such as the internal combustion engines of the late 20th century, which had vacuum advance as mechanical feedback mechanism. However, mechanical feedback devices were gradually replaced by electronic engine management systems once small, robust and powerful single-chip microcontrollers became affordable. These examples show the wide range and importance of feedback controller devices utilised in mechanical engineering.

Furthermore, feedback exists in *nature*. Generically speaking, biological systems contain many types of regulatory circuits, among them positive and negative feedback cycles. The purpose of those cycles is that in biological systems, such as organisms, ecosystems and the biosphere, most parameters must stay within narrow boundaries around a certain optimal level under certain environmental conditions; this is called homeostasis (see Section 3.3). An example is keeping the pH level of water reservoirs (an indication how acidic or basic a substance is) at a specific value, which allows certain flora and fauna to flourish; the chemical composition of a water reservoir acts like a buffer so that a certain pH level will be maintained. In other biological systems, such as the human body, the value of the parameter to maintain is recorded by a reception system and conveyed to a regulation module via a transmission channel. An example of that type of negative feedback is the reaction activated by heat receptors in the skin. A sudden increase in temperature will trigger a neuro-physiological signal, in turn activating a muscle contraction. Another example of biological feedback, but in this case positive feedback, happens at the onset of contractions during childbirth. When contractions occur, oxytocin is released into the body stimulating more contractions; this hormone helps to relax, to reduce blood pressure and cortisol levels and to increases pain thresholds, among other effects. Thus, the result is an increased amplitude and frequency of contractions. In general for biological systems, the negative feedback loops tend to slow down a process, while positive feedback loops have a tendency to accelerate it.

Moreover, feedback can also be found in *economics and finance*. A most famous example is the stock market, which has both positive and negative feedback mechanisms. This is due to both cognitive and emotional factors that belong to the field of behavioural economics. In the example of the stock exchange, the following well-known applications of feedback can be identified:

* When stocks are rising (a bull market), the belief that further rises are probable gives investors an incentive to buy (positive feedback); however, the increased price of the shares and the knowledge that there must be a peak after which the market will fall, ends up deterring buyers at the same time (negative feedback).
* Once the market begins to fall (a bear market), some investors may expect further losses and refrain from buying (positive feedback), but others may buy because stocks become more and more a bargain (negative feedback).

The existence of negative and positive feedback mechanisms make the stock market prone to hunting (oscillating). Well-known investor George Soros [1987] described the workings of feedback in the financial markets based on those self-reinforcing effects of market sentiment and developed an investment theory based on those principles. However, the more traditional economic equilibrium model of supply and demand supports only ideal linear negative feedback and was heavily criticised by Ormerod [1994] in his book 'The Death of Economics', which in turn was criticised by traditional economists. The discussions are a reflection of the changing perspective as economists started to recognise that non-linear feedback processes might apply to financial markets (non-linear behaviour is the topic of Chapter 9). Hence, feedback mechanisms play an important role in the understanding of phenomena in economics, as also exemplified in recent economic crises.

The principles of feedback processes have also been applied to the domain of *organisations and management*. For example, as an organisation seeks improving its performance, feedback from customers (in the form of sales, complaints, etc.) assists it in improving quality of products and components as well as adjusting organisational processes and structures. Within organisations, feedback is used as well: performance measurement of organisational units, 360-degree feedback, etc. Most particularly, feedback constitutes an essential part of the movement of cybernetic management and other approaches related to systems theories, such as the design of organisations (for both, see Section 11.3). Hence, feedback as control mechanism has found its way in a wide range of practices in organisations and management.

6.4 Feedforward as Control Mechanism

Whereas the emphasis of feedback as control mechanism in all its applications is on reacting to already existing deviations in the output of a process, feedforward is a term describing a control mechanism that reacts to changes in its environment, usually to maintain some desired state of the process, before the actual primary process takes place. The mechanism of feedforward control can be illustrated by comparing it with a familiar feedback process for the cruise control of a car. When in use, the cruise control enables a car to maintain a steady speed. When an uphill stretch of road is encountered, the car slows down below the set speed; this speed error causes the engine throttle to be opened further, bringing the car back to its original speed (this is feedback). In contrast, feedforward control would in some way 'predict' the slowing down of the car. For example, it could measure the slope of the road and, upon encountering a hill, would open up the throttle by a certain amount, anticipating the extra load on the engine. By using feedforward, the car does not have to slow down at all for the correction to come into play. However, other factors than the slope of the hill and the throttle setting influence the speed of the car: air temperature, pressure, fuel composition, wind speed, etc. Just setting the throttle based on a function of the slope may not result in the constant speed being maintained. Since there is no comparison between the output variable, speed, and the input variable, it is not possible to resolve this problem only with feedforward as control mechanism. A control process that exhibits feedforward behaviour responds to a measured disturbance before the actual primary process takes place in a pre-defined way.

Looking at the general concept of feedforward for the transformation process, this type of control embeds the concept of intervening prior to the execution of the primary process in a downstream direction, when following the flowing element (see Figure 6.4). The intervention is performed either on the flowing element itself or on the primary process or on the system of resources. In the case that a deviation from the desired value of the specific parameter is being identified an intervention will follow to compensate for this deviation. Feedforward control is about preparing flowing elements for the primary process or adjusting the primary process to handle the flowing elements. Similar to feedback, the control mechanism of feedforward

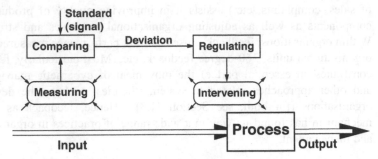

Figure 6.4 *Feedforward. Generic representation showing that a measurement taken from the properties of the input results in an intervention downstream. The regulatory mechanism depends on a model connecting the deviation to the intervention.*

encompasses the activities of measuring, comparing, assigning and, if necessary, intervening. But such measurements and interventions lead to four specific cases of feedforward, see Box 6.2. These four cases demonstrate that the exact configuration of feedforward depends on the type of disturbance occurring in the input prior to the transformation process.

Because of the response being known in advance, some prerequisites come along with a feedforward control scheme:

- the disturbance must be measurable;
- the effect of the disturbance on the output of the system must be known;
- the time it takes for the disturbance to affect the output must be longer than the time it takes the feedforward control loop to affect the output.

If these conditions are met, feedforward might be extremely effective. Feedforward control will respond more quickly to known and measurable disturbances, but will not cope with novel disturbances. For example, the sight of food triggers an anticipatory salivary flow as a form of preparing the human body for eating and digesting food. However, food disguised as another shape will not set off this reaction. In contrast to feedforward only dealing with known disturbances, feedback deals with any deviation from nominal process behaviour; however, feedback requires that the measured variable of the process (in most cases, the output) reacts to the disturbance in order to notice a deviation. Such a response by feedback is always subject to a delay in time, caused by the execution of the primary process and the time to intervene. In that sense, the response by feedforward is only driven by the time lapse between the occurrence of the deviation and the intervention, and not by the execution (time) of the primary process. Feedforward control can be exemplified by learned responses to known cues; applications of feedforward control can be found in control theory (with early contributions by Lefkowitz [1966] and Morgan Jr. [1964]) , physiology (with the earliest contribution by MacKay [1966]) and computing.

The two types of control, feedback and feedforward, are not mutually exclusive. For example, feedforward control could be combined with feedback to allow quick responses with feedforward to defined deviations

Box 6.2:   Basic Types of Feedforward

1ST CASE – INTERVENTION AFTER MEASUREMENT OF INPUT

In this case of feedforward, the properties of the input are measured and translated into an intervention downstream. For example, this type of feedforward occurs if a supermarket would measure the number of customers entering the store and adjust the number of cash registers that are open to the expected queues. To be effective, this case requires that the time delay between measurement and intervention suffices not to disrupt the primary process and the output.

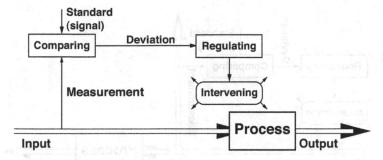

2ND CASE – DISRUPTIONS AS DIFFERENT INPUT

In this particular case, the disruptions enter the main flow of elements, and create an amended input for the primary process – note that the disruption is viewed as a different input rather than a property of the flow as in the first case. An example of this type of disruption would be rush-orders for a company that will go through the same primary process as the regular orders; the tuning of the capacity of the primary process, or alternatively prioritising orders, would constitute possible interventions (note that rush orders are measured separately in this example). To be effective, this case also requires that the time delay between measurement and intervention suffices not to disrupt the primary process and the output.

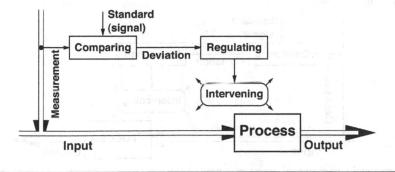

Box 6.2 (CONTINUED)

3ᴿᴰ CASE – DISCHARGE OF INPUT FLOW

Furthermore, a discharge of the input flow to the environment might lead to disruptions. In this case, feedforward control measures this emission and adjusts the input of flowing elements through an intervention. For example, if leakages appear in a pipeline and it is measured how much leakage occurs, feedforward of this type will result in adjusting the flow. If the discharge is harmful to the environment, e.g. leakage of oil and chemicals, then the intervention should result in shutting down the input.

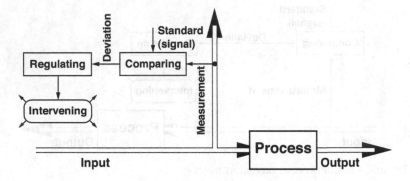

4ᵀᴴ CASE – INTERVENTION BASE ON A QUALITATIVE MEASUREMENT INPUT

The fourth case happens when the measurement leads to an intervention downstream based a qualitative description of the input. In the case of the supermarket, that would mean not only measuring the number of people entering the store, but also their reasons for entering. If they enter the store looking for a quick ready-made meal for the night or if they have extensive shopping lists, that would determine how the supermarket would adjust the number of cash registers. Typically, this type of control anticipates on other than quantitative measurements and results in fine-tuning of the parameters for the process.

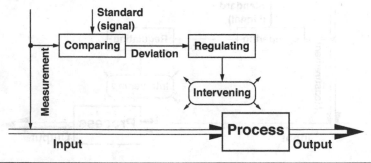

and with the feedback cleaning up for any error in the predetermined adjustment made by feedforward control. Or the feedforward acts as a complementary control mechanism, as shown for the case of hydro-electric power generation in Box 6.D. Most importantly, feedforward does not have the stability problems that feedback may have. As mentioned, feedforward needs to have a pre-calibrated cause-effect relationship, where feedback does respond to any deviation despite its cause. Another way of saying would be that feedforward control applies to measurable disturbances with known effects and feedback control reacts to any disturbances but with delay. For this reason, feedforward complements feedback in most cases.

6.5 Completing Deficiencies

The fourth concept for control mechanisms is the concept of completing deficiencies. This control mechanism compares aspects of the output with standards and then aims at recovering the deficiencies; this mechanism goes back to the feedback amplifier invented by Harold Stephen Black in 1927 [Black 1977]. Most importantly, this control mechanism does not lead to an

---

BOX 6.D: FEEDFORWARD – CASE HYDRO-ELECTRIC POWER GENERATION

For the case of hydro-electric power generation, see Boxes 6.A–6.C, it could also be that an additional influx of water could lead to overcapacity in the power generation. A potential intervention might be diverting the additional potential for power to another distribution channel; if that is done in advance then it would be called feedforward. Note in the figure how the feedback is used in conjunction with directing (Box 6.B) and feedback (Box 6.C). For example, the Paraguayan side of the Itaipu Dam (Box 6.A) uses this principle for 'selling' its overcapacity to the Brazilian distribution network.

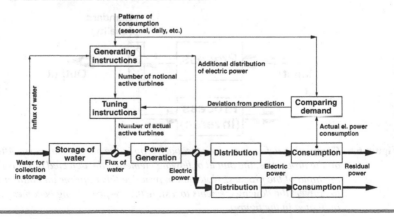

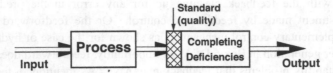

Figure 6.5 *Completing deficiencies by an additional process. After the primary process, a check on properties will reveal deficiencies, which will be completed; note that this is an additional process to the primary process.*

intervention of the primary process or an adjustment of the input (flowing elements), but recuperates the flowing elements as output to ensure that they comply with the target state.

To this purpose, the control process of completing deficiencies comes into action after the primary process. After the transformation process, the flowing elements are being measured in terms of relevant aspects. In the case that a deviation to the normative values of these parameters is ascertained, not a feedback loop comes into action but an intervention takes place by completing the deficiencies of the flowing elements. An example is when a product is made and a part is missing, the adding of the missing part is completing the deficiency; in the case of feedback, the process or the input would be adjusted, ensuring that the next products will be complete (and the deficient product will be simply discarded). This process of completing the deficiencies should be viewed as an additional process to the primary process. Its value resides in the objective to reduce the defects in the overall output given the characteristics of the primary process. The (sub)process of completing implies correcting until the desired value has been reached. Principally, the recovery from deficiencies takes two forms:

- By completing deficiencies as an additional process (see Figure 6.5). This type of correction can be characterised as feedforward after the transformation process has taken place before other processes happen; however, it does not generate an intervention, like the feedforward processes in Section 6.4. An example of this type of control is when cars

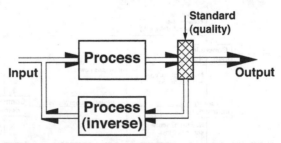

Figure 6.6 *Completing deficiencies by a feedback loop in the primary process. After a check on the properties, the output of flowing elements (the defected ones) is fed back to the input of the process. Note that in practice this requires an additional process to convert the flowing elements to a state that re-processing becomes possible (as indicated in the figure).*

Box 6.E: Completing – Hydro-Electric Power Generation

It might be that in the case of hydro-electric power generation, see Boxes 6.A–6.D, the actual demand for electric power outstrips the power generated. That means that either the actual power consumption should be reduced of additional electric power generated. In practice, such additional power could come from gas turbines that use traditional fossil fuel (oil or gas). That could be seen as completing the deficiency in the output of power generation (see figure below).

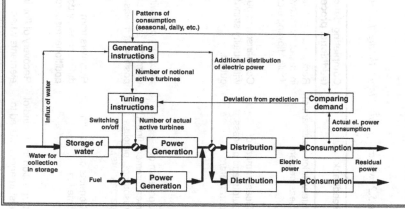

---

have been assembled, they are checked and then missing parts added or faulty parts replaced; that is necessary as the next stage, driving the car, can generally not happen without all parts installed in the car or with parts being defected.

- By a feedback loop into the process or the input of the process (Figure 6.6). After a qualitative measurement of the properties of the input, the flowing elements are fed back into the primary process. In practice, that means that the input has to be converted to possess properties that make this type of recycling possible; for example, a product has to be dismounted before putting back together again during the primary process.

Completing deficiencies is a control process that consists of the following activities: measuring and adjusting or reversing. Typically, the intervention takes place on the flowing element only; see Box 6.E for the example throughout the chapter. However, interventions to prevent these deficiencies recurring again require feedback mechanisms.

6.6   Application of Control Mechanisms

The application of the control mechanisms presented in this chapter (directing, feedback, feedforward and completing deficiencies) depends on the required state of the output and the capability to intervene, see Table 6.1.

Table 6.1 Overview of control processes. This overview takes Figure 5.2 as starting point (input, throughput, output, resources). The applications for the control mechanisms are merely indicative.

	Directing	Feedforward	Feedback	Completing deficiencies (additional process)	Completing deficiencies (loop)
Principle	'One-time' intervention.	Intervention downstream of measurement.	Intervention upstream of measurement.	Recuperating deficiencies in output by recovering 'faulty' flowing elements.	Recuperating deficiencies in output by sending them through primary process again.
Measurement	No measurement.	Measurement of input: 1. Flowing elements. 2. Additional input as disruption. 3. Discharge of flowing element(s). 4. Quality of input.	Measurement of output and process parameters.	Comparison of output with quality standard for output.	Comparison of output with quality standard for output.
Intervention	Only based on changes in standard.	Downstream of measurement (input, resources, process parameters, except Case 3).	Upstream of measurement (input, resources, process parameters).	Recovery of flowing elements at or near position of measurement.	Inversion of flowing elements and processing them again.
Characteristics	Lack of monitoring output or state of resources.	Prevents output to be outside set boundaries.	Corrects output with time lag.	Recovers from deficiencies by using additional resources.	Recovers from deficiencies by using additional resources (loop).
Application	Limited to situations where output is predictable once standard is set (causal model for intervention).	Application limited to causal model for known potential disturbances.	Correction for all kind of disturbances related to aspect of measurement (see also Figure 2.2).	Recovery of flowing elements when corrections by feedback are difficult.	Recovery of flowing elements when corrections by feedback are impossible.

Even then, that may leave options open for the design of a control system since each control mechanism has its own characteristics; even a combination integrated into one design of a control system might be necessary to meet a wide range of disruptions (see Boxes 6.A–6.E). But not only does the design for a definite control mechanism depend on the effectiveness, the resources for the control process itself need consideration as well. For example, if the use of additional resources should be limited, completing deficiencies is a less preferable option. Hence, the choice which control mechanism(s) to deploy and how to design a control system depends on the capabilities for effective interventions and the resources needed to exert the interventions.

When looking at the mechanism of directing again, first of all, it can be considered an activity of control, particularly present in feedforward and feedback. Somehow, standards from the environment have to be converted into directives for control, this will also be discussed in Chapter 7; the basic mechanism for directing does that. An example is a deadline for a project. For managing individual project activities, directing, called project planning in this case, transforms the deadline into a project plan. And it is that project plan that allows monitoring activities and forms the base for interventions. This means that directing may be used in connection with other control mechanisms to be fully effective. Applying directing only is also possible and is appropriate for single events or situations when there is at least a considerable amount of time between every intervention and when the state of the output is predictable and acceptable within set boundaries; actually, any change in the intervention by directing depends only on changes in the standard.

Feedforward control is preferable in case the input as flowing elements needs adjusting in any sense to make it suitable for the transformation process; the same applies to tuning the resources and the process parameters. However, this type of control requires a causal model for associating a deviation (mostly the measurement) with an intervention. For example, the flowing element needs to have a certain temperature before the actual transformation takes place; think about the stages in heat treatment of metal products. In this particular case, the causal model is that metal being at a certain temperature before the next step in a heat treatment results in certain properties; if the product needs specific properties then the metal product should be at a specified temperature before the next step in the heat treatment. In case of a deviation of the normative temperature, the actual temperature has to be corrected to enable the transformation process to be successful. Applying feedforward seems necessary in case the flowing element is rather 'raw' (or 'coarse') and needs to be adjusted in any sense prior to the transformation process. This principle can also be applied when the deviation measured is relative easy to compensate by adjusting the primary process. A case in point is when a teacher adjusts the teaching methods depending on the level of proficiency in a class. Another reason for applying feedforward can be that, when for example the products are very expensive, one could not permit one product to show any malfunction; a quality check of parts before assembling is a case in point. This is related to the fact that correction in the case of feedforward takes place before the actual transformation process. Theoretically, the advantage of this control mechanism is that not a single flowing element has to be discarded. In all cases, the adequate application of feedforward depends on the capability to exert an intervention prior to the execution of the transformation process; or, even better, the principle of feedforward results in an intervention downstream relative to the point of measurement.

Where feedforward measures and intervenes prior to the transformation process, feedback operates where a measured aspect of the output shows a deviation from the normative value; the correction always takes place after the execution of the actual transformation process. Practically, that means that at least part of the output has to be scrapped. This is the case when cooking and the food does not taste at all good. The prepared dish is thrown away and a new preparation starts; however, this will only be successful if there is some understanding about the relationship between the deviation and what to change. Feedback is likely to be applied in case the to be transformed aspect(s) of the flowing element are difficult to measure prior to the transformation process or intervention at that stage proves difficult. Feedback can be used well when the relationship between effect and cause has not yet been clarified. The principle of feedback implies an intervention upstream relative to the point of measurement.

The fourth control mechanism for reaching a target state of the output is completing the deficiencies. Completing seems to be a rather good solution in case of a one-time executed transformation process and when the objective is to achieve output without deficiencies for the overall process. Another situation leading to completing can be that the aspect is difficult to monitor and would require much effort and investments. Taking the example of preparing a dish again, simply adding salt might resolve that it did not taste very well. Also when the rate of occurrence of rejected flowing elements is low, it could be preferable to apply completing the deficiencies. Nevertheless, the principle of completing deficiencies applies to output of the primary process and requires additional resources for the activity of completion or inversion.

The order of the presentation of control mechanisms in this section reflects their capability to predict outcomes on before hand and their capability to correct for unknown sources of deviation. Take directing, which entirely depends on the capability to predict in advance the outcome of the directives given the process. On the other hand, one does find feedback and completing the deficiencies where a causal relationship between the source of deviation, i.e. the cause, and the effect do not have to be known; these control mechanisms simply measure the outcome and intervene. They rely on a model relating intervention to deviations but the relationship does not have to be one-on-one (for example, control mechanisms based on fuzzy logic). Although the capabilities for intervention do quite differ, all control mechanisms need some kind of understanding for relating deviations to interventions.

6.7 Echelons of Control

So far, the control mechanisms, introduced in this chapter, have been treated more or less as single processes leading to an intervention. But in practice, one control process, which does it all, hardly exists and combinations of control mechanisms constitute practice in all kinds of domains. That is demonstrated by the case of hydro-electric power generation in Box 6.E. Also, the co-existence of positive and negative feedback in biological systems has been mentioned in this respect. Moreover, the control mechanisms complement each other. Feedforward control might act on the input of flowing elements entering the process, whereas at the same time feedback control corrects for deviations occurring during the execution of the primary process. The design and implementation of control processes may employ the different control mechanisms to supplement and to complement each other.

Furthermore, a primary process consists mostly of cascading activities, each having their own characteristics and therefore, requiring control mechanisms that fit with the change of state of flowing elements and the capabilities for intervention. Output of one activity might be the input for the next step; on the output of the process feedback might serve as an adequate solution, while the input of the next step might have a feedforward control to correct for any remaining disturbances. Hence, control processes are also linked to each other by the subsequent steps in a primary process.

The cascade of activities with their own control mechanisms leads to a situation that either all control process act independently or the need arises for an overarching control mechanism (or mechanisms). These echelons of control, overarching control mechanisms, act by using the same mechanisms as those for the individual activities (see Figure 6.7); each higher level of control interacts with a number of control processes at a lower level. These echelons should also prevent that the individual control mechanisms contradict each other; the higher levels of control mechanisms might integrate different

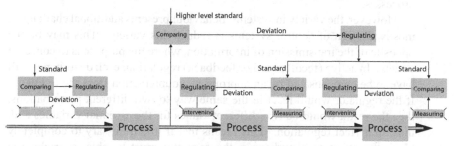

Figure 6.7 *Symbolic representation of echelons of control. Higher levels of control supersede the control mechanisms at lower levels making sure that control is adapted to overall process rather than the optimisation of individual subprocesses.*

aspects, whenever that is possible (remember that distinct types of aspects are hard to compare and to integrate, see Chapter 2).

Therefore, control mechanisms complement each other for a single transformation process, link to different activities in a primary process and might have echelons of control. The application and design of a specific control process for a specific primary process does not only depend on the effectiveness of an individual control mechanisms but also on how several can be combined to achieve overall effectiveness.

6.8   Law of Requisite Variety

This necessity for a wide range of complementary control mechanisms comes also into play when considering the phenomenon of variety. Suppose that a complex system has to be controlled: it has a high variety in elements and relationships, and this variety cannot be ignored. A complex system, looked upon from the content and internal structure, has many interrelationships between all the elements and then most likely covering a broad range of aspects. Ecological systems represent such systems, many elements of various types constitute the whole and it has been difficult or even impossible duplicating it because of its variety and complexity (think about the dome Biosphere II, see Section 11.2). Even then in a controlled environment it appears difficult to predict the behaviour. For the sake of control, a complex system must be represented in a homeostatic causal model and itself capable of maintaining its state. However, simple control mechanisms may have insufficient variety to cope with the variety of perturbations from within the system and from the environment. The notion: 'Give me a simple control system; one that cannot go wrong' underestimates the complexity of variety. Building further on this argument, control or regulation is most fundamentally formulated as a reduction of variety: perturbations with high variety affect the system's internal state, which should be kept as close as possible to the target state and, therefore, exhibit a low variety. In this sense, control mechanisms aim at preventing the impact of the variety of disturbances present in the environment on variations in the state of output, systems of resources or processes.

However, the variety in systems' behaviour presents additional challenges; mostly the aim of control systems is reduce this variety. This may be the opposite of the transmission of information, where the purpose is to conserve variety. In active (feedforward or feedback) regulation, each disturbance will have to be compensated by an appropriate counteraction from the regulator. If the regulator would react in the same way to two different perturbations, then the result would be two different values for the essential variables and, thus, imperfect regulation. This means that if it is necessary to completely block the effect of disturbances, the regulator must be able to produce at least as many counteractions as there are disturbances. Therefore, the variety of the regulator must be at least as great as the variety of disturbances.

Ashby [1956] has called this principle the law of requisite variety: in active regulation only variety can counteract variety. It leads to the somewhat counter-intuitive observation that the regulator must have a sufficiently large variety of actions in order to ensure a sufficiently small variety of outcomes in the essential variables. This principle embedded in the law of requisite variety has important implications for practical situations: since the variety of perturbations a system (or process) can potentially be confronted with is unlimited, the designer should always try maximise its internal variety (or diversity), so as to be optimally prepared for any foreseeable or unforeseeable contingency.

Ashby's law of requisite variety can be seen as an application of the principle of selective variety (the larger the number of states a system goes through, the more likely that one of these states will be retained). However, a frequently cited stronger formulation of Ashby's law, 'the variety in the control system must be equal to or larger than the variety of the perturbations in order to achieve control' does not hold in general, according to Heylighen and Joslyn [2001]. The underlying 'only variety can destroy variety' assumption contradicts with another principle, called the principle of asymmetric transitions [Heylighen 1991], which tells that transitions from unstable states to a stable state is possible but the converse is not likely (think about systems maintaining homeostasis). The principle implies that a 'spontaneous' decrease of variety is possible. An example of this the principle of asymmetric transitions is a bacterium searching for nutrition and avoiding poisons; a bacterium has a minimal variety of only two interventions: increase or decrease the rate of random movements. Each position after a random movement is a new state and searching the space for states leads to it find a favourable set of positions. Its random movements are normally sufficient to find a favourable situation, thus escaping all dangers. This demonstrates that this bacterium with only two possible interventions is capable of coping with a quite complex environment, with many different types of perturbations and opportunities. This example also shows that not necessarily many interventions are necessary to counteract variety. This implies that the interventions of a composite control system (through complementary control mechanisms and echelons of control) might have few interventions at is disposal to respond to a complex environment but that the sensing of the environment (measurements) should at least address that variety and complexity.

For both the understanding and design of control mechanisms, Ashby's law is perhaps the most famous principle of cybernetics. It found its way into many applications, such as electronic control systems and design of computer systems and software. Many did build on the law of requisite variety for their own purposes; for example, Beer [1979, p. 286] restated the law as 'variety absorbs variety'. Whereas the core of this law takes many forms, it depends on simple principle: a control system or controller can only model or control something to the extent that it has sufficient internal variety to represent it.

For instance, in order to make a choice between two interventions, the control processes must be able to assert at least two possibilities, which requires distinguishing at least one characteristic of the primary process or the flowing elements. From the perspectives of comparing alternative interventions, the quantity of variety that the control process encompasses provides an upper bound for the quantity of variety in the process or flowing elements it can control or model.

From this point of view, the blackbox approach (Sections 3.2 and 5.7), facilitates dealing with variety; according to Beer [1959, p. 50], a suitable blackbox model will contain enough information to handle the variety of a complex system [*ibid.*, pp. 52 ff., 76 ff.]. This has much to do with ignoring the internal structure as additional variable(s): the control mechanisms focus only on the external effects of the interventions. The blackbox approach has been introduced as reducing the information to these external effects, allowing an overall view on the process and system. In this sense, the blackbox approach only works when the internal structure of the system and the process possesses the capability to handle the potential disturbance; when this exceeds the capability an intervention in the structure becomes necessary.

Particularly, this is the case for organisations when their environment changes. In such cases homeostasis, balancing the internal structure and the environment, may not fit as model for change when talking about dynamically changing environments in which less time remains to implement gradual changes. If the structures for increasing the so-called complexity handling capability [Boswijk 1992, p. 101], the ability of an organisation to deal with the imposed complexity by its environment, exist, homeostasis does not drive adaptation; his proposition seems an extension of the law of requisite variety to the changing and modification of structures of firms. Companies that do not employ processes and structures to cope with the imposed complexity are forced to exert severe interventions heavily drawing on resources within and in reach of the company, decreasing the chances of survival. To increase their complexity handling capability, organisational entities might decrease their internal complexity through redefining their organisational structures and their product structure (products seen as output to the environment and fulfilling the function of an organisation as a system, according to Applied Systems Theory). The effect of these internal measurements seems limited; an organisation may win more by learning to increase its base of capabilities for dealing with the imposed complexity of the competitive landscape on which it operates [Dekkers, 2005]. Hence, in any case, organisations might improve their complexity handling capability by modifying their organisational structure to comply with the law of requisite variety.

6.9   Summary

Thus, control is all about interventions in the flowing elements, the primary process or the system of resources (consistent with the definition and scope of a process in Chapter 5). If one wants to reach certain objectives or better states of the flowing elements as output, interventions in processes (by tuning parameters) will correct deviations happening during the execution of primary processes. Even in its simplest form, control acts on processes and flowing elements for standards to be maintained, whether it concerns technical processes or organisational processes or any other process; those standards are derived from objectives for the system and they define the states of the flowing elements or resources.

For the control mechanisms four basic principles to exert interventions are at hand with their own advantages and consequences:

- Directing: this control mechanism converts standards into directives without any measurement of the state of the flowing elements, the process or the system of resources.
- Feedforward control: this control mechanism measures the input and intervenes in the inflow of elements, in the parameters for the process or in the system of resources (the intervention takes place downstream).
- Feedback control: this control mechanism intervenes based on the measurement of the output and the intervention takes place upstream in the input of flowing elements, the parameters for the process or the system of resources.
- Completing deficiencies: this control mechanism checks the state of the output and corrects any deficiencies as a complementary primary process or as feedback loop (inverse process) for the flowing elements.

The first three control mechanisms rely on the availability of a model to relate settings or deviations to an intervention, whereas the fourth one corrects deficiencies in the output.

For the effective deployment of control mechanisms, the law of requisite variety simply means that a flexible control system with many options is better able to cope with variety in change. One that is tightly optimised for an initial set of conditions might be more efficient whilst those conditions prevail but fail totally should conditions change. In its original setting of control theory, Ashby's law of requisite variety concerns controllers trying to keep a system stable. The more options the control process has, the better able it is to deal with fluctuations in the flowing elements, the system of resources and the process. For organisations that means they may have to improve their complexity handling capability by modifying their organisational structure to match the variety imposed by the environment. Variety in input, resources and processes can only be dealt with by variety of interventions.

References

Ashby, W. R. (1956). An introduction to cybernetics. New York: J. Wiley.
Beer, S. (1959). Cybernetics and Management. New York: Wiley.
Beer, S. (1979). The Heart of Enterprise. Chichester: Wiley & Sons.

Black, H. S. (1977). Inventing the negative feedback amplifier: Six years of persistent search helped the author conceive the idea "in a flash" aboard the old Lackawanna Ferry. IEEE Spectrum Magazine, 14(12), 55–60.

Bogdanow, A. (1928). Allgemeine Organisationslehre Tektologie (Vol. II). Berlin: Organisation Verlagsgeschellshaft (S. Hirzel).

Boswijk, H. K. (1992). Complexiteit in evolutionair and organisatorisch perspectief, het zoeken naar balans tussen vermogens en uitdagingen. Rotterdam: Erasmus Universiteits Drukkerij.

Dekkers, R. (2005). (R)Evolution, Organizations and the Dynamics of the Environment. New York: Springer.

Dekkers, R. (2015, 3–5 August). On the Origins and Applications of the Steady-State Model. Paper presented at the 23rd International Conference on Production Research, Manilla.

Farcot, J. (1873). Le servo-moteur ou moteur-asservi. Paris: J. Baudry.

Heylighen, F. (1991, Nov.). The Principle of Asymmetric Transitions. Principia Cybernetica Web.

Heylighen, F., & Joslyn, C. (2001, 31st August). The Law of Requisite Variety. Principia Cybernetica Web. Retrieved from http://pespmc1.vub.ac.be/REQVAR. html

Iancu, Ş. (2009). Ştefan Odobleja – The Main Romanian Forerunner of the Cybernetics. Annals of the Academy of Romanian Scientists – Science and Technology of Information, 2(1), 41–58.

Lefkowitz, I. (1966). Multilevel Approach Applied to Control System Design. Journal of Fluids Engineering, 88(2), 392–398. doi: 10.1115/1.3645868

MacKay, D. M. (1966). Cerebral Organization and the Conscious Control of Action. In J. C. Eccles (Ed.), Brain and Conscious Experience (pp. 422–445): Springer.

Maxwell, J. C. (1968). On Governors. Proceedings of the Royal Society of London, 16, 270–283.

Minorsky, N. (1922). Directional stability of automatically steered bodies. Journal of American Society of Naval Engineers, 42(2), 280–309.

Morgan Jr., B. (1964). The synthesis of linear multivariable systems by state-variable feedback. IEEE Transactions on Automatic Control, 9(4), 405–411. doi: 10.1109/ TAC.1964.1105733

Ormerod, P. (1994). The Death of Economics. New York: St. Martin's Press.

Soros, G. (1987). The Alchemy of Finance: Reading the mind of the Market. New York: Simon and Schuster.

White, W. H. (1900). A Manual of Naval Architecture: For Use of Officers of the Royal Navy, Officers of the Mercantile Marine, Yachtsmen, Shipowners, and Shipbuilders. London: John Murray.

7 Steady-State Model

The steady-state model adds the control for the system boundary, or more precisely the process boundary, in addition to the control processes of the previous chapter, such as feedback and feedforward. The two previous chapters have emphasised the (primary) process and its control, based on the premise that a process constitutes an interaction between flowing elements (as system) and resources (as an other system). In addition, systems operate in relation to their environment and interact with other elements from that environment. The processes that occur in systems convert input, consisting of flowing elements, into output by changing the state of these flowing elements. The resources that execute that conversion might have a limited capability for dealing with variations in input and throughput. To this purpose, it becomes necessary that systems of resources have capabilities to deal with the varieties in input and output, as the steady-state model will describe for the processes.

The steady-state model builds on the concept of homeostasis and cybernetic principles in the writings of E. J. Miller and Rice [1967] and Emery and Trist [1969], and the conceptualisation of living systems [James Grier Miller, 1965a, b] in which the mechanisms of control from the previous chapter are integrated (see Dekkers [2015] for a full description and how it related to Beer's [1972] viable system model). These writers apply these principles to organisational, technological and biological systems; the consequences of such thinking for organisms and organisations are found in Chapter 8, which is about autopoiesis. Maintaining homeostasis applies to fourth level of the systems hierarchy of Boulding and higher (see Section 3.5); from that level on, systems are called open systems, which interact with their environment, and it is this interaction that calls for the need of boundary control.

To this purpose, section 7.1 will expand on the implications of boundary control, following the concept of Miller and Rice. They discuss the concept of boundary zones for organisational systems; however, this section will also talk about the validity for technical systems and biological systems. Chapter 8 will deal with the concept of autopoiesis, which application resembles in some aspects the steady-state model, but has a limited application to biological and organisational systems; the current chapter focuses on open systems in general. The treatment of the three boundary zones is the subject of Sections 7.2–7.4. For each of the boundary zones, the main concepts are elaborated. Finally, Section 7.5 integrates the separate boundary zones into one model and also discusses the limitations of the steady-state model.

7.1 Boundary Control

The behaviour of processes and systems is bound by how the throughput and control mechanisms deal with variations imposed through interactions with

© Springer International Publishing AG 2017
R. Dekkers, *Applied Systems Theory*,
DOI 10.1007/978-3-319-57526-1_7

the environment. For example, a banking process only allows authorised users to enter the system and discards any other entities through an electronic signature or password or any other form authentication. This process executed by computer systems identifies authorised users and ideally prevents unauthorised customers and others to enter and use data for other purposes. This way it secures the input of data by users and the output of an accurate and authenticated report of banking accounts and it prevents unauthorised access to data recorded in the system. These actions take place principally before and after entering and converting data, which is the primary process for a bank; hence, the authentication of users happens in the boundary zone relative to the entering and converting of data. This so-called boundary control acts on the primary process itself and the internal control processes, and at the same time serves as an intermediary between the environment and the primary process.

Most processes are in some measure self-regulating in the sense that the nature or structure of the processes imposes limitations and constraints on the associated processes and systems of resources. Thus, a given activity (whether part or no part of a process) is regulated by preceding and succeeding activities. An example is the capacity of a process, which is mostly determined by the availability of the system of resources; hence, the output of a process is limited by the capabilities of the system of resources. Regulatory activities that relate a set of processes to its environment, from the perspective of the allocated resources, occur at the boundary of the process and its resources, and they control the input and output for the process to maintain a steady state. Before expanding on the steady-state model, this section will first expand on the concept of steady state, boundary zones and the particular phenomenon heterostasis as key concepts for self-regulation.

Steady State

When all variables in a system are balanced to the point where no change is occurring, the system is said to be in static equilibrium. In such a case, internal processes do not take place since no deviation activates control processes to preserve a specific state. In practice, that seems hardly the case. Perturbations enter the process or affect the system of resources and activate process within the system boundary to maintain a balance with the environment. Even when everything seems in balance, it may denote a situation in which forces dynamically interact to maintain a point of equilibrium. The human body when standing up is a case in point; muscular actions ensure a posture that may look static to an external observer. A dynamic (steady-state) equilibrium exists when the system components are in a state of change, but at least one variable stays within a specified range. Whereas static equilibrium implies no changes at all taking place, dynamic equilibrium indicates equilibrium between at least two system or process variables.

In the case of dynamic equilibrium, it is necessary for many systems to maintain their equilibrium in changing environments or in response to disturbances, otherwise they cannot function properly or their goals cannot be attained. In living systems, the process of self-maintenance or 'homeostasis' proves essential to ensure their survival and sustained viability. The term homeostasis is referred to by Flood and Carson [1993] as a process by which a system preserves its existence through the maintenance of its dynamic equilibrium. By some this equilibrium is termed 'homeostatic equilibrium' [e.g. van Gigch, 1978]. Even when a mature organism as an open system appears to be unchanged over a period of time, there is a continuous exchange and replacement of matter, energy, and information between the system and the environment. Homeostasis is not only one of the most important properties of any living organism, but is also readily applicable to human or work organisations treated as open systems. For example, organisations need to recruit new employees to replace those who retire; they also need raw materials, energy and information for use in their processes and operations to maintain a steady state. In fact, an organisation that appears externally static and unchanged to external observers is internally in a state of flux, in a state of dynamic equilibrium (as with most open systems). Maintaining equilibrium constitutes a major activity for open systems.

Another significant aspect of an open system in a state of dynamic equilibrium is that it relies at least on feedback mechanisms to remain in this specific state. Based on the systems hierarchy of Boulding (see Section 3.5), which classifies the system according to its complexity, it is not surprising to find that properties exhibited by systems lower in the hierarchy are also found in those higher in the hierarchy because the latter are built on the former. Therefore, a system that is classified as an open system would possess all the qualities that belong to the system at a cybernetic (or self-regulated systems) level, see Box 7.1; open systems are found at the fourth level, whereas many of the principles are derived from cybernetic systems, the third level of Boulding, including feedback mechanisms. The behaviour of open systems is determined, to a great extent, by the feedback mechanisms present in them. Negative feedback reduces or eliminates the system's deviation from a given standard, so a negative feedback mechanism tends to neutralise the effect of disturbance from the environment. Positive feedback amplifies or accentuates change, which leads to a continuous divergence from the starting state. Positive feedback works together with negative feedback in living systems (e.g. in organisms), and organisations too; both types of feedback are present during adaptation even though the net result might be positive. However, the operation of positive feedback alone will eventually result in the system's disintegration or collapse. Negative feedback plays the key role in the ability of open systems to achieve a steady state, or homeostasis; therefore, negative feedback mechanisms are inherent to systems achieving and maintaining a steady-state. This means that at least negative feedback is needed to maintain equilibrium; positive feedback without negative feedback will lead to moving away from a stable but dynamic state. The other control mechanisms in Chapter 6 as singular mechanisms are less effective in ensuring that a steady state will be maintained. Directing assumes that a one-time intervention will lead to the stable state and does not verify it. Similarly, feedforward can only act on correcting the input, the system of resources and the process, but does not measure the output. And, completing deficiencies only corrects the output of the process but does not address the causes. Thus, negative feedback is a necessary condition for maintaining dynamic equilibrium by an open system in its interaction with the environment.

Boundary Zones

Any open system, read a system of resources in the spirit of Chapter 5, interacts with its environment and mostly through a conversion process. Take (living) cells, for example. Living systems require the continuous uptake of energy and nutriments from their environment, to excrete and to react in specific

---

Box 7.1:   KEY CONCEPTS FOR STEADY STATE

STATIC HOMEOSTATIC EQUILIBRIUM

Static homeostatic equilibrium refers to a steady-state situation with no dynamic events acting on the system of resources. By definition, in the case of static equilibrium there is balance, but no change, disturbance or event.

DYNAMIC HOMEOSTATIC EQUILIBRIUM

In contrast, dynamic homeostatic equilibrium occurs when perturbations or changes act on processes or systems of resources causing a temporary deviation from the equilibrium. By activating internal processes in response to that perturbation or change, the process or system of resources tries to regain its point of equilibrium with the environment. Sometimes, this concept is also used for the situation where a system loses its equilibrium and finds a new state of balance (for example, when a glass of water is tipped over and comes to a rest at the counter).

SELF-REGULATION

Although self-regulation has many connotations, in the context of Applied Systems Theory, it refers to the capability of systems of resources or processes to maintain a (fixed) state. This state might be subject to external influences or perturbations; when these occur internal process within the system of resources ensure appropriate responses to maintain that state. Self-regulation might cover simple control processes, such the ones in Chapter 6, or complex interactions with the environment, as shown in this chapter.

ways. Therefore, cells – just like all other biological systems – have to be regarded as open systems that are characterised by inputs and outputs and a transition. These open systems are never in a static equilibrium but always in a steady state as a dynamic equilibrium. As long as we do not know what happens in the transitional element (in this case the cell) it can, according to system theories, be regarded as a blackbox. The relation between input and output characterises a flow of (coded) information through the system. A physical or chemical energy may influence the system through the input and cause certain changes that may again have an influence on other systems or system elements via the output. From a cybernetic point of view, neither the inner structure of the transitional element nor the form of the energy is of importance. The input and output as events in the boundary of a system and the connection between both is decisive for determining the behaviour of an open system.

Therefore, the boundary of a system of activities implies both a discontinuity and the interpolation of a region of control [Miller and Rice, 1967, p. 9]. Difficulties arise if a boundary is imposed at a point in the process, which does not satisfy these two criteria of the boundary of an activity system. Unless there is a discontinuity, there can be no boundary to separate a system from its environment and thus no distinction in which activities are carried out within the supposed system and that are insulated from other activities 'outside'. Such a discontinuity happens, for instance, when somebody wants to use a copier or printer. The paper arrives in packs from its manufacturing process and needs to be unwrapped and stored in the device before the actual copying or printing starts; the packs represent the disconnection between the manufacturing process of papers and the use of papers for printing or copying. A different example of discontinuity is imports at the level of a national economic system; whereas these imports are a necessity for factories, acquiring goods as imports constitutes a discontinuity. The second criterion for the boundary zone implies that the discontinuity leads to regulatory activities of some sort. The discontinuity results in a mismatch between the two systems of processes, regulatory mechanisms aim at achieving equilibrium so that outputs of one process match with the input requirements for the next process. A case in point is a water collection and irrigation system in agriculture that ensures relatively constant water supply even though it might rain at very unpredictable times. But also raising import levies on goods in the context of a national economic system constitutes a regulatory mechanism. Hence, boundary zones indicate both a discontinuity and the presence of regulatory activities of some kind.

Regulation in the boundary zones itself can be analysed as an input-conversion-output process. Input activities are the collection of data from measurement or other observation, conversion activities the comparison of these data with objectives or standards of performance and output activities the decisions to stop or modify the process or to pass the product. An example is the inspection processes of goods at manufacturing enterprises;

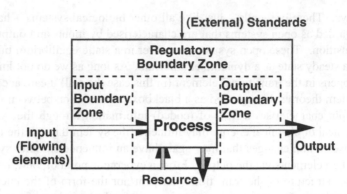

Figure 7.1 *The three boundary zones of a process. The boundary zones for the input and the output interact directly with environment through the flowing elements. The regulatory boundary zone converts standards imposed externally on the system to the internal control processes.*

raw materials are tested before being accepted and products inspected before being dispatched. In larger factories, inspection processes are also positioned between departments. Provided the inspection occurs at boundaries between the enterprise and its environment or between distinct constituent systems of the total enterprise, there are few problems with regard to boundaries. Coordination problems at the boundary increase when considering the (organisational) effect of introducing continuous automatic controls, particularly those that incorporate feedback and other self-correcting mechanisms. Consequently, these automated regulatory activities eliminate time lapses between checks for one system of operating activities and the next system. Hence, the design of regulation in the boundary zones should allow effective interventions that compensate the impact of discontinuities.

The boundary zones can be divided into three zones, the input boundary zone, the output boundary zone and the regulatory boundary zone (see Figure 7.1). Each zone fulfils a specific function. The input boundary zone acts on the inflowing elements and their properties so that the primary process at the heart of processing will operate within its boundaries. An example of this is the filtering of air through respiration systems before it enters the lungs. The output zone operates the other way around: output of the process is converted to be suitable for entering the environment. When software generates a report and completes the data with other information, such as headings, tables and figures, the output is ready for use by an operator or manager. The regulatory zone translates external standards and information from the environment to operational directives for the control processes. All three zones constitute the boundary zone of a process and its system of resources.

Heterostasis

Through the three boundary zones, processes and systems of resources maintain homeostasis, a property of an open system; especially living organisms regulate their internal structure and state to maintain a stable, constant condition, by means of multiple dynamic equilibrium adjustments, controlled by interrelated regulation mechanisms. The term is most often used in the sense of biological homeostasis. Multicellular organisms require a homeostatic internal environment, in order to live; many environmentalists believe this principle also applies to the external environment. Many ecological, biological, and social systems are homeostatic. They oppose change in favour of maintaining equilibrium. If the system does not succeed in re-establishing its balance with the environment, it may ultimately lead the system to stop functioning; the extinction of species is a case in point. Complex systems, such as the human body, must strive for homeostasis to maintain stability and to survive. Each of the three boundary zones contributes to that purpose of maintaining the steady state for open systems.

These open systems do not only have to endure to survive; they must adapt themselves and evolve with modifications that fit to the dynamics of the environment. The now widely accepted concept of complex adaptive systems, see Chapter 9, was first suggested by Selye's [1976] adaptation syndrome, which emphasises, among other things, the positive role of inflammation in striving for homeostasis. This is observed in the self-limiting illnesses and many febrile conditions, where complementary and alternative treatment may be supportive, or enabling the self-healing process, as long as the effort is within the vital capabilities of the patient. Selye also coined the term heterostasis to describe the potential of healing, such as inducement of a febrile response by Echinacin in herbal medicine and constitutional hydrotherapy in naturopathy. Klopf [1972] developed a basis for learning in artificial neurons based on a biological principle for neuronal learning called heterostasis. He states that organisms are not hiding in the environment; on the contrary, they are trying to minimise action and change. In general, organisms actively seek stimulation. Heterostasis is the seeking of this maximum stimulation. For example, all parts of the brain are independently seeking positive stimulation (or 'pleasure') and avoiding negative stimulation (or 'pain'). Emotion provides the sense (a measure) of what the organism needs, whereas cognition provides the means for achieving those needs. The concept of heterostasis is currently applied to the body's endurance when taking out organs and conducting an external treatment before placing them back; that (temporary) change in internal structure forces the system to function at a different point of dynamic equilibrium than usual. Heterostasis implies that open systems might temporarily operate at other points of dynamic equilibrium than the optimal point of homeostasis.

7.2 Input Boundary Zone

One of the three boundary zones to maintain homeostasis is the input boundary zone. This constitutes the zone before the actual primary process where interventions are exerted on the properties, i.e. the quality, and the

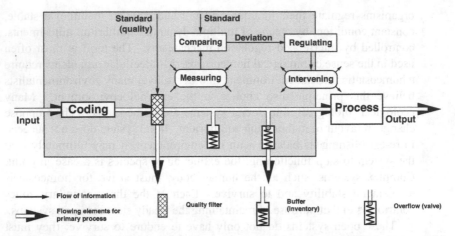

Figure 7.2 *Boundary control at input zone. Elements flowing into the boundary zone will be coded first before a check on appropriate properties. After this check the combined activities of feedforward control, the input buffer and the overflow valve prevent the input exceeding the capability of the primary process.*

quantity of the input as flowing elements. By doing so, the process ensures that the input matches the properties of the flowing elements as input with the capability of the transformation process; the same applies to the quantity of the input. The generic interventions of the boundary input zone are shown in Figure 6.2 and explained in the subsections.

Coding

Often the flowing elements as input do not have the properties for the actual transformation process to identify them or the properties for further processing. For example, an order is sent to an English company written in the Chinese language. After receipt, the translation of the document into a native language constitutes coding as well as the identification of which products and services the order refers to (e.g. article numbers). But also chewing food before swallowing to facilitate digestion is an example of coding. Thus, coding also requires determining which aspects are crucial to the further processing; this means that coding is related to the aspects for modelling (see Sections 3.1 and 3.4). The principles of coding go back to Shannon's [1948] communication theory; his proposition that still reverberates today in many disciplines, including systems theories. Coding constitutes a necessity for either identification during later activities in the steady-state model or for matching properties of the input to the capability of the transformation.

Generically speaking, only after coding, it becomes possible to activate control processes and other activities that make the flowing elements suitable for downstream processes. Without the identification of the flowing elements needed for further processing, a quality check becomes difficult to perform. Even the coding as a conversion process could be looked at from this perspective; unpacking paper from a box to put into a large volume copier makes it possible to control the flow of paper into the core processing unit of that copier. Internal standards are mostly linked to classification of the flowing elements. For example, in case of an organisation, the goods receipt for stationary and raw materials for the primary process differ substantially and have very different standards to adhere to. For the transformation process itself the input needs to be coded before passing through a filter for acceptance.

Quality Filter Input

Next, the filter of acceptance, called the quality filter, checks the coded input against standards for its properties. If the qualitative standards of defined aspects are not met, then the flowing elements should be brought up to standards; however, generally, this takes place outside the system boundary and is symbolised by the discarding of the elements that do not meet those standards (system refers to the system of resources). Under normal circumstances, not all properties of the flowing elements undergo a quality check, but only those that represent 'critical' properties to the transformation process. Taking the case of imports of goods again. It might be that the imported goods are inadequately labelled or packaged according to regulations; in this case, the goods are refused by the customs and returned to producer in the country of origin, where they can be repacked, re-used or discarded. The packaging and labelling represents only one specific aspect of the imported goods. The customs officers do not check other properties, such as geometry, which may be important for another manufacturing company using the imported goods. When the quality of the input meets the standards, the flowing elements can be processed further. The quality filter checks against standards as set by higher echelons of control and it accepts or rejects the flowing elements for the actual transformation process.

Control Mechanism (Feedforward)

After the acceptance of the quality in the input boundary zone, the control mechanism of feedforward acts on 'quantitative' features of the flowing elements. Quantitative features refer to the parameters and facets of these flowing elements (see Section 2.2 and Figure 2.2) that trigger control mechanisms to adhere to standards with respect to performance. In this sense, the quality filter corrects only the 'feasibility' of flowing elements for entering the transformation process; but it does not check against the actual performance (for example, the capacity of the transformation process as measured through the timely delivery of output). If the control of quantity is placed before the quality check, it may occur that defected flowing elements are accounted for when exerting intervention; in that case the intervention may even be based on inaccurate parameters. To avoid this systematic error, the control mechanisms for quantitative aspects (or parameters) are position after the coding and the quality filter.

As the control mechanism feedforward measures the quantitative aspects of the flow and exerts an intervention. The intervention leads to a correction before the transformation process takes place or to adjustment of the transformation process. This could be the case when the influx of orders is measured and depending on volume of these orders the number of workers is adjusted. Or it could be that the parameters for printing are adjusted when using certain types of paper (many inkjet desktop printers offer that change of settings for printing). Therefore, feedforward ensures the more quantitative interventions in the flowing elements as part of the input boundary zone before the transformation takes place.

Input Buffer

The input buffer corrects for differences in supply of flowing elements by the environment and the capability (or capacity) of the transformation process. When the inflow is too much or irregular in comparison to the capability of the transformation process, the excess of flowing elements is buffered. At moments that the supply undercuts the capability of the transformation process, the input buffer supplies the deficiencies in the influx depending on its own capacity. However, the input buffer does not necessarily have to be positioned away from the flowing elements; for example, in the case of a 'first-in-first-out' approach all flowing elements will enter the buffer. An example is queuing in a post office; people enter the post office, take their position at the end of the queue and wait for their turn. As a different approach, in the case of a production line, the excess of goods might be put into a separate storage area until the influx of elements is insufficient for the transformation process; then the excess of goods will be put back to the further processing. Such an approach is only possible when flowing elements do not deteriorate with regard to their properties. Both examples show that the input buffer corrects for irregularities in the influx of flowing elements.

Overflow (Valve)

When the system of resources cannot cope with the supply of flowing elements, then the abundant flowing elements will be discarded into the environment. This acts as a last resort in case the feedforward loop could not anticipate sufficiently and the input buffer has reached its maximum capacity. An example of this is a parking garage when it is full; cars wanting to park are turned away and have to find another possibility for parking or change destination. This overflow of flowing elements makes it possible that the transformation process operates within set limits with respect to the aspect of the control mechanism.

The three mechanisms – the feedforward loop as control mechanism, the input buffer and the overflow valve – have similar contributions to the input for the actual transformation process. The feedforward loop controls the flux of flowing elements (by linking it the capability of the transformation process), whereas the input buffer and valve react beyond the deviations that the feedforward control mechanism handles. It could be that not all three are necessary or useful in certain cases. In the example of serving coffee, in which the serving itself is the transformation process, it makes no sense to have a buffer of filled coffee cups, because the drink will cool down and become distasteful. Hence, the feedforward control mechanism, the input buffer and the overflow valve have distinct and yet complementary functions for the input boundary zone for the primary process.

7.3 Output Boundary Zone

Similar to the input boundary zone, the transformation process needs additional activities to ensure that its output matches with what is needed by the environment and acceptable to it, see Sections 6.3–6.7. Sometimes, control processes for the primary process suffice to generate output that meets standards set by the environment; in other cases, additional activities in the boundary zone are needed to achieve an output that conforms to those standards. In addition, a primary process might also consist of processes at a lower level of detail that need individual control loops. This might evoke the necessity to deploy higher internal echelons of control to avoid contradictory interventions and to achieve overall objectives for the primary process. Because of meeting overall objectives, activities in the boundary zones are mostly connected to internal control processes as well as higher external echelons of control for interaction with the environment.

Hence, the output boundary zone, positioned after the primary transformation process, regulates the transfer from flowing elements into the environment. The boundary zone at the output-side of the transformation process (see Figure 6.3) constitutes of similar processes found in the input zone (Figure 6.2); their specific use in the output boundary zone will be elaborated in the following subsections.

Control Mechanisms (Feedback and Completing Deficiencies)

In the output boundary zone, there are principally two control mechanisms of the ones discussed in Chapter 6 present. The first one is the feedback control mechanism that measures the flowing elements against standards set by the environment and intervenes in the primary process or its input or the system of resources. For instance, it could be that the output of a factory lags behind on the delivery schedule for orders; the intervention could be either increasing the capacity of the primary process or reducing the input of orders, resulting in less materials and components entering the actual transformation process. A quality filter on the output side supplements the feedback control. This filter measures the properties of the flowing elements that come out

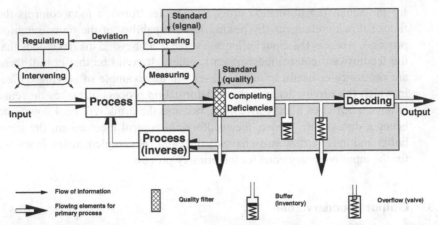

Figure 7.3 *Boundary control at output zone. Elements flowing through the boundary zone will be checked on appropriate properties and this might result in discarding them or completing the deficiencies; note that in case the deficient elements are fed back into the primary process an inverse process takes place, which might be part of the boundary zone of the system of resources or not. The combined activities of feedback control, the output buffer and the overflow valve prevent the output exceeding the capability of the environment. The final step is decoding the flowing elements for the transition to the environment.*

of the process, compares them with the qualitative standards and exerts an intervention; this intervention is either discarding the flowing elements or adding and replacing deficiencies in the flowing elements or recuperating the flowing elements by using an inverse process and passing on the flowing elements back to the primary process. In the case of a factory, it could be that some products as output are discarded because they are beyond repair, other products might just need a few components to be added before they can be shipped to customers, such as adding user instructions, and some products might have to dismantled and the components can be used again for making the products. Both control mechanisms, feedback and completing deficiencies, have a different function and exert different interventions.

It should be noted that the filter of acceptance, called the quality filter, checks the coded output against qualitative standards of pre-defined aspects. If the quality of the output meets the standards, the flowing elements can be further processed and delivered to the environment. When the qualitative standards of the aspect are not met, then the flowing elements should be brought up to standards. However, this recuperation might take place outside the boundaries of the system of resources in general and is then symbolised by the discarding of the elements that do not meet those standards. It is also possible that the recovery takes place within the boundaries of the system of resources. Whether the recuperation takes place within the system or outside it depends on the processes needed for converting the defected output into flowing elements that can be processed again by the system of resources for the primary processes.

Output Buffer

Similar to the buffer in the input boundary zone, the output buffer corrects for differences in supply by the transformation process and the capability (or capacity) of the environment to absorb the process' output. When the outflow is too much or irregular in comparison to the capability of the environment, the excess of flowing elements is buffered. At moments that the supply undercuts the capability of the environment, the output buffer supplies the shortages in the flowing elements depending on its own capacity. An example of this is a petrochemical plant. Switching on and off capacity or even tuning capacity is often lengthy and difficult. To match demand with capacity of the plant, it uses storage of finished products as output buffer, for example by using oil tanks in depots. Thus, the function of output buffer is mediating between variances of the process and the intake of flowing elements by the environment of the system of resources.

Overflow (Valve)

Again similar to the input boundary zone, if the environment of the system of resources cannot cope with the supply of flowing elements and the output buffer has reached its maximum capacity, then the abundant flowing elements will be discarded into the environment (the valve). This is a very different process from the quality filter directly after the primary process. The overflow valve acts as a 'safety' measure to prevent an overload of flowing elements to the environment by discarding acceptable products from a quality perspective. Factories might use such a mechanism when they get rid of products they cannot sell anymore (products at the end of the life-cycle or competition from more attractive products). However, companies discard those products sometimes by selling them through different distribution channels at reduced prices; in such cases this could not be considered as part of the overflow, but a regular output with different qualities only. Therefore, the valve represents a principal safeguard against overloading the environment with regard to the regular output.

Also similar to the activities in the boundary zone for the input, the three mechanisms – the feedback loop and quality filter as control mechanisms, the output buffer and the overflow valve – have similar contributions but yet distinct functions for the flow of elements into the environment. The feedback loop controls the flux (by linking it the capability of the transformation process for defined aspects), whereas the output buffer and overflow valve react beyond the deviations that the feedback loop handles. It could be that not all three are necessary or useful in certain cases. In the example of preparing and serving dinners à la carte, there will be no buffer of ready since the taste will detoriate and customers expect freshly prepared meals. Again, the design and the set-up of the boundary zone depend much on the characteristics of the primary process and the requirements set by the environment.

Decoding

Decoding happens after the completion of the process and adapts the flowing elements to the environment. For example, exhaust gasses of a car are processed through a catalyser before streaming into the environment, that way reducing the output of CO and NO_2. Again, this process of decoding stems from Shannon's [1948] communication theory. It can be a simple process of labelling to more complex steps of identification or even shaping the flowing elements. An example of the latter is the adding of a scent to natural gas to make it detectable when leakages occur in the distribution systems. It should be kept in mind that decoding principally does not change the properties of the flowing elements.

7.4   Regulatory Boundary Zone

During the discussion of the two boundary zones and the preceding chapter about control mechanisms the standards for comparison with measurements were more or less treated as coming from external to the system of resources and its related processes. These standards will vary between precisely defined and vague. In addition, the environment will set the standards but not relate or assign them directly to a specific control mechanism. Therefore, the system has to convert these external requirements into operational standards that will allow the system of resources to fulfil its function within the larger whole. For tuning the system's internal control processes to the environment, there is a regulatory boundary zone, see Figure 7.4. This zone deploys the initiating process and the evaluation process for maintaining the function of the output and homeostasis for the system of resources.

Initiating Process

The process that transfers relatively vague requirements into standards or otherwise adapts them for internal control mechanisms is called the initiating process. Through initiation the standard as imposed by the environment is transformed into standards suitable for use by the internal control process. An example of a standard is the reliability of delivery for customer orders by a company; the key performance indicator itself is not directly usable for the control process but needs to be translated into acceptable queue lengths before the primary process (feedforward) and internal performance evaluation (feedback). Within this line of thought, the initiating function bridges the transformation process with its environment for the interpretation of externally generated standards and is considered as part of the boundary of the processes and the system of resources.

Ideally, this conversion from external standards to internal standards (or directives because of the similarities with the control process of directing) requires no elaboration. Coding, akin Shannon's [1948] communication theory, might occur to internal standards to make them fit to the internal

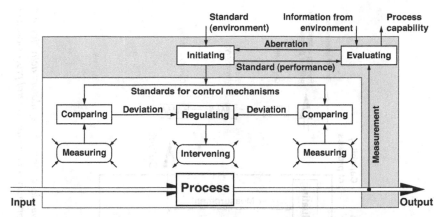

Figure 7.4 *Regulatory boundary control (partly derived from Bogart [1980, p. 238]). Standards from higher levels are converted into standards for the control mechanisms by the initiation process. The evaluation process compares internal measurements with the standard set by the initiation process and the information obtained from the environment. The evaluation process passes on aberrations to the initiating process and information about the system of resources' capability to the environment.*

control mechanisms. However, the coding is generally not necessary for the link to the evaluation process and is not displayed in Figure 7.4.

Evaluation Process

The process of evaluation consists of obtaining information, comparing that to the standard and informing the initiating process about aberrations. This is necessary because a standard makes less sense when no regular check takes place on its validity. For example, changes in sales and lead-times of parts do directly effect requirements for minimum levels of inventory. Hence, changes in the environment or within the system may induce updating of the standard. In this respect, it is absolutely necessary to evaluate the standard and to revise it when deviations show up.

One source of information for the evaluation process comes from internal mechanisms for measuring the output. These measurements are not necessarily linked to the internal control mechanisms. An example is measuring how many individual customers (output) were served in a restaurant, whereas internal control mechanisms might focus on the individual dishes (each customer might order one or more dishes). Once the information is obtained, it is compared with the standard as issued by the initiating process. On assessment, when aberrations occur, these will lead to the initiating process to issue revised standards for internal control mechanisms.

In addition, as a second source, information from the environment may influence that evaluation process. Take for example, changing weather conditions and the effect on the human body. In such a case the homeostasis becomes influenced through the perception of the external conditions; for

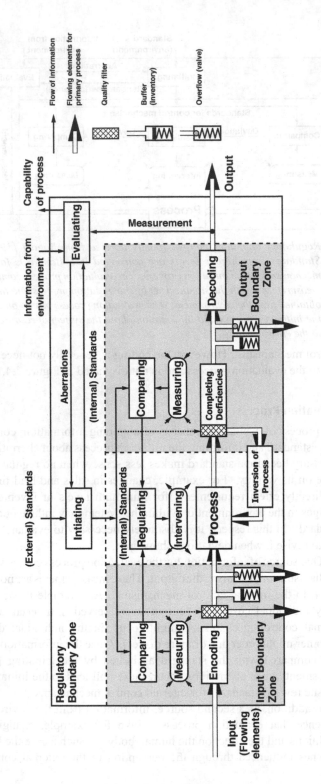

Figure 7.5 *Steady-state model. This generic model provides a complete overview of all processes in the boundary zones, the regulatory activities and the control mechanisms. For reasons of simplification the resources needed for all processes (primary process and control processes) and activities (coding, quality filters, buffers, overflow valves and decoding) have been omitted.*

example, long periods of rainy weather may make people more pessimistic. Therefore, the evaluation process should not only account for the internal information about the performance process but also consider the (relevant) information it obtained from the environment.

Moreover, the evaluation process might conclude that the internal process is not capable of maintaining homeostasis; in that case, a signal is sent to the environment about the capability of the process (and the capability of the system of resources, if applicable). For example, when experiencing fever, the human body sends signals, such as a rise of temperature, transpiration and expressions of pain, to the environment. Another case is a factory unable to cope with orders or with structural issues affecting the output; the environment (management) may have to make a decision for expanding resources. This signal that the process and the system of resources is not capable anymore of sustaining static or dynamic equilibrium is informing the environment that the capability for self-regulation (see Box 7.1) is exceeded.

7.5 Limitations of Steady-State Model

These two control processes in the regulatory boundary zone (see Figure 7.4), initiating and evaluating, could concern quantitative standards as well as qualitative standards based on information of the environment. The initiating process should also consider for which of the aspects to generate the operational standards. Following the thoughts of Sections 2.3 and 2.4, this implies that the steady-state model with its control mechanisms applies to only one aspect system at a time. For example, such a situation occurs when the processing of orders does take structurally two weeks rather than the standard of one week. To perform the evaluation, information from the environment helps to assess the standards or creates the need for adaptation. Sales growth might end up in increasing the levels of inventory to allow the same degree of service levels for delivery to customers. If the change of standards affects the performance to the environment, a signal will be generated to inform the environment of the changing capability of the transformation process. But how does that work if the quality of products is put into the equation? It will be difficult to compare the quality of delivery with the quantity of delivery; in case of conflicts, which one takes precedence for decision-making? Hence, the processes as depicted in the boundary zones as basis for the steady state model principally apply to one defined aspect.

In addition to the limitation of being focused on one defined aspect, the steady-state model (see Figure 7.5) only applies to recurrent processes. Through the control processes it will adapt to changes in its environment and the adaptation is limited to scope of the primary process, the capabilities of the system of resources and the possibilities for interventions by control mechanisms. Do we want to go beyond the existing capabilities of a process, then we have to introduce a new internal (and if necessary an external) structure for the system of resources. When the system of resources and its processes proves incapable of coping with the perturbations imposed by the environment it means that the limits of the existing capability have been reached (thus triggering a signal from the evaluating process to the environment).

The maintenance of homeostatic equilibrium depends strongly on the capabilities of resources; however, the system of resources might also operate at another equilibrium caused by stimuli. The concept of heterostasis denotes such states that might even be beneficial to the system of resources. At best, the representation in the steady-state model would indicate a temporary setting of standards operating at different levels (aligned with stimuli). Therefore, the steady-state model does not incorporate the concept of heterostasis adequately.

One cause might be that it should be noted that the steady-state model only covers one aspect of a system. This has much to do with the conversion from external standards to internal directives for the control mechanisms. The limitation that the steady-state model applies to one aspect brings up the question how to handle situations with multiple aspects, when strictly following the definitions of basic concepts as introduced in Chapter 2 and the modelling principles in Chapter 3. As outlined for modelling in Section 3.4, Applied Systems Theory indicates: 'create or design a new model for each aspect', which aligns with the definition of a system. This means that multiple steady-state models may be needed for describing separate control processes for each aspect; each steady-state model offers a different view on the control processes. This is not far from reality in some cases. Just look at how organisations divide the financial responsibilities and the responsibilities for quality and logistics; this notion suggests that different steady-state models for different aspects are commonly embedded in organisational structures. In addition, a comparison between aspects remains in this view always subjective; for example, one individual manager will value some issues higher than peers do and weigh them different in a specific situation. The implication is that steady-state models apply to only one aspect for solving problems adequately and also for practical reasons of valuating and weighting different aspects.

Another limitation of the steady-state models occurs when the environment has no longer the need for the output of the transformation process; the standards will reduce to 'zero'. Theoretically, this leads to a standstill of the system even though (people and) resources within the system can still fulfil their functions. In such a case, the system looses its right of existence but will enhance itself after exploration of other needs and goals that the same resources can fulfil. The only way for the system to sustain is either an external intervention, especially for technical systems, or adaptation, especially for living systems and organisations. Both the external intervention and the adaptation call for re-arrangements in the structure of the system of resources considered (although it should be noted that for living systems this depends on the occurrence of mutations). These topics and related processes will be discussed in Chapters 8–10.

7.6 Summary

The steady-state model adds the control for the system boundary to the control processes for the primary process (directing, feedback, feedforward, completing deficiencies), which were presented in Chapter 6. The steady-state model has three boundary zones: the input boundary zone, the output boundary zone and the regulatory boundary zone for the system of resources. These three zones ensure that the system of resources is capable of taking in flowing elements from the environment and transferring the transformed elements to the environment (that means after the primary process) in a controlled way.

In the input boundary zone the regulatory activities ensure the suitability of the influx of flowing elements for the transformation process. First the input is coded before it is checked on its qualitative properties against standards. Then, an input buffer smooth out irregularities in the influx of flowing elements, whereas an overflow valve assures that the flux of flowing elements aligns with the capabilities of the transformation process. Hence, the input zone aims at offering the flowing elements to the primary process that it is capable of transforming, both in a quantitative and qualitative manner (depending on the aspect considered).

Conversely, the boundary zone at the output of the transformation process ensures the transition from the flowing elements to the environment. As part of this zone, the feedback control mechanism checks the flowing elements against standards. In addition, if necessary, a quality filter as control mechanism, discards the flowing elements into the environment, completes the deficiencies or feeds them back into the transformation process. An output buffer and an overflow valve smooth the irregularities between the supply of the transformation process and the absorption by the environment. Finally, the step of decoding makes the flowing suitable for the environment.

For aligning the control mechanisms of the system of resources to the environment, the steady-state model deploys the initiating process and the evaluation process in the regulatory boundary zone. Through the initiating process the standard, imposed by the environment, transforms into directives suitable for use by the 'internal' control processes. By evaluating the aberrations (as input from the evaluation process) and the external standards from the environment, the initiating process issues new or revised directives. These appraised standards and the performance of the process combined with information from the environment constitute the input for the evaluation process; this process informs the initiating process about deviations and also relays the capability of the system of resources to remain within its own standards to the environment (which is called self-regulation).

The steady-state model (resulting from the three boundary zones and the control mechanisms for the primary process) applies to only one (selected) aspect and aims at maintaining homeostasis. The limitation that the model covers only one aspect indicates that principally for each different aspect a separate steady-state model should be developed. Furthermore, the maintaining of a homeostatic equilibrium means also it depends strongly on the need of the environment, i.e. the continual existence of the function it fulfils. When this function is no longer necessary, the system of resources with its process is no longer able to produce output.

References

Beer, S. (1972). Brain of the Firm - the Managerial Cybernetics of Organization. Chichester: John Wiley & Sons.

Bogart, D. H. (1980). Feedback, feedforward, and feedwithin: Strategic information in systems. Behavioral Science, 25(4), 237–249. doi: 10.1002/bs.3830250402

Dekkers, R. (2015, 3–5 August). On the Origins and Applications of the Steady-State Model. Paper presented at the 23rd International Conference on Production Research, Manilla.

Emery, F. E., & Trist, E. A. (1969). Socio-technical Systems/Systems Thinking. London: Penguin.

Flood, R. L., & Carson, E. R. (1993). Dealing with Complexity: An Introduction to the Theory and Application of Systems Science. New York: Plenum Press.

Klopf, A. H. (1972). Brain Function and Adaptive Systems - A Heterostatic Theory (AFCRL-72-0164). Bedford, MA.

Miller, E. J., & Rice, A. K. (1967). Systems of Organization: The Control of Task and Sentient Boundaries. London: Tavistock Institute.

Miller, J. G. (1965a). Living systems: Basic concepts. Behavioral Science, 10(3), 193–237. doi: 10.1002/bs.3830100302

Miller, J. G. (1965b). Living systems: Structure and process. Behavioral Science, 10(4), 337–379. doi: 10.1002/bs.3830100402

Selye, H. (1973). Homeosatis and Heterostasis. Perspectives in Biology and Medicine, 16(3), 441–445.

Shannon, C. E. (1948). A Mathematical Theory of Communication. The Bell System Technical Journal, 27(3), 379–423. doi: 10.1145/584091.584093

van Gigch, J. (1978). Applied General System Theory. New York: Harper & Row.

8   Autopoietic Systems

Whereas the previous chapters have mostly focused on cybernetic systems cumulating in the steady-state model, this chapter moves to more recent development in systems theories, particularly the theory of autopoiesis. Autopoiesis literally means 'autocreation' (from the Greek: auto - αυτο for self- and poiesis - ποιησις for creation or production) and expresses a fundamental complementarity between the structure and the function of a system. Originally, autopoiesis was formulated as an alternative to Darwinian ecology theory [Hernes and Bakken, 2003, p. 1512]; now it is seen as being complementary to existing models for evolution and change. It presents a different way of looking at how entities interact with their environment and how they evolve.

Thus, the theory of autopoiesis has found its way into explanations for biological evolution, interactions between humans and organisational development. Especially four fields have adapted the concepts of autopoiesis: computing (McMullin [2013] is a case in point), social systems in general (see Jones [2014], economics (for example, Valentinov [2015]) and organisational science (for instance, Demetis and Lee [2016], though their position is contested by others, such as Mingers [in press]). At the moment, there seems a revival of academic interest into this topic to deploy these principles to a larger variety of phenomena observed in nature and science. Many of the topics presented in this chapter and Chapter 9 have interrelations and, therefore, are mostly used in combination to explain these phenomena that otherwise seem difficult to comprehend. For part, these conceptualisations for autopoiesis constitute qualitative approaches and they often link the autopoietic principles to complex adaptive systems (see Chapter 9). Although autopoiesis appeals as a concept for explaining phenomena, it has proven difficult to connect to practice and implement it; related concepts, such as self-organisation, serve sometimes better as description for the same phenomena.

For this purpose, this chapter will take a broad view on autopoiesis and relate it to different disciplines for explanation and application. Section 8.1 will shortly describe the concept of autopoiesis as a different way of looking at systems from both a closed systems view and an open systems view. Section 8.2 pays attention to three main principles of autopoiesis that govern the development of systems. This results in Section 8.3 discussing the interaction of autopoietic systems with their environment. Section 8.4 explores perception and cognition. A slight different conceptualisation is presented in Section 7.5: allopoiesis for systems that do not posses all characteristics of autopoietic systems.

© Springer International Publishing AG 2017
R. Dekkers, *Applied Systems Theory*,
DOI 10.1007/978-3-319-57526-1_8

8.1   Autopoiesis

Morgan [1997, pp. 253–258, 413–414] points to the theory offered by Maturana and Varela [1980] to explain evolutionary processes: autopoiesis, the ability to self-create or self-renew through a closed system of relations, whether that concerns living organisms or possibly organisations and society. In this view, living systems engage in circular patterns of interaction whereby change in one element of the system is coupled with changes elsewhere, setting up continuous patterns of interaction that are always self-referential. Thus, a system enters only interactions that are specified by its external structure; this structure is directly related to its internal structure. This means that a system's interaction with its environment is a reflection and part of its own structure. It also implies that a (living) system interacts with its environment in a way that facilitates its own self-production; this way, the environment becomes really a part of itself. These implications of the concept of autopoiesis mean that it needs a further explanation.

When the term autopoiesis was originally introduced by Chilean biologists Francisco Varela and Humberto Maturana in 1973, they described it as:

'An autopoietic machine is a machine organized (defined as a unity) as a network of processes of production (transformation and destruction) of components which: (i) through their interactions and transformations continuously regenerate and realize the network of processes (relations) that produced them; and (ii) constitute it (the machine) as a concrete unity in space in which they (the components) exist by specifying the topological domain of its realization as such a network.' [Maturana and Varela, 1973, p. 78]

'(...) the space defined by an autopoietic system is self-contained and cannot be described by using dimensions that define another space. When we refer to our interactions with a concrete autopoietic system, however, we project this system on the space of our manipulations and make a description of this projection.' [Maturana and Varela, 1973, p. 89]

The most used example of an autopoietic system is the biological cell (by the way, one of the entities that motivated these Chilean scientists to define autopoiesis). For example, the eukaryotic cell is made up of various biochemical components, such as nucleic acids and proteins, and is organised into bounded structures, such as the cell nucleus, various organelles, a cell membrane and cytoskeleton. These structures of resources, based on an external flow of molecules and energy (the flowing elements), produce other elements, which, in turn, continue to maintain the bounded structure that gives rise to these elements. Examples of those elements are chromosomes and cell membranes that are created during and after a division of single cell. During the study of biological entities, Maturena and Varela [1980] arrived at some fundamental notions[1]:

---

[1]   The word 'component' has been replaced with 'element' for overall consistency throughout the book. Element may mean subsystem in specific contexts, but this

- Individual systems are characterised by their autonomy. Even when they are part of organisms or populations and when they undergo environmental influences, the individual entities remain internally closed and self-defined.
- Living systems consist of elements with different properties. These elements and their interaction with adjacent elements determine the total behaviour of living systems.
- All explanations and descriptions of these living systems are generated by observers external to the system. Such an observer will denote the entities and the environment in which they exist. Elements within the system do not possess this capability of observation and will only react to behaviour of other constituent elements.
- An observer can describe the objectives and functions of elements present in the system; the living system itself is incapable of making these observations. Only the interactions with adjacent internal elements can be observed.

The development of autopoiesis can be seen as a reaction to the cybernetic movements within the general systems theory (see also Chapter 1) and aims at explaining the unique features of biological systems and entities.

8.2 Principles of Autopoiesis

In the context of unique features of biological and social systems, the first of the three principles for the theory of autopoiesis is the possibility for self-reproduction by systems. This concept of self-reproduction is also used by the attempts of Kauffman [1993, pp. 298–341] to explain the origin of life when he relates it to the concept of autocatalytic sets. The concept of autocatalytic sets builds on the combinatorial consequences of polymer chemistry. As the maximum length in a system of polymers increases, the number of reactions by which polymers can interconvert rises faster than the number of polymers present. Then, a sufficiently complex set of polymers has very many potential reactions leading to the synthesis of any of these polymers. Consequently, for many possible distributions of catalytic capacity for those reactions among the same set of polymers, autocatalytic sets will emerge, such as peptides stepping up to deoxyribonucleic acid (known as DNA). The hypothesis proposed by Kauffman is that life is a collective emergent property of complex polymer systems; this seems likely to give an answer to the critical question why free-living systems exhibit a minimal complexity. The self-reproduction principle of autopoiesis tells that the structure of all components and processes together produce the same components and processes to ensure the continuity of the living system [Maturana and Varela, 1980]. This principle creates an autonomous, self-producing entity.

Furthermore, as a second principle out of three, autopoietic systems are structurally closed which does not imply that no interaction takes place with

also depends on the aggregation stratum.

the environment. For example, living entities feed themselves through input taken from the environment. This type of input and other relationships with the environment do not trigger changes in the living entity and generally support the continuity of the system through recurrent processes. However, within the definitions of autopoiesis, perturbations lead to disruptions within the system. For example, disruptions and irregularities in the food supply lead to reactions within the living system to cope with the changes in the environment through mutations in self-production; these are the well-known adaptation processes by species as observed in nature. However, from an autopoietic perspective, the environment has little influence on the actual responses by the system and the consequences for the (internal) structure and, and vice versa, because of being structurally closed.

The connection to the environment is called the structural coupling, the third principle. Structural coupling is the term for structure-determined (and structure-determining) interaction of a given system with either its environment or another system; this should be viewed particularly through the lens of the external structure (see Section 2.2). An example is the vision of living systems; one set of interactions takes place through eye contact between human beings. Both the possibilities and limitations of eye contact determine the effectiveness of this visual interaction as communication. Structural coupling between living systems facilitates the realisation of autopoiesis. When the homeostasis created by the organisational closure of the systems can no longer be maintained, disintegration occurs, which leads van der Vaart [2002, p. 5] to the statement: *Autopoiesis is all or nothing!*

Using the three principles – self-production, being structurally closed and structural coupling – an autopoietic system can be defined as a composed unit with a network of components (see Figure 8.1) that
- through interactions repeatedly completes the production process of components by which the self-production sustains,
- and that realises a unit for self-production in a space in which the system exists by creating and specifying boundaries in which only components are allowed that participate in the realisation of the production process.

When reading this definition think about cells and organisms as examples of autopoietic systems. Also, the Earth can be viewed as an autopoietic system, referred to as Gaia; think about the exchange with its environment (energy in the form of solar energy and radiation), how the internal structure responds to these exchanges and the internal processes that seem almost independent from the environment (for example, the movement of the tectonic plates). Hence, there are many examples of autopoietic systems when considering the definition.

8.3   Autopoiesis and Self-Organisation

Particularly, with autopoietic systems being structurally closed, self-organisation has a prominent position in maintaining these entities. Self-

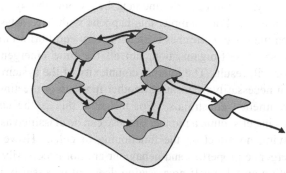

Figure 8.1 *Symbolic representation of an autopoietic system. Internal components interact with the environment and have internal relationships that constitute the structure of the system. At the same time, the boundary of the system defines the internal structure as being closed and it defines the interaction with the environment through its external structure.*

organisation can be seen as a process in which the internal structure of a, normally an autopoietic, system increases in complexity without being directed or controlled by an external source. Self-organising systems typically display emergent properties (see Section 2.2) as a result of transitions in states. The concept was first noted in physics, such as convection cells in gravity fields and spontaneous magnetisation. Self-organisation is also relevant in chemistry, where it has often been taken as being synonymous with self-assembly (defined as a reversible process in which the pre-existing parts of a pre-existing system form structures of patterns, as is the case for some complex polymers, see also the previous section). Furthermore, the concept of self-organisation is essential to the description of biological systems, from the subcellular to the ecosystem level. 'Self-organising' behaviour also appears in many other disciplines, both in the natural sciences and the social sciences, such as economics or anthropology. It has also been observed in mathematical systems such as cellular automata, which are grid of cells that have a finite set of states. All these examples and domains show that self-organisation constitutes a basic process that allows systems of resources, particularly autopoietic systems, to adapt to stimuli by the environment without the guidance of an external resource (which does not mean that they do not interact with the environment).

This process of adaptation might sometimes lead to the notion that self-organisation is mixed with that of the related concept of emergence. However, there may be instances of self-organisation without emergence and emergence without self-organisation. In this respect, it is useful to distinguish between emergent properties and emergent behaviour. Emergent properties might already appear when elements of a system are put together. For example, a car you can drive but you cannot do the driving without the parts assembled. In such cases, emergent properties do not result from self-organisation, but from the whole having properties that the constituent elements do not have (a bolt for the engine is necessary, but itself does not make the engine rotate).

Contrastingly, in the case of emergent behaviour, the system responds to external stimuli. For example, this happens with heterostasis (see Section 7.1) when the system searches for a temporary equilibrium in a poised state. In the case of self-organisation, not all the time emergent properties or behaviour will result. The internal complexity of the system might increase without it necessarily performing another function or meeting another scope of environmental constraints. For instance, this is the case for random mutations in cells; quite a number of these can be classified as neutral genetic drift, having no effect on the functioning of cells. However, this means that emergence properties and behaviour are not necessarily related to self-organisation and that self-organisation does not necessarily imply emergent properties or emergent behaviour.

The concept of self-organisation is a process where a system searches for internal adaptation to external events (or stimuli) and to that purpose usually relies on three basic mechanisms [Halley and Winkler, 2006, p. 12]:

- Positive feedback amplifies certain deviations rather than damping them, giving a greatly increased output to any input change from flowing elements or resources. Often amplifying deviations is undesirable and is countered by negative feedback measures, leading in complex systems to a mix of feedback influences. However, positive feedback might ensure a fast transition between a detrimental state and target state. This can be seen in evolution where fitness of species operates in this way, success breeds more success. The interactions of these types of feedback lead to self-limiting systems and often to cycles and oscillations in nature.

- Negative feedback maintains equilibrium and counteracts positive feedback when reaching new states. Whereas most self-organising systems use positive feedback to reach new states, for such systems both negative and positive feedback are indispensable. Camazine et al. [2001, p. 489] point out that negative feedback often takes the form of regulation, competition, reduction or saturation. For example, in the ants' nest negative feedback dominates when there is competition among food sources, the food source is fully consumed, too many ants are feeding from a food source, there are not enough food sources in a particular area, lack of space or any other similar event that overtakes the positive feedback processes of the ants' nest. Consequently, the ants are forced to hunt for other food sources and commence the feeding cycle again. An another example used in biology is the case of pillar formation in termite nests [e.g. ibid., 402]. In this event, negative feedback takes over when there is no more material in the area close to the formation of these types of pillars. It has also been observed that there seems to be a certain type of competition among termites building other pillars in the same area. This pattern of competition is recognised as negative feedback. So negative feedback complements positive feedback for reaching new states as well as maintaining equilibrium.

- Multiple interactions enable self-organising systems to respond to external events and stimuli. These responses rely on internal processes, multiple interactions through the external structure and passing of information among subsystems and elements; in this respect, information can also appear in physical or chemical form, such as physiological reactions, e.g. pheromones. As Fuchs [2003, p. 161] points out, '... all self-organising systems are information-generating systems'. Self-organisation takes place in systems with multiple active interactions among many subsystems and elements (sometimes these are called actors when referring to social systems). Because there are many, often identical, subsystems and elements, there is no requirement for a single one to carry out a series of connected processes. These sequences of processes result in the emergence of new structures.

So, self-organising systems use information of some kind as events through multiple interactions to trigger internal and external processes of positive and negative feedback that ultimately establish new structures within the system by going through transitions of states using outcomes of multiple interactions.

Self-organised Criticality

During those transitions of states self-organised criticality might occur. This is a property of (classes of) dynamic systems, which have a critical point as an attractor, mostly a phase transition. An attractor is a set of states of a dynamic physical system toward which that system tends to evolve, regardless of the starting conditions of the system. An example is a pile of sand. By continuously adding new grains of sand to a small pile of sand, the formation of small local avalanches starts. The small local avalanches decrease the local slopes whenever they become to steep. Perturbing the system, the small sandpiles, provoked by avalanches, create still greater sandpiles and eventually we end up with only one big sandpile. At some point (the transient point) this pile ceases to grow. The (global) average slope has reached a steady state corresponding to the angle of repose that the sandpile cannot exceed no matter how much sand is added. That means that the pile has reached a statistically stationary state and additional grains of sand will ultimately fall off the pile. This particular state reflects self-organised criticality and the pile of sand is just one of many examples.

This phase transition takes place without the need to tune parameters to precise values. Contrary to intuition, parameters themselves tend to gravitate towards a dynamic equilibrium. This happens in work groups that have to perform complex tasks; a work division will occur over time, whereas different groups might create different divisions, though these have similarities as well. The concept of self-organised criticality implies that larger interactive components (in terms of autopoiesis subsystems or elements in Applied Systems Theory) will self-organise into a critical state. Once in the critical state perturbations result in a chain of events among the elements and subsystems, which can affect a number of elements within the system.

Hence, it is the perturbations that lead the self-organising system towards the critical state or the state of the phase transition.

Self-organisation versus Entropy

The idea of self-organisation challenges an earlier paradigm of ever-decreasing order that was based on a philosophical generalisation from the second law of thermodynamics. Each system that has a number of states (structure, properties of elements and relationships) has also a likelihood that a particular state may exist. The concept of 'entropy' expresses that measure of the statistical 'disorder'. That also means that the higher the 'entropy', the less likely that a discrete state will occur. In other words, entropy is an expression of randomness. A practical example is a black marble in a box full with white marbles. In the first scenario the black marble is put into a large box and the box is shaken intensely. Because the box is large, there are many possible places inside the box where the black marble could be, so the black marble in the box has high entropy (many possible states). The second scenario is repeating this experiment with a small box. In the case of a small box, the black marble has low entropy (more limited number of states). Entropy plays a large role in self-organising systems.

In open systems, it is the flow of elements and energy through the system that allows the system to self-organise and to exchange entropy with the environment. This is the basis of the theory of dissipative structures by Nicolis and Prigogine [1977]. Dissipative structures are not limited to living things, such as cells, organisms, trees, internal organs and people, but might also include some non-living structures. For example, a whirlpool is a dissipative structure requiring a continuous flow of matter and energy to sustain its form. At the same time, its entropy is transmitted to its environment as it seeks to reduce (read: stabilise) itself. At the same, such a dissipative structure displays self-organisation that can only occur far away from (thermodynamic) equilibrium. Since closed systems cannot decrease their entropy, only open systems can exhibit self-organisation.

However, such open systems can gain macroscopic order while increasing their overall entropy. Specifically, a few of a system's macroscopic degrees of freedom can become more structured at the expense of microscopic disorder. In many cases of biological self-assembly, for instance metabolism, the increasing entropy of small molecules more than compensates for the increasing organisation of large molecules; this is especially the case for water. At the level of whole organisms and over longer time scales, biological systems are open systems taking in input from the environment and discarding waste into it (as output). In economics, such a concept exists under the label: externalities. These externalities are costs or benefits that result from an activity or transaction and that affects an otherwise uninvolved agent (system, subsystem or element) who did not choose to incur that cost or benefit. Some examples of these externalities are pollution, dumping of toxic waste and labour conditions that they not allow employees to sustain themselves reasonably. Therefore, the concept of entropy entails looking at a more aggregated level of order while considering effects at a lower level of detail.

Autopoietic Aspects of Self-Organisation

The concept of autopoiesis is linked to self-organisation and to self-assembly. Self-assembly is a spontaneous process in which elements (or subsystems), either separate or linked, spontaneously form ordered aggregates. This process of self-assembly searches for equilibrium and happens in nature, in biological systems (created systems) and human engineered systems (including organisations). It leads to an increased complexity of the internal structure to respond to organisational and externally imposed constraints. Both self-organisation and self-assembly happen from within the system. Also, autopoiesis is based on reproduction leading to mutations as a result from internal reproduction processes, self-assembly being one of them, and happens as well from within. During these reproductions and self-assembly errors might happen during mutations, so-called point mutations, which lead to changes in reproduced elements and indirectly to changed structures. Therefore, observed mutations of autopoietic systems can be the result of self-production or of errors – the point mutations; when detecting those mutations, their origin might not necessarily be clear.

The new internal structure of autopoietic systems might possibly induce changed behaviour, either by constraints of functions or by new functions. It is written as 'might' since some mutations do not yield any effect themselves. An example is the use of a calculator for the addition of two figures instead of adding them up manually. However, if this was done electronically in a spreadsheet and stored at a server centrally, then others in an organisation might use those data for other purposes resulting in new patterns of interaction within the organisation. Therefore, self-organisation during these (point) mutations leads to integration, which in turn might end up in new structures. These adapted structures imply changes in processes and therefore in behaviour, which is either covering a different range of constraints or generating new functions; selectional forces in the environment determine the viability of these mutations as interaction between the autopoietic system and the environment.

8.4   Interaction with Environment

For the interaction with the environment, the structural determinism of autopoietic systems and the related principle of structurally closed entities constitute the most important principles for selectional processes, such as evolution. The structural determinism tells that all changes are embedded in the structure of the entity itself. Each change of a composed unit is a change of the structure moulded by the properties of the elements and subsystems in that very same structure. A case in point is the construction of eyes that is very similar across a wide range of (related) species and serves the same functions. A change will occur as a reaction to the internal dynamics of the system or the interaction with the environment and, even then, the internal relations of components shape the change rather than the environment dictates the internal adaptations.

This view of structural determinism of autopoietic systems does not imply that autopoietic systems are isolated. These types of open systems interact with the environment through a continuous pattern that has principally no end or beginning since it is a closed loop of interaction. For instance, living beings absorb flowing elements as food from the environment and have to continue to do that; and if they cease to exist themselves, they have passed that on to the next generation. However, the theory of autopoiesis includes that systems can be recognised as having environments but insists that relations with any environment are determined internally by their structure. This is because the boundary of the system consists of elements and subsystems generated by the interactions of internal elements and subsystems of the system with the environment. Relations and interaction with adjacent elements and subsystems in the boundary maintain the boundary. Without these elements and subsystems, the autopoietic system does not sustain self-referring processes for retaining its autonomy and steady state.

The structural coupling governs by which interactions a element or subsystem of an autopoietic system is influenced. When interactions initiate changes in the structure and composition, the structure is called plastic. Through repeated interaction and initiations, selection of subsequent structures happens by the environment. In a way, the environment and the plasticity of the structure drive the selection by its own elements, subsystems and internal relationships. The environment does not determine the internal adaptations! Therefore, autopoietic systems are interactively open and structurally closed [van der Vaart, 2002, p. 11].

Structural coupling in biological systems arises as a result of the plasticity of their internal structures and the plasticity of the interaction with the environment. As suggested above, autopoietic systems are structurally determined – how they respond to environmental events – and what events they respond to is something that is determined by their structure at a given moment in time (the same goes for the interaction with the environment). Since a plastic structure is one that can be affected by outside events, it can be perturbed. So, structure determines what an autopoietic system does, and when the structure of such a system changes, what it does is likely to change in a manner determined by its system's structure. The similar reasoning applies to the interaction with the environment. Over time, the structure of both system and interaction with the environment change as a result of mutual non-destructive perturbations. An autopoietic system strives to respond to events in its environment in an appropriate manner; minimally, the system seeks not to be destroyed.

8.5 Perception and Cognition

One of the foremost reasons for the conceptualisation of autopoiesis stems from the quest for the nature of perception and cognition in the interaction of systems with the environment. Perception and cognition derive from the internal processing of stimuli in the interaction with the environment, consistent with the concept of autopoiesis. To exist, continuously interactions should be repeated since the structural coupling exists; in this sense, cognition represents gathering knowledge about all effective interaction for sustainability, particularly for living systems.

Learning as a process of cognition originates in the properties of self-reference of the system. When learning exceeds the level of direct interaction and moves towards orientation in the common domain of two autopoietic systems, communication is established. When descriptions lead to being observer of its own behaviour, self-conscience arises. Hence, the composition of a system related to an external point of reference defines the identity of an autopoietic system [van der Vaart, 2002, pp. 7, 24–25]. The identity is strongly related to the composition of the entity, changes in the composition lead to a changed identity; through self-reference autopoietic systems seek to maintain their identity unless perturbations provoke adaptations. These notions lead Mingers [1995] to connect the theory of autopoiesis to the systems hierarchy of Boulding (see Figure 8.2); see Section 3.5 for this hierarchy. The higher the system is positioned at the levels in the hierarchy of Boulding, the more pronounced and complex the perception and cognition processes.

An example of the application of perception and cognition processes is the 'living company'. Connecting it to the capabilities of autopoietic systems, de Geus [1999, p. 111] points to learning as characteristic for organisations (similar to learning by organisms). Learning becomes possible through self-cognition as typical for the higher levels of the systems hierarchy of Boulding. Following the ideas laid down by de Geus, it was Senge [1992]

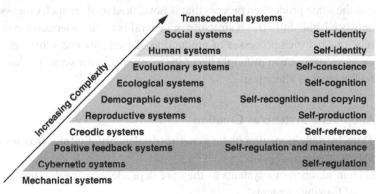

Figure 8.2 *Connection of the theory of autopoiesis to the systems hierarchy of Boulding, based on Mingers [1995]. The interaction with the environment changes because of the model of reference to the environment differs for each level; thus, this induces different internal processes at the higher levels of the systems hierarchy.*

who has expanded these theories and gained recognition about the importance of the continuous process of gathering understanding about the interaction between organisation and environment. Therefore, the conceptualisation of the learning organisation originates in the principles of autopoiesis.

8.6 Allopoietic Systems

However, organisations are an example of entities that do not self-produce by generating offspring. In such a case, we talk about allopoiesis; this is the process whereby a system produces a mutation of itself rather than replicating its own structure in a new system. Social organisations, such as manufacturing companies or political parties, are examples of allopoietic systems. In such a case, the entities that are produced consist of other elements or subsystems than those required for producing them. Some say that even reproduction in biology is allopoietic to some extent because offspring is materially distinct from the parent organisms and might even occupy different spaces. In that sense, reproduction is not equal to self-production.

Allopoiesis is neither a separate concept nor even a background for the articulation of an autopoietic system, but rather an ideal construction of a non-autopoietic system, while sharing some of its characteristics, see Figure 8.3. The main difference between autopoietic and allopoietic systems appears to be the difference in their structures. Whereas the former have a structure that is defined by the relationships between processes of production of elements and subsystems, the latter thus have a structure that is defined by the spatial relations between elements and subsystems [Maturana and Varela, 1980, pp. 79-80]. Therefore, autopoiesis and allopoiesis belong to different domains, namely, the domain of the 'concatenation of processes' and the domain of the 'concatenation' of components that participate in one and the same production process that is not linked to other such processes in a network [ibid., 80]. But if structure in general is to be understood in terms of relations between processes of production of elements and subsystems, then the term structure in principle becomes inapplicable for system's description on the level of allopoiesis.

Allopoietic Systems as Creation

Thus, by definition, allopoietic systems (drawn partially from [Maturana and Varela, 1980, p. 81]) are:

(1) non-autonomous systems as they are dependent on the continuous influx of flowing elements;

(2) systems without individuality because their identity depends on the observer (note the parallel with de Geus' [1999, p. 99 ff.] proposition of 'persona');

(3) systems which are not unities because they do not have self-defined boundaries but rather have boundaries defined externally [allopoietic systems are created or emerge in creation];

(4) systems with inputs and outputs which can be perturbed by external events.

However, to these distinctive characteristics one more could be added, which, although of crucial importance, still remains without articulation. Zeleny [1981, p. 96] remarks that autopoiesis and allopoiesis should be treated as inseparable concepts in the same way in which 'organisation is inseparable from structure'. He further concluded that while an allopoietic system may emerge out of chaos and disorder, or out of 'non-systems', autopoietic systems cannot but emerge from another system. This is based on the argument that processes producing elements or subsystems must exist before the process of their linking together can take place. Such considerations gave Zeleny certain grounds to question the possibility that autopoietic systems operate on an essentially unordered environment of components. He was rather inclined to accept the other alternative, namely, that an autopoietic system acts 'upon an already ordered, structured milieu, favourable to its enhancement and maintenance' [ibid., p. 95]. Hence, autopoietic systems might emerge rather than be created and allopoietic systems are created.

According to Misheva [2001], this shows that the domain that the term allopoiesis was supposed to designate still remains undefined. Because an allopoietic system is created by an external entity (or entities), its lacks the links between independent self-production processes as autopoietic systems do have. Also, for autopoietic systems maintaining homeostasis is an artificial construct. Thus, allopoietic systems lack the ability to maintain certain critical systemic variables within unchangeable limits; there is some parallel with the capability for self-regulation as mentioned in Section 7.4. In this sense, allopoietic systems lack permanence and sooner or later vanish without leaving any trace of their existence. Think about manufacturing, such as early iron furnaces in the German region of North Rhine-Westphalia.

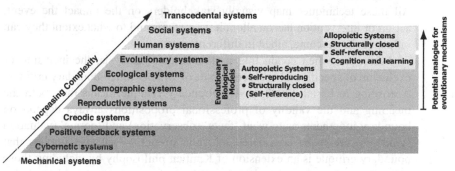

Figure 8.3 *Connection of the theories of autopoiesis and allopoiesis to the systems hierarchy of Boulding, according to Dekkers [2005, p. 148]. Allopoietic systems have a different domain of application than autopoietic systems, particularly for cognition and learning.*

Except the artefacts, remains of building and traces in the soil, almost all traces of its organisation have disappeared. However, in human beings there are still traces of the origins in the RNA and DNA, dating back 85,000 years and more. In the view of Misheva, this is so because allopoietic systems lack an essential 'device' and are basically non-homeostatic systems; even when they briefly attain the autopoietic structure, they lose it and revert to their previous state [Zeleny, 1981, p. 93]. Hence, the theory of autopoietic systems does not explain, observe, or describe life or living systems, but rather only the 'device' that makes a living system a homeostatic machine capable of maintaining its own structure under some strictly defined range of conditions; this is commensurate with the definitions. In this respect, one could say that even allopoietic systems are dependent on an external entity to remain purposeful.

Stakeholders and Boundary Critique

If allopoietic systems are dependent on an external entity, or may be more than one, then the purpose is defined by these entities; this investigation into the purpose of an allopoietic system is called stakeholders' analysis. A stakeholder is an individual or group that is going to be affected by an event, action or intervention. That means that stakeholders are not limited to proprietors (or in business administration, shareholders) but also other that might be experiencing an impact. An example of that is when a new bypass is build around a village those near to new route might be experiencing noise pollution. There different techniques for assessing the impact of events, actions or interventions, for example [Bourne and Weaver, 2010, p. 101]:

- The influence-interest grid.
- The power-impact grid.
- The power-interest grid.
- The three-dimensional grouping of power, interest and attitude.
- The stakeholder circle.

All these techniques map visually stakeholders on the impact the event, action or intervention has on allopoietic systems and to what extent they can influence the outcome, albeit in different ways.

Moreover, when conducting a stakeholders' analysis, the inclusion or exclusion of specific stakeholders is part of what is called boundary critique. According to Ulrich [2002, p. 41] boundary critique means that 'both the meaning and the validity of professional propositions always depend on boundary judgments as to what 'facts' (observation) and 'norms' (valuation standards) are to be considered relevant' or not. That also implies that boundary critique is an extension of Kantian philosophy by including those actors that are sources of motivation, sources of power, sources of knowledge and sources of legitimation [Ulrich, 2000, p. 258]. Hence, the definition and existence of allopoietic systems relies on the inclusion of specific external stakeholders.

8.7   Social Systems as Autopoietic Systems

Despite the ambiguity of allopoietic systems being autonomous reproducing entities and at the same time dependent on external entities, the study of social systems as being autopoietic or allopoietic systems offers some interesting perspectives. However, traditionally, the focus is on autopoietic concepts rather than allopoietic considerations. This perspective reverberates when according to Luhmann [1986, p. 186] autopoiesis is 'a theory of self-referential systems, to be applied to observing systems as well'. This links social autopoiesis theory to second-order cybernetics, as expressed by von Foerster [1981], Geyer and van der Zouwen [1978] among others. The second-order cybernetics evoke that an observer becomes also subject of study; that means that for the individual researcher that a study becomes just as much a question of self-observation as observation of the social system. Luhmann [1986, p. 187] remarks: 'To combine these two distinctions (between autopoiesis and observation, and between external observation and self-observation, our inclusion) is one of the unsolved tasks in systems theory'. Essentially, this suggests that an observer examining a social system constitutes an autopoietic system in its own right; when gathering information about social systems, we cannot avoid collecting information about ourselves. Luhmann [ibid., p. 188] points out that in order to solve this ambiguous problem (paradox) a sort of exchange between external observation and self-observation is required.

Moreover, another property of autopoietic systems, the conceptual pairings (normatively closed and cognitively open), makes it possible for a social system to be simultaneously self-producing in terms of social norms and to still maintain the capability of learning through the cognitive openness of the system. Luhmann [1986, p. 183] states: 'the concept of autopoietic closure has to be understood as the recursively closed organisation of an open system'. The point is the extent to which normative closure and cognitive openness exists in a specific system. According to Luhmann [ibid., p. 186], it is communication that constitutes the evolutionary potential for the structure of systems able to 'maintain closure under the condition of openness'. Even if the system is closed normatively, it does not follow that it is not subject to influences from the environment (not the universe, see Section 2.1 and 8.2). An autopoietic system is cognitively open, and, therefore, can both influence other systems and at the same time learn and adapt to the environment.

However, there is no agreement as to whether social systems can be truly regarded as autopoietic systems (or allopoietic systems for that matter). The works of de Geus [1999], Luhmann [e.g. 1986] and Robb [1989] argue in favour of the contention that the theory of autopoiesis can be adapted to social systems. However, Maturana [1980], Varela [1981] and Mingers [1989, p. 175] cast more doubts about the fruitfulness of this analogy. From the perspective of systems theories and the systems hierarchy of Boulding (Section 3.5), autopoietic processes and systems can be used as metaphor, but that implies not as identical in social systems and organisations. Particularly, the cognitive and knowledge-based facets of the principles of autopoiesis at the cell level can be adapted for the purpose of acquiring knowledge of social processes in organisations regarded as social systems. For example, this is what Senge (1992) has done when introducing the concept of the learning organisation. This interpretation is similar to Luhmann's point of view [1986, p. 173]. Luhmann's application of the autopoiesis theory can be used to describe, explain and possibly predicate change or lack of change in social systems. But this also means that autopoiesis for social systems is an evolutionary model and relies on mechanisms derived from evolutionary biology, for example.

8.8 Summary

Thus, the theory of autopoiesis provides further insight into the behaviour of systems in addition to the more cybernetic views that have dominated the previous chapters. Principally, it tells that next generations of autopoietic systems build on the elements and structures of previous generations. Such is the case in evolutionary biology for offspring. Autopoiesis implies also that these systems are self-referential in their interaction with the environment; only that what can be perceived acts as stimulus for activities in the system and for the next generation. However, it is also a very difficult theory to apply to systems, because the observers themselves have cognitive limitations, too. Therefore, the principles of autopoietic theory serve as explanation for phenomena at higher levels of the systems hierarchy of Boulding, but should applied with reservations.

A special class of autopoietic systems are allopoietic systems – by some considered an opposite to autopoietic systems. These systems do not self-produce as autopoietic systems do, but are 'created' or emerge from systems external to them. The dependency on the external systems for justifying its existence means also that it depends on the perceived need of the output or function by the external actors. Therefore, adaptations of an allopoietic system are also driven by external enactment, while at the same time building on the self-referential aspects. A conversion of an allopoietic system relies on its extant subsystems and elements and in that sense follows autopoietic principles. Only when further adaptations are not possible the allopoietic systems become extinct and in the best case, elements or subsystems are re-used by external (perhaps, different) actors.

The application of autopoiesis to social systems proves not only difficult but also contentious. Social systems could be viewed from both an autopoietic perspective and an allopoietic view. The latter implies an account of stakeholders and their influence on the purpose of the system. Including or excluding particular groups of stakeholders may influence the outcomes of analysing a situation, advancing a solution and societal progress. Other aspects of autopoietic theories lead to more philosophical views on interaction between systems and actors, where the capability of self-reflection takes a paramount role.

References

Bourne, L., & Weaver, P. (2010). Mapping Stakeholders. In E. Chinyio & P. Olomolaiye (Eds.), Construction Stakeholder Management (pp. 99–120). Oxford: Wiley-Blackwell.

Camazine, S., Deneubourg, J.-L., Franks, N. R., Sneyd, J., Theraulaz, G., & Bonabeau, E. (2001). Self-Organization in Biological Systems. Princeton, NJ: Princeton University Press.

de Geus, A. (1999). The Living Company. London: Nicholas Brealy Publishing.

Dekkers, R. (2005). (R)Evolution, Organizations and the Dynamics of the Environment. New York: Springer.

Demetis, D. S., & Lee, A. S. (2016). Crafting theory to satisfy the requirements of systems science. Information and Organization, 26(4), 116–126. doi: 10.1016/j.infoandorg.2016.09.002

Fuchs, C. (2003). Structuration Theory and Self-Organization. Systemic Practice and Action Research, 16(2), 133–167. doi: 10.1023/A:1022889627100

Geyer, R. F., & van der Zouwen, J. (1978). Sociocybernetics: An actor-oriented social systems approach. Boston: Martinus Nijhoff Social Sciences Division.

Halley, J. D., & Winkler, D. A. (2008). Consistent concepts of self-organization and self-assembly. Complexity, 14(2), 10–17. doi: 10.1002/cplx.20235

Hernes, T., & Bakken, T. (2003). Implications of Self-Reference: Niklas Luhmann's Autopoiesis and Organization Theory. Organization Studies, 24(9), 1511–1535. doi: 10.1177/0170840603249007

Jones, P. H. (2014). Systemic Design Principles for Complex Social Systems. In G. S. Metcalf (Ed.), Social Systems and Design (pp. 91–128). Tokyo: Springer Japan.

Kauffman, S. A. (1993). The Origins of Order: Self-Organization and Selection in Evolution. New York: Oxford University Press.

Luhmann, N. (1986). The autopoiesis of social systems. In F. Geyer & J. van der Zouwen (Eds.), Sociocybernetic Paradoxes: Observation, Control and Evolution of Self-steering Systems (pp. 172–192). London: Sage Publications.

Maturana, H. R., & Varela, F. J. (1973). Autopoiesis and Cognition – The Realization of Living. Dordrecht: D. Reidl.

Maturana, H. R., & Varela, F. J. (1980). Autopoiesis and Cognition. London: Reidl.

McMullin, B. (2013). Computational Autopoiesis. In W. Dubitzky, O. Wolkenhauer, K.-H. Cho, & H. Yokota (Eds.), Encyclopedia of Systems Biology (pp. 461–464). New York, NY: Springer New York.

Mingers, J. (1995). Self-Producing Systems: Implications and Application of Autopoiesis. New York: Plenum Press.

Mingers, J. (in press). Back to the future: A critique of Demetis and Lee's "Crafting theory to satisfy the requirements of systems science". Information and Organization. doi: 10.1016/j.infoandorg.2017.01.003

Misheva, V. (2001, 25–26 January). Systems Theory from a Gender Perspective. Paper presented at the Annual Meeting of the Swedish Sociological Association, Uppsala.

Morgan, G. (1997). Images of organization. Thousand Oaks, CA: Sage Publications.

Nicolis, G., & Prigogine, I. (1977). Self-organization in nonequilibrium systems : from dissipative structures to order through fluctuations. New York: Wiley.

Robb, F. F. (1989). Cybernetics and Suprahuman Autopoietic Systems. Systems Practice, 2(1), 47–74. doi: 10.1007/BF01061617

Senge, P. M. (1992). The Fifth Discipline. Kent: Century Business.

Ulrich, W. (2000). Reflective Practice in the Civil Society: The contribution of critically systemic thinking. Reflective Practice, 1(2), 247–268. doi: 10.1080/713693151

Ulrich, W. (2002). Boundary Critique. In H. G. Daellenbach & R. R. Flood (Eds.), The Informed Student Guide to Management Science (pp. 41–42). London: Thomson Learning.

Valentinov, V. (2015). From equilibrium to autopoiesis: A Luhmannian reading of Veblenian evolutionary economics. Economic Systems, 39(1), 143–155. doi: 10.1016/j.ecosys.2014.10.004

van der Vaart, R. (2002). Autopoiesis! Zin of Onzin voor organisaties? Delft.

Varela, F. (1981). Describing the logic of the living: The adequacy and limitations of the idea of autopoiesis. In F. Varela (Ed.), Autopoiesis: A theory of living organization (pp. 36–48). New York: North Holland.

von Foerster, H. (1981). Observing systems. Seaside, CA: Intersystems.

Zeleny, M. (1981). "What is autopoiesis." Autopoiesis: a theory of living organization. New York: Elsevier.

9 Complex Adaptive Systems

The previous chapter has already shown that the deterministic view of the earlier chapters remains insufficient to address the complexity at the higher levels of the systems hierarchy of Boulding. Building on this notion, the domain of complex adaptive systems believes that the dynamics of complex systems are founded on universal principles that may be used to describe disparate problems ranging from particle physics to the economics of societies, such as in the work of Kauffman [1993]. In addition, the development of complexity science offers a shift in scientific approach with the potential to profoundly affect business, organisations, institutions and government about the effectiveness of change and interventions. The science of complexity science strives to uncover the underlying principles and emergent behaviour of complex systems that are poorly described by deterministic approaches. Generally speaking complex systems are composed of numerous, varied, simultaneously interacting agents (elements or subsystems in terms of Applied Systems Theory). The goal of complexity science is to understand these complex systems – what 'rules' govern their behaviour, how they adapt to change, learn efficiently and optimise their own behaviour.

This chapter intends to provide an introduction to the principles of complex adaptive systems but not go into detail for every application, method and theoretical concept; it will restrict itself to some main principles. To this purpose, Section 9.1 gives a brief introduction how concept of complex adaptive systems can be understood and to what domains it has been applied (some of the descriptions also appear in Dekkers [2005]). The attributes of complex adaptive systems constitute Section 9.2. Central to the understanding of the behaviour is the process of adaptation on fitness landscapes, which appears in Section 9.3. Akin the development in autopoiesis, the concept of self-organisation has been linked to complex adaptive systems; Section 9.4 gives a short overview on that matter and also discusses dissipative structures. Section 9.5 covers recursive behaviour. Finally, Section 9.6 relates complex adaptive systems to connectivity, one of the attributes of complex adaptive systems, for human-influenced networks.

9.1 Dimensions of Complexity

Adaptive behaviour, see Section 5.2 for introducing related processes, becomes complex when many elements or subsystems interact. However, we refer to complex adaptive systems, when exact behaviour is difficult to predict, although the behaviour of subsystems or elements may be known. A typical example is the weather system; the generic behaviour of tornadoes is known but their emergence, their exact behaviour and the path they follow are more difficult to predict. Hence, the behaviours of complex adaptive systems that

© Springer International Publishing AG 2017
R. Dekkers, *Applied Systems Theory*,
DOI 10.1007/978-3-319-57526-1_9

emerge as a consequence of non-linear spatial-temporal interactions among a large number of elements or subsystems are ubiquitous in nature. Further examples of complex adaptive systems that exist in nature include immune systems, multicellular organisms, nervous systems, ecologies and societies. Examples of synthetic (man-made) complex adaptive systems include parallel and distributed computing systems, large-scale communication networks, artificial neural networks, large software systems and economies; note that these could also be called allopoietic systems (see Section 8.6). Therefore, some will say that complex adaptive systems are and have become part of our real life, and thus pose challenges for those interacting with them.

These challenges have attracted researchers from a number of disparate areas to study the behaviour and application of complex adaptive systems. Their interest goes to the control, coordination, communication, adaptation, learning, evolutionary structures and processes in complex adaptive systems. Of particular interest are algorithmic, information processing and theoretical conceptualisations of complex adaptive systems. The study of these systems is found in computer networks and applications, information theory, artificial intelligence, cognitive science, neuroscience, psychology, sociology, control theory, economics, mathematics, physics, evolutionary biology and engineering among others. The resulting tools for analysis of complex adaptive systems have found applications in many areas of science and even the humanities. Computational experiments and simulations used in this strand of research have led to the development of mathematical and computational techniques; these are equally applicable to the design of distributed control systems based on the model of a complex system composed of multiple, autonomous, intelligent agents that are competing and cooperating in the context of the whole system's environment (note that this chapter is limited to some basic principles of complex adaptive systems). These examples of studies into the behaviour and application of theories for complex adaptive systems demonstrate the wide range of domains and the vast extent of tools that are being developed.

Nevertheless, the term complexity carries some ambiguity. This is mostly due to complexity being understood in two distinct ways:

1. As an expression of structure, mostly internally oriented, either as part of networks or as an individual system.
2. As an expression of emergence, more rooted in new behaviour and complexity imposed by the environment.

The first interpretation of complexity, internal complexity, can be seen as a design parameter, even though not sufficiently defined in cybernetic approaches. Returning to the basic definitions in Section 2.1, internal complexity simply means a large number of elements with a large number of interrelationships. The intricate interrelationships of elements within a complex system might give rise to multiple chains of dependencies. An example is an economic system of a country with many firms, government agencies, education institutes, etc. being dependent on each other. However, this means that a system cannot be reduced to simple understanding, but it does not mean it cannot be understood or its behaviour cannot be predicted. Contrastingly, the second meaning of complexity is related to new behaviour that could not directly be foreseen. To cope with emergence, different entities might develop different types of complexity handling capability [Boswijk, 1992, p. 100]; that means building on existing capabilities for new situations or incorporating new knowledge for creating new capabilities to cope. Under these conditions, balance will hardly be achieved, only paradigms that address the dynamics of networks constituting of agents and the environment will elect for elaboration within the context of complex adaptive systems. In an organisational context, complexity provides an explanatory framework of how agents behave, how individuals and agents interact, relate and evolve within a larger (social) ecosystem. Complexity as emergence also explains why interventions may have unanticipated consequences [Buchanan, 2004]. Whereas an economy is an internally complex system, it also has emergent behaviour that may result from external interaction, such as its resilience to recover from global economic crises. It is the second meaning of complexity – emergence of new behaviour and imposed complexity by the environment – that is taken to the forefront in the remainder of this chapter.

9.2   Attributes of Complex Adaptive Systems

Thus emergent behaviour, in response to the events in the environment and as a result of dependencies internally and externally, arises because complex adaptive systems constitute of agents that link together and that do form a network (note that the concepts for complex adaptive systems refer to agents instead of elements and subsystems). The actions and strategies of one agent depend on the actions of the other agents it relates to in the system. That differs slightly from the basic concepts as introduced in Chapter 2 because the individual elements of a system do not necessarily display individual behaviour. Because of individual behaviour of agents, the intricacies of complex adaptive systems reside in three unique attributes: distributed control, connectivity and co-evolution.

Distributed Control

The need for distributed control in complex adaptive systems came about because of the limitations of the hierarchic approaches (i.e. the traditional control mechanisms from Chapter 6). With the proliferation of the network paradigm the hierarchical approach towards control has lost its charm and attention in science. Inspired by the Zeitgeist of the late 1980s, the trend of decentralisation and the postulation of non-hierarchical, participative and distributed control in society and organisations also penetrated complexity science [Malik, 1992]. Starting with the works of the Santa Fe Institute in the early 1980s, the paradigm of self-organisation emerged and opened a new branch in the explanation and control of complexity [Jost, 2004]. With the increasing number of elements in artificial systems – turning them into net-like entities – their control became increasingly complex. This made the deterministic, top-down approach to systems control inefficient, if not impossible, especially against the background of a highly dynamic environment.

From the perspective of cybernetic control we would try to achieve control by building on lower levels of control and for higher levels of control impose echelons of control (see Section 6.7); in situations of interacting agents these echelons turn out to have moderate effects. An example is traffic control in case of congestion on highways; despite being able to regulate a local traffic jam, flows of cars on highways to avoid traffic jams are more difficult to regulate from a central governance point of view. Hence, complex adaptive systems rely on distributed control within the network. Alternatively, the interaction of distributed control from agents leads to dynamics that cannot be understood by deterministic behaviour but require modelling, analysis and design allowing for decision-making at lower levels than the level of the system as a total; this applies to technical systems, biological systems, organisations and society.

Connectivity

This implies that the ways in which the agents in a system connect and relate to one another is critical to the survival of the system, because it is from these connections that patterns are formed and that the interventions by feedback are disseminated; this is the second attribute of complex adaptive systems. These systems are made up of interdependent interacting parts. The relationships between the agents are generally more important than the agents themselves for understanding the behaviour of the whole system. In the case of an economy that puts the emphasis on how firms interact with each other; for example, the buyer-supplier relationships in a specific economic sector. If the onus were on the firm itself, their classification, size, revenue and profit would be of interest. To understand complex adaptive systems, the researcher would rather look at how companies compete, how they collaborate for delivering products and services, and how they disseminate new products and services (the latter is often called technology diffusion). Hence, the interrelations as being interconnected determine patterns that are formed and the related feedback mechanisms (both positive and negative, see Section 6.3).

The interconnectedness between agents in a complex adaptive system gives rise to non-linearity in behaviour. The concept of non-linearity is presented under very many different names: synergy, linkages, network effects, complementarity, superadditivity, etc. For example, an organisation is made up of a network of interconnections at many different levels: teams, projects, divisions, functions and business units. In terms of management science, these 'agents' create synergy when collaborating. The interactions, not only internally but also externally, mean that it is often difficult to find the boundaries of organisations and for that reason many organisations are larger than their assets and resources (incl. human resources). Just think of the alliances and partnerships and the multiple formal and informal arrangements that 'internal agents' have with other organisations. In this example if a supplier improves its performance, then expectations of the focal firm for all its suppliers might increase forcing the other suppliers to become better. As shown by the example, a change in one element of the organisation as a system of resources might cause changes in a second element and other elements that might come back to the first element and other elements than affected in the first instance. That implies that potentially every element of such a system is linked to every other element in one way or another. In the case of negative synergies there are terms such as diseconomies of scale, overcrowding or negative spillover effects, subadditivity, etc. An example of negative synergy is if centralised coordination in an organisation reduces flexibility to meet more local demand. Whether it concerns positive synergy or negative synergy we can no longer think in terms of simple cause effect relationships (A causes B) but of patterns of relationships, evolution and the emergence of novelty. Thus, the second attribute of complex adaptive systems, the connections between agents, has a profound effect on the behaviour of the individual agents and the complex adaptive system as a total.

Co-Evolution

The third attribute of complex adaptive systems, co-evolution, stems from the argument that complex adaptive systems exist within their own environment and at the same time they are also part of that very same environment. Therefore, as their environment changes, they need to change to ensure a best fit (that requires them to be autopoietic or allopoietic systems, see Chapter 8). Because they are part of their environment, when they change, they change their environment, and as it has changed they need to change again, and so it goes on as a constant process. Co-evolution relates to the two-way interplay between the system and aspects of its environment and can occur in various forms. First, the system can affect its environment, by changing the adaptive pressures on itself (e.g. by moving around, digging holes), thus the environment should not be regarded a static 'object'. Second, changes in this environment can affect the system (for example, weather changes) leading to adaptation or changes in behaviour. However, most forms of co-evolution will occur with respect to other systems. Those interactions encountered can be with systems of its class or other classes, and be either competitive or cooperative. In evolutionary biology, economics and game theory the dynamics of such interactions lead to situations where no agent can improve its fitness so long as the others continue with their current strategy. This leads to repetitive behaviours (cyclic attractors), where stability is at best temporary due to the multiplicity of influences.

All these influences may be multi-faceted, i.e. they can affect several interacting values or properties of each system. A particular set of relevant interactions relates to symbiotic co-evolution, where a number of organisms have a mutual 'arrangement' such that the net benefit to all exceeds their individual costs (a win-win situation). This particular type of co-evolution takes many forms, including symbiotic, social and ecological networks. In general, symbiotic co-evolution will be present to some degree in any complex system, including economic, technological and cultural ones. Such co-evolving systems naturally adjust their parameters to maximise overall group fitness. However, the recognition of co-evolving systems (or entities) does not necessarily mean beneficial relationships for both types of entities. Think about parasitism and mutualism, where one entity benefits more from the 'relationship' than the other one. Hence, symbiotic relationships can take the form of beneficial or detrimental effects for one of the agents.

At the level of biophysical co-evolution – akin fifth to eighth level of the systems hierarchy of Boulding (see Section 3.5) – biological organisms interact with the world's physical resources and with each other to ensure continued existence of species (or evolution). As one part of those interactions, matter is transformed into life and conversely (upon death) life is transformed back into matter (recycling is endemic to this process). This interplay between the physical and biological realms leads to dynamic evolutionary processes in which the interaction between organisms and species and other organisms and species plays a key role.

The next level, socio-biological co-evolution, comparable to the ninth and tenth level of the systems hierarchy of Boulding, adds the interplay between individual life forms and others life forms of the same class, the 'collective' behaviour. Hence, co-evolution becomes more complex and dynamic whereby many types of social structures become possible, including the move between individual and social forms. In this context, animal societies can take innumerable complex forms, as well as much human behaviour, especially for day-to-day social interactions – which take place largely unconsciously and are automatically constrained by cultural norms. This co-evolution between life and society has been described by Wilson [1999], who is known for his propagation of the concept of sociobiology. This level of socio-biological co-evolution relates to the maintenance of an autopoietic socio-ecosystem (akin to the structure of a multicellular organism, but rather less constrained), the emergence of societies and ecosystems.

The level of mythological-social co-evolution, akin the eleventh level of the systems hierarchy of Boulding, is generally thought to apply only to humans, who uniquely have the ability to generate abstract ideas, non-material concepts like mathematics, philosophy, ethics and politics. Thus we have co-evolution between our social level and a mythological world of the imagination (yet very real to all the people involved), which takes place along many separate abstract dimensions. It is here that language and symbolism come to the fore and the autopoiesis here takes place largely in these realms; think about the concept of memes by Dawkins [1989, p. 192] as concepts, ideas and artefacts that are diffused in society similar to how genes are generating offspring. This autopoietic level maintains 'culture' itself, in the form of self-sustaining and self-reinforcing 'belief systems' or philosophical 'world views'. That means that this level may be more subject to perturbation and systemic disintegration by outside influences (e.g. extinction or transformation of human societies or civilisations – compared to either ecosystems or individual species).

9.3 Fitness Landscapes

These levels of the systems hierarchy of Boulding – the fourth to the eleventh – describe the search of systems or their entities for fitness through physical fitness and fitness in the social domain; note that the focus of the current section will be on biological systems mostly. To this purpose, one could view the adaptation as a process in which agents in complex adaptive systems search for a fit between traits and selectional forces from the environment. According to Colby [1996], natural selection may not lead a population to a state in which it possesses the optimal set of traits (or properties, see Section 2.1). These traits are found in specific forms of genes, which are called alleles; for example, the gene for eye colour has an allele for blue eye colour and an allele for brown eye, etc. In any population, a certain combination of possible alleles exists that would produce the optimal set of traits (the global optimum); alleles are two or more alternative forms of genes caused by mutation of an original set of genes. However, there are also other sets of alleles present that would yield a population almost as adapted (local optima). In social systems, the concept of genes for alleles might have to be replaced with memes; Dawkins^^ [1989, p. 192] has proposed memes, as unit for imitation and recombination. Memes constitute elements of a culture or system of behaviour that may be considered to be passed from one individual to another by non-genetic means, especially imitation. Dawkins extends this concept to a wide variety of topics like ideas, artefacts, including people, products, books, behaviours, routines, knowledge, science, religion, art, rituals, institutions, and politics. The search for optima of traits (alleles) finds place in so-called fitness landscapes; these imaginary landscapes contain the combinations of alleles and also have optima. Transition from a local optimum to another optimum may be hindered or forbidden because the population would have to pass through less adaptive states to make the transition to optimal states.

Wright's Adaptive Landscape

One of the tenets of searching for fitness is that natural selection only works by bringing populations to the nearest optimal point of traits. This theory is referred to as Sewall Wright's adaptive landscape [Wright, 1982], even though others have shown that, mathematically, his landscapes do not exist as envisioned. The fitness of species develops by moving from one to another selective peak (see Figure 9.1). As Wright wonders, which force will act against the pressure of selection moving species from one peak to another one? According to him [ibid., pp. 8–9], an effective process for shifting the balance involves three phases: first, extensive local differentiation, with wide stochastic variability in each locality; second, occasional crossing of a saddle leading to a higher selective peak under mass selection; and third, excess proliferation of, and dispersion from, those local populations in which a peak-shift has occurred, leading to occupation of the superior peak as a whole. Not only do the species adapt to the landscape, the surface of fitness values also mutate when environmental conditions change. With changing conditions, the location of the species follows the movement of the peak if the change does not lead to extinction. That might lead to old adaptations are being lost as new ones are acquired. That means that populations, as systems, move from one peak to another but not necessarily the optimal one and that they follow the dynamics of the fitness landscapes (emerging peaks and changes in existing peaks) by adaptations.

Referring to the development of species, Worden [1995] encourages the application of genetic algorithms – as a combinatory method for optimisation – for mimicking natural selection with limitations in terms of speed of development. He maintains that the development of species through the deployment of genetic algorithms is subject to the speed limits he proposes. The adaptations to the environment, homeostatic development, inhibit moving the development of a species to the next successful set of adaptation, akin a ball rolling across hilly terrain. Then species are always trying to reach the adaptive peaks of the landscape and are continually modified in response to the shifting of the peaks. In this perspective, Gould [1980, p. 129] refers to the metaphor of Galton's polyhedron borrowed by St. George Mivart to describe the evolution of organisms and the state of equilibria. Prothero [1992] uses the metaphor of the polyhedron, which can roll rapidly

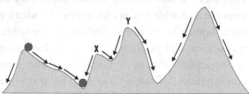

Figure 9.1 *Representation of Wright's adaptive landscape [adapted from Heylighen, 2000]. Species move from one peak to another (represented by the circle, being at a fitness peak, a stable position, and a saddle, representing a stable point with low fitness). During the move they might experience reduced fitness before reaching the more adaptive saddle. The ball will move most likely from the saddle to the nearest, highest fitness peak (on the left) or reach point X because each adaptive step yields higher fitness than moving left. Point X and point Y indicate stable positions of high fitness. A species at point X will have to move through a saddle with lower fitness before reaching the higher fitness peak Y.*

over from face to face, but resists change when it is sitting on one of its stable faces. Change only occurs when the threshold necessary to tip it over has been exceeded, and then the polyhedron will resist further change until that threshold is once again reached. Between stable states (the faces), the transitions are very rapid. The practical implication is that a lot of energy and momentum is needed to initiate a change process but once the system approaches a possible stable state in relation to the environment, the intended change will unfold very fast; furthermore, because of step-wise change in a given fitness landscape, populations are limited by this landscape in how quick they adapt.

Random Fitness Landscapes

Building on the notions of Wright's adaptive landscape and how populations can move from one optimum to other ones, Kauffman [1993, 1995] has extended the concept of the fitness landscapes to explain the diversity of life. His two publications cover mostly the same ground and provide different texts for explaining some phenomena during the evolution of life on earth. The theory of his fitness landscapes directly connects to the impact of natural selection. The framework of adaptation on rugged landscapes [Kauffman, 1993, pp. 36–67] applies to adaptive evolution in sequence spaces, i.e. spaces in which the successive steps depend on the preceding steps (the underlying mechanism for mutations to appear in a population), akin principles of autopoiesis.

To explain the thoughts of Kauffman about the fitness landscape, this section captures the text for explaining the simpler N-model before moving on to the more complex NK-model. In both models, fitness is expressed as height, a measure for expressing the fitness of a genotype. An example of this is the darkening of the peppered moth population, which took place when the soot-darkened trees resulting from heavy industry made light-coloured moths easier targets for hungry birds [Max, 2001]. According to Colby [1996], Kettlewell found that dark moths constituted less than 2% of the population before 1848. The frequency of the dark morph increased in the years following. By 1898, 95% of the moths in Manchester and other highly industrialised areas were of the dark type. Their frequency was less in rural areas. The moth population changed from mostly light coloured moths to mostly dark coloured moths, determined primarily by a single gene. This example shows that for the changing environment (industrialisation) dark moths have a higher fitness than light-coloured moths. Kauffman [1993, pp. 36-67] builds on this phenomenon by attributing fitter genotypes with have a higher number of heights than less fit genotypes.

The N-model uses this premise of attributing fitter genotypes to have a higher number for height than less fit genotypes, when using this model for evolutionary biology (height represents fitness). As an example, consider a genotype with only four genes, each having two possible alleles: *1* and *0* (i.e. a Boolean representation of the state of each gene; think as one of the genes the moth being dark or light-coloured). That results in 16 possible genotypes, from which each is a unique combination of the different states of the four genes. If these are mapped in a so-called hypercube, each node corresponds to one of the 16 possibilities (see Figure 8.2); a hypercube is used in mathematics to represent multiple dimensions, in this case four because of the four components of a gene. Each node on the hypercube (called vertex) differs only by one mutation from the neighbouring ones, representing the step of a single mutation; that means that each mutation is independent from the state of the other components of the gene. Each genotype is arbitrarily assigned to a fitness value ranked from worst (i.e. *1*) to best (i.e. *16*). An adaptive walk might start at any vertex and move to vertices that have higher fitness values. An adaptive walk ends at a local optimum, a vertex that has a higher fitness value than all its one-mutant neighbours but that is not necessarily the highest optimum. Figure 9.2 shows that there are three local optima at which adaptive walks may end; only one of these three is the global optimum. This means that in the N-model the search for fitness might end at either a local or a global optimum; moving from one local optimum to another optimum with a higher value means following a pathway in which losing fitness occurs before an optimum is reached.

For finding optima, Kauffman [1993, p. 39] refers to Gillespie [1984], who has shown that the search by an adapting population corresponds to an adaptive walk. If the population begins at the less fit allele, a single mutant will eventually encounter the fitter allele. Either this mutant dies out before leaving offspring or a few of the fitter mutants are produced. Once the number of fitter mutants produced is sufficient to reduce the chance of fluctuation

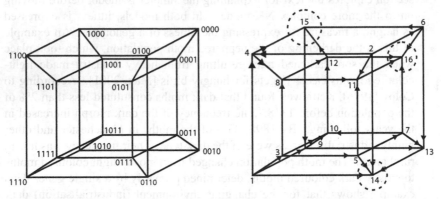

Figure 9.2 *The N-model (derived from Kauffman [1993, p. 38]). The 16 possible combinations of four digits are arranged as vertices on a four-dimensional Boolean hypercube. Each combination connects to its four one-mutant neighbours, accessible by changing a single digit from 1 to 0 or from 0 to 1. The hypercube on the left represents this four-dimensional space. In the hypercube on the right-hand side, each combination has been assigned, at random, a rank-order fitness, ranging from the worst (1) to the best (16). Arrows from the less fit to the more fit show directions of such moves between adjacent positions. Combinations fitter than all one-mutant neighbours are local optima (three in this case).*

leading to their death, the fitter type rapidly takes over the entire population. Thus the entire population moves to the fitter genotype. Gillespie has shown that the entire adaptive process can be treated as a continuous-time, discrete-state process of transitions (that is called a Markov chain in mathematics). Hence, in practice, that implies that finding an optimum set of traits follows a pathway of distinctive steps for mutations; however, each individual mutation represents a chance event.

This notion of pathways building on individual mutations being a chance event implies that evolution on landscapes that are random is a search by chance, as Kauffman [1995, pp. 166-167] states. In fact, finding the global peak by searching uphill on random landscapes is totally useless; the entire space of possibilities has to be searched to find the local optima as well as the global optimum. However, from any initial point on a random landscape, adaptive walks reach local peaks after some number of steps. Additionally, no matter, where an adaptive walks starts, if the population is allowed to walk only uphill, it can reach only an infinitesimal fraction of the local peaks. But in reality, the fitness landscapes that underlie the mutation steps of gradual evolutionary pathways are correlated and local peaks do often have similar heights. Hence, reaching a local peak, which might or might not be the global optimum, will generally end the search on a random landscape.

Rugged Fitness Landscapes

In reality, random fitness landscapes hardly occur. Each component of a system does not exist on its own; therefore, the notion of the random landscape has limited meaning for the evolution of life forms. For example, in the case of genes, some do have correlations to others, e.g. the hierarchy of genes, or sets of genes exist all contributing to particular appendages, organelles, etc. (in terms of systems theories, elements, subsystems and even aspectsystems). As Kauffman [1995, p. 170] notes: the fitness contribution of one allele of one gene may depend in complex ways on the alleles of other genes. This is often referred to as epistatic coupling or epistatic interactions for the components of a system. Rugged fitness landscapes are those landscapes in which the fitness of one component of a system depends on that part and upon K other parts among the N present in the landscape [Kauffman, 1993, p. 40]. That model that expresses the interdependency of components and the fitness landscapes being less random is called the NK-model.

This NK-model offers further insight in the mechanisms of evolution and selection, from here on applied to genes. Let us look at an organism with N gene loci, each with two alleles (two possibilities), *1* and *0*. The parameter K stands for the average number of other loci that epistatically affect the fitness contribution of each locus; that means that they between the components there are connections that determine an individual component's fitness (for example, such is the case for an organisation with its products having a relationship between quality and price). Hence, the fitness contribution of the allele at the ith locus depends on itself (whether it is *1* or *0*) and on the other alleles, *1* or *0*, at *K* other loci, hence upon *K+1* alleles. The number of combinations of these alleles is just 2^{K+1}. The example of Kauffman selects at random a different fitness contribution to each of the 2^{K+1} combinations from a uniform distribution between *0.0* and *1.0* (this pattern has been followed in Figure 9.3 where three optima appear); note that this does not work if the fitness does not have a random distribution. The fitness of one entire genotype can be calculated as the average of all of the loci. When this example is visualised in a hypercube, it becomes clear that again more than one local optimum exists. It should be noted that, normally, these fitness landscapes based on epistatic coupling are more rugged in comparison to the random fitness landscapes.

For the study of evolutionary biological phenomena a number of conclusions can be drawn by deploying the NK-model. Kauffman [1995, p. 161] states that biologists hardly know what such fitness landscapes look like or how successful a search process is as a function of landscape structure (this is most likely not restricted to species). Nevertheless, these (imaginary) landscapes for selection of species may vary from smooth, single-peaked to rugged, multi-peaked landscapes. During evolution species search these landscapes using mutation, recombination and selection, a process for which

1	2	3	4	w_1 (.4)	w_2 (.3)	w_3 (.2)	w_4 (.1)	W
0	0	0	0	0.6	0.4	1.0	0.3	0.59
0	0	0	1	0.8	0.5	0.6	0.2	0.61
0	0	1	0	0.1	0.7	0.1	0.8	0.35
0	1	0	0	0.2	0.5	0.8	1.0	0.49
1	0	0	0	0.9	0.6	0.8	0.7	0.77
0	0	1	1	0.9	0.5	0.2	0.8	0.63
0	1	0	1	0.8	0.2	0.9	0.7	0.63
1	0	0	1	0.6	0.1	1.0	0.3	0.50
0	1	1	1	0.4	0.7	0.8	0.5	0.58
1	1	0	1	0.4	0.7	1.0	0.5	0.62
1	0	1	1	0.4	0.1	0.6	1.0	0.41
1	1	1	1	0.7	0.3	0.8	0.2	0.55
1	1	1	0	0.2	0.4	0.2	0.1	0.25
1	1	0	0	0.1	0.1	0.4	0.5	0.20
1	0	1	0	0.5	0.8	0.7	0.2	0.60
0	1	1	0	0.3	0.8	0.3	0.6	0.48

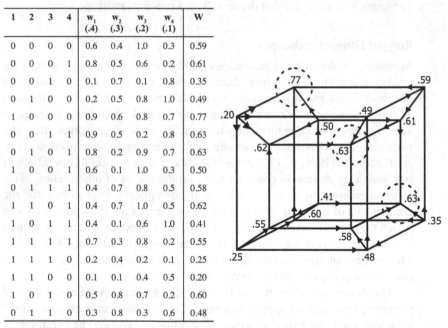

Figure 9.3 *NK-model derived from Kauffman [1993, p. 42] and Figure 9.2. The assignment of the fitness values to each of the four components. These fitness values then assign fitness to each of the $2^3=8$ possible genotypes as the weighted value of the fitness contributions of the four components. The figure depicts the fitness landscape on the four-dimensional Boolean cube corresponding to the fitness values of the 16 genotypes. The coupling leads to a smoother landscape with relatively lower peaks compared to Figure 9.2.*

the NK-model provides insights into particular phenomena accompanying the adaptive walk:

- On smooth surfaces and rugged surfaces of the fitness landscape, the search process may fail to find the high peaks [Kauffman, 1995, p. 161].
- The search process on rugged and random landscapes might be facilitated by multi-steps and long jumps [Kaufmann, 1993]. While these multi-step and long jumps are effective search strategies for those landscapes, they also increase the risk that local peaks or a global optimum may be missed.
- When on smooth surfaces the high peaks are found by a population, mutations might cause the complexity error catastrophe [Kauffman, 1993, p. 96; Kauffman, 1995, p. 161, 183–184]. Normally, when a species or population reaches a peak in the landscape, smooth or rugged, it remains stable at the peak. Through higher mutation rates, the population might increase by number and diversity, causing a greater area of spread. This spread might extend so far from the peak itself that part of the population starts the search for new peaks. Kauffman labels this phenomenon the complexity error catastrophe, which indicates that the mutation rate exceeds the equilibrium force of remaining at the peak.
- On random landscapes, finding the global peak by searching uphill is useless, the entire space of possibilities needs to be searched [Kauffman, 1995, p. 168]. At the same time, wherever the adaptive walk starts, only a fraction of the local peaks will be reached.
- When the population climbs higher to a local or global peak, it becomes exponentially harder to find the direction uphill [Kauffman, 1995, p. 178, 193–194]. As complexity increases, meant as number and diversity, blind long jumps become a more wasteful strategy, even on the best of landscapes [Kauffman, 1993, p. 74].
- Fitness can increase more rapidly near peaks when mutation and selection are joined by recombination [Kauffman, 1995, p. 182]. This covers both local and large-scale features of the fitness landscape.
- Complex artefacts or real organisms never find the global optimum of the fixed or adapting landscape [Kauffman, 1993, pp. 77–78].
- A breakdown of populations in patches enhances adaptability of species and populations, especially in changing landscapes [Kauffman, 1995, p. 263].
- Mass extinction [Kauffman, 1993, p. 78].

From these points it follows that adaptive evolution is a search process by populations – driven by mutation, recombination and selection – on fixed or deforming fitness landscapes. Nevertheless, the dynamics of the environment are driven in his view by the shape and the dynamics of the landscape. In the theories for the N-model and the NK-model, the landscape is more or less fixed by the values assigned to specific genotypes. The populations move around in this landscape and cause the dynamics. If the fitness landscapes are shifting, then the deformation of the landscape needs to be explained qualitatively, hence the shifting dynamics of the landscape are less present in the models and mathematical approaches underpinning this theory; other theories, such as adaptive dynamics [Geritz et al., 1997; Meszéna et al., 2001], are more adequate for the dynamics of evolving species influencing the fitness landscape. The search in these (semi-)static fitness landscapes of the NK-model is directed to finding fitness peaks, and to which mutation types (one-step, multi-step or long jumps) fit best to the shape of the fitness landscape.

Co-Evolution and NK-model

Kauffman [1993, pp. 243–245] extends the NK-model to co-evolution by adding the constraint that each trait in species 1 depends epistatically on K traits internally and on C traits in species 2, the so-called NK[C]-model. More generally, in an ecosystem with S species, each trait in a species will depend on K traits internally and on C traits in each of the S_i among the S species with which it interacts. Therefore, if one species adapts, it also changes the fitness of other species and deforms their landscapes in the NK[C]-model. This does not only apply to biology but also to organisational constructs, such as supply chains [Dekkers, 2009; Surana et al., 2005]. Therefore, the NK[C]-model offers an explanation for not only the development of individual systems and populations but also for the interconnection between systems and populations.

The coupling of the fitness landscapes affects the search for increased fitness [Kauffman, 1993, pp. 252–253]. When a new link is introduced (i.e. increasing K) the genetic locus spreads throughout a population in three ways: (a) the new epistatic link, when it forms, causes the genotype to be fitter, (b) the new epistatic link is near neutral and spreads through the population by random drift, and (c) the new link not only has a direct effect on the fitness of the current genotype but also increases the inclusive fitness of the individual and its genetic descendants. It suggests that optimisation in co-evolutionary dynamics becomes possible by optimisation mechanisms that search for optimal traits in relation to the coupled traits (we could view the industrial development of the Pearl River Delta as a complex adaptive system [Noori and Lee, 2002; The Economist, 2002]). The second option for a network consists of increasing its reach, which compares to increasing the number of species S. When this happens the waiting time to encounter a new equilibrium increases, the mean fitness of the co-evolving partners decreases [McKelvey, 1999, p. 312], and the fluctuations in fitness of the co-evolving partners increase dramatically. The increase of agents may lead to a new optimisation in traits and coupled traits but only after going through a period of instability.

9.4 Self-Organisation by Complex Adaptive Systems

Because of these alternating phases of stability and instability, the behaviour of complex adaptive systems has been linked to self-organisation; for example,

Kauffman [1993, pp. 567–568] connects the NK-model to the concept of self-organisation. Note that the concept of self-organisation already appeared in Section 8.3; in the context of this section it should be noted that the concept of self-organisation of systems arose from the study by Nicolis and Prigogine [1977] into thermodynamic systems for from equilibrium. These concepts of self-organisation have drawn the attention of researchers in many fields of science [e.g. Mikulecky, 1995]. The interests of all are directed to the explanation of emergent behaviour and the establishment of patterns that cannot be explained only by the actions of agents or reduced to the agents' behaviour [Stacey, 1996, p. 63]. The explanations for the emergence of behaviour and patterns vary; some of these explanations for complex adaptive systems will follow now.

Simple Rules and Complex Behaviour

One explanation for those complex adaptive systems is that simple rules might lead to complex behaviour. The famous and often quoted example concerns the flocking behaviour of birds, originating in the work of Reynolds [1987]. The simulation of Boids, an artificial creature, in computer applications shows the complex behavioural pattern of flocks that emerges when a Boid adjust its behaviour by only looking at its neighbour's position and speed. With no more than these rules, Boids flock, fly around obstacles and regroup. Another famous example is the Game of Life [Conway, 1970]. In this game, a grid of cells progresses from an initial state to form complex patterns; each cell can either be 'alive' or 'dead' and each can change its state from round to round. At the start of every new round, simple rules are concurrently applied to each cell to decide whether it is alive or dead:

1. Any live cell with fewer than two live neighbours dies, as if caused by under-population.
2. Any live cell with two or three live neighbours lives on to the next generation.
3. Any live cell with more than three live neighbours dies, as if by overcrowding.
4. Any dead cell with exactly three live neighbours becomes a live cell, as if by reproduction.

Those two examples of experiments point to simple rules that cause complex patterns of behaviour, however, they limitedly explain how complex adaptive systems respond to changes in their environment.

Attractors

A second explanation for complex behaviour arrives through the insights of attractors [Kauffman, 1993, pp. 178–179; Morgan, 1997, p. 264; Stacey, 1996, pp. 58–60]. Attractors are set of physical properties toward which systems evolve regardless of their initial state. Note the parallel with equifinality (Section 3.3); however, the difference is that a system might have attractors consisting of a finite set of states, a curve or a manifold. A pendulum is an example of such a system that moves between two states (a state of balance with no movement and a state where the velocity is reaching its maximum). Even if there is a perturbation the system will return to the states associated with attractors. In the case of the pendulum, a force that is exerted will make it eventually move between the two states again. Hence, attractors are foremost representing 'predictable' behaviour that nevertheless may be quite complex.

When the control parameters in a deterministic non-linear feedback network are tuned up (e.g. when information or energy flows are increased), the behaviour of the network follows a potential bifurcating path in which it continues to display regular, stable patterns but they become increasingly complicated. A critical level of the control parameter moves the system in a state between stability and instability. Sensitivity of these patterns to initial conditions, tiny deviations, might result in vast differences in the subsequent behaviour of the system; this is the so-called butterfly effect [Lorenz, 1963]. This means that dependent on the initial conditions, a complex adaptive system may end up in different attractors. An attractor does not mean bifurcation; the latter is related to evolutionary biology and means that two states co-exist 'next to each other', i.e. the branching of a species into subspecies. Attractors only indicate points of stability with high or low dimensionality (states) to which the behaviour of a system evolves. Low dimensionality of attractors is mostly related to more orderly behaviour [Kauffman, 1993, p. 179]. Thus, the existence of attractors as a state of systems between stability and instability causes systems to evolve towards these states, though dependent on how these attractors are related to flows of energy and information.

Dissipative Structures

The theories of Prigogine and Nicolis [1977] expand this matter of energy and information flows by introducing dissipative systems, whereby away from the point of homeostasis a temporary and complex order is maintained. Only a part of the exchange with the environment, such as the energy flow, sustains the order, most of it dissipates to the environment. At the end, a new state arises in which the internal complexity has been increased, and new structures and behaviour do emerge. Examples of dissipative systems are cyclones, hurricanes and living organisms in distress; these are instances of systems with a dynamic regime that nevertheless seems to be in reproducible steady state. These reproducible states may be reached by natural progression or by artifice. An example of the latter are Bénard cells; these occur in a horizontal layer of fluid on a plane heated from below, in which the fluid develops a regular pattern of convection cells. Thus, dissipative structures, naturally occurring or artificially created, are another source for complex behaviour of systems.

Edge-of-Chaos

Another explanation for complex behaviour of complex adaptive systems arrives from so-called chaos theory. Four particular states arise when the NK-model from Section 9.3 is more closely looked at for the principles of self-organisation [Kauffman, 1993, pp. 191–203]. First, at $K=1$, the orderly regime appears, in which independent subsystems function as largely isolated islands with minimal interaction. At $K=2$, the network is at the edge-of-chaos, the ordered regime rules at maximum capacity but chaos is about to emerge. At values ranging from $K=2$ to $K=5$ the transition to chaos appears although indications are that this transition happens already before $K=3$. From $K>5$, the network of subsystems displays chaotic behaviour, meaning unpredictable behaviour. All these four types of behaviour related to the value of K indicate that the behaviour of systems strongly varies according to the connectivity, one of the principle attributes of complex adaptive systems.

Using this notion Kauffman [1993, p. 198; 1995, p. 91] claims that a position in the ordered regime near the state of chaos affords the best mixture of stability and flexibility. Such a state optimises the performance of the complexity of connected tasks and optimises evolvability of complex adaptive systems. Although Kauffman's models merely generate semi-static fitness landscapes, they imply the similarity between structures within forms of life. The developmental pathways embedded in the fitness landscapes and the principles of self-organisation are bound in the search for optimisation. Hence, the resemblance in existing forms of complex adaptive systems is no matter of chance, but a result from previous mutations and developments at the edge-of-chaos (a mixture of stability and flexibility).

9.5 Recursive Behaviour

The behaviour related to state between stability and flexibility can also be found in multi-agent systems; increasingly these systems are being used and designed for solving problems in a variety of complex and dynamic domains; the purpose of these multi-agent systems is mostly simulating and controlling complex adaptive systems. A multi-agent system is a computer application composed of multiple interacting intelligent agents within an environment. Those multi-agent systems can be used to solve problems that are difficult or impossible for an individual agent or a monolithic system to solve. The intelligence of agents may include some methodical, functional, procedural or algorithmic search, retrieval and processing approach. Typical applications are software engineering, collaborative networks and factory automation. Although there is considerable overlap, a multi-agent system is different from an agent-based model. The goal of an agent-based model is to search for explanatory insight into the collective behaviour of agents obeying simple rules rather than in solving specific practical or engineering problems. Some examples of the use of agent-based models are epidemics and economic analysis. The terminology of agent-based models tends to be used more often in the sciences and multi-agent systems in engineering and technology.

Effective agent learning in such domains raises some of most fundamental challenges for agent in these systems, whether it concerns agent-based models or multi-agent systems. Agents are autonomous but might lack a global view of the system; in addition, there is no designated controlling agent, hence, agents exert responses locally. Typical agents are software agents, human, teams (or groups) and autonomous equipment, such as robots. An agent may often need to model the behaviour of other agents, learn and adapt from its interactions, negotiate with other agents, and so on. The typical assumption in most of the studies on learning is that the data is uniformly distributed and available to agents. However, in practice, not all data are available to all agents, not even to similar agents, and not uniformly distributed either. Data are often available at progressively more detail, in similar patterns and recursively made available, though chaotic and random at times. For instance, almost all biological systems contain self-similar structures that are made through recurrent processes, while many physical systems contain a form of functional self-similarity that owes its richness to recursion. Therefore, this indicates that recursion plays a key role in the behaviour of complex adaptive systems and that the modification of recursive behaviour plays a role in evolutionary processes.

If the tenet of complex adaptive systems is recursive behaviour, it would be expected that behaviour of the holistic system (complex adaptive system)

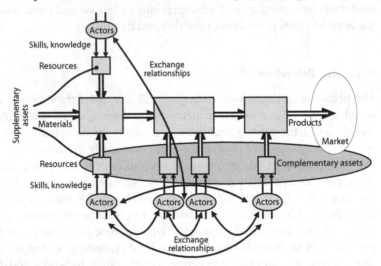

Figure 9.4 *Collaboration model for the value chain [Dekkers, 2005, p. 330]. Vertical collaboration (which is the horizontal dimension in the figure) indicates the capability of actors to manage the supply chain. Horizontal collaboration (vertical dimension in figure) contributes to the dynamics of the network by recombining resources or creating substitution.*

is fairly predictable. However, as it turns out, the human mind, economic markets, network data, agent behaviour, internet browsing behaviour and nature create enormously complex behaviour that is much richer than the behaviour of the individual constituting elements. Complex systems with emergent properties can often be viewed as highly parallel collections of similar elements. A parallel system is inherently more efficient than a sequential system, since tasks can be performed simultaneously and more readily via specialisation. Parallel systems that are redundant have fault tolerance and subtle variation among the parts of a parallel system allows for multiple problem solutions to be attempted simultaneously. Hence, complex adaptive systems are characterised by agents operating in parallel that results in emergent and complex behaviour, although each constituent agent displays mostly recursive behaviour, while not necessarily behaving identical to a similar agent or similar agents.

9.6 Connectivity in Human-influenced Networks

In addition to emergent behaviour resulting from connectivity, human-influenced complex networks, e.g. the World-Wide Web and human acquaintance networks, have common properties, which are hardly compatible with existing cybernetic approaches. Since that incompatibility has been recognised, increased efforts have been dedicated to identify other measures of complex (enterprise) networks [Fricker, 1996]. Therefore, this section looks more closely at interaction in networks.

The lack of network-orientation within traditional systems theories becomes obvious, considering that nowadays most companies are viewed as being part of networks, such as supply chains and service providers. More specifically, the network-orientation requires managing both the relationships between agents in networks (sometimes called actors) and the delivery of products and services. Such a model combines three parts of networks: resources, activities (processes), and actors; see Figure 9.4 for the model by Dekkers [2005, p. 330] and Dekkers and Kühnle [2012, p. 1095]. In this model, companies ensure value innovation spanning the entire value chain and the integration of skills and knowledge for meeting performance requirements through vertical collaboration; these complementary resources not only cover manufacturing and the supply chain but may also include product innovation and new product development. By horizontal collaboration, actors will increase the chance for achieving economics of scale for products or their components (which could include process innovation); that calls on supplementary resources [Das and Teng, 2000, p. 49], which should lead to synergy. Both vertical collaboration and horizontal collaboration allow companies to deploy effective resources for both product and process innovation and meet criteria of sustained competitive advantage. But ultimately, this model combines the linear cybernetic process for delivery of processes and services and the agent approaches from complex adaptive systems for managing the interrelationships between resources and agents.

For collaborative networks between firms and for human acquaintance networks, there is a property that relates path length to size of the network. This so-called small-world property, the most known of specific properties for networks, states that the average path length in the network is small relative to the system size [Milgram, 1967]. This phenomenon has already been scientifically studied more than three decades ago, long preceding its notoriety. For example, the phrase *six degrees of separation* [Guare, 1990], another popular slogan depicting the small-world phenomenon, can be traced back to Milgram's 1967 experiment. Most famously, the actor Kevin Bacon became an unwitting part of networks with small-world properties, when a couple of decades ago some college students, scheming to get on Jon Stewart's show on MTV, seemingly decided that '6 degrees of Kevin Bacon' sounded enough like '6 degrees of separation' that it must imply that Kevin Bacon was the centre of the acting universe. Hence, the small-world property has been related to the potential of connectivity for human-influenced networks and become attached to complex behaviour.

Another notable property of complex networks is clustering, i.e., the increased probability that pairs of nodes with a common neighbour are also connected. That becomes apparent when a critical fraction of nodes (or links) is removed, then the network becomes fragmented into small, disconnected clusters. In mathematical terms this is called percolation theory. Clustering means also that sometimes nodes start to act as hub; these hubs connect to multiple nodes that have multiple connections. The further away a node is from a hub the lower the number of connections generally. In addition, an other approach to small-world networks was formulated by Watts and Strogatz [1998] that has helped network science become a medium of expression for numerous physicists, mathematicians, computer scientists and many others; mainly because it confirmed not only small-world properties but also clustering as essential trait of networks. In that sense, clustering facilitates the connectivity of the human-influenced networks.

Perhaps, the most important property of human-influenced networks is the distribution of degrees, i.e. the distribution of the number of links the nodes have. One way of looking at a network is that each node (or element) has a probability to be connected to another node; such networks are called random networks. It has been shown that several real world networks have scale-free distributions, often in the form of a power law. In these networks, a huge number of nodes have only one or two neighbours, while a couple of them are massively connected. Many networks are thought to be scale-free, including the World-Wide Web, biological networks and social networks. Take for example a social network in which nodes are people and links are acquaintance relationships between people. In those networks people tend to form communities, i.e. small groups in which everyone knows everyone (subsystems). Moreover, the members of a community also have a few acquaintance relationships to people outside that community. However, some people are connected to a large number of communities (for example, celebrities and politicians), the earlier mentioned small-world property. Mostly, the behaviour of a randomly distributed network and a scale free network differs. The connectivity of a randomly distributed network decays steadily as nodes fail, slowly a into smaller, separate domains that are unable to interact. In contrast, scale-free networks may show almost no degradation as randomly nodes fail. With their well-connected nodes, which are statistically unlikely to fail under random conditions, connectivity in the network is maintained. The distribution of these links among the elements of a scale free system means that relationships are not randomly or evenly distributed, generally speaking; that means that scale free networks are more robust towards perturbations.

These three specific properties – small world property, clustering, distribution of degrees – hardly appear in the original systems theories; however, they explain how agents as systems in networks as systems behave and how networks might be structured to be more robust.

9.7 Summary

The concepts and theories for complex adaptive systems aim at explaining non-linear behaviour. This type of behaviour cannot directly be explained by the behaviour of individual elements or subsystems (sometimes called agents). In addition to the concepts of autopoiesis, the theories about complex adaptive systems state that these have many autonomous entities, that they are able to respond to external changes and that they form self-maintaining systems with internal pathways for feedback. The concepts related to complex adaptive systems aim at explaining the behaviour of such systems, called non-linear behaviour, that cannot directly be explained as a result of the behaviour of individual entities of this system.

One of the mechanisms to explain that non-linear behaviour is that simple rules for the interaction between entities in a complex adaptive system could lead to complex patterns. A famous example is flocking of birds; computerised simulations that use rules such as maintaining a certain distance to the nearest-by entity result in patterns that look very similar to the patterns of flight by a large group of birds. However, such patterns might only appear in homogeneous and regular environments.

Another mechanism for explaining the non-linear behaviour is the search by complex adaptive systems for optimal points on a fitness landscapes. These landscapes can be imagined as real-life landscapes with rugged areas, hilly areas and flats. Complex adaptive systems seek out peaks on these landscapes, which might be either local peaks or global peaks. Moving from one peak to an other (higher) one follows pathways that will lead to passing through sub-optimal points, such as depressions and valleys in the landscape. These pathways can be circumvented if a complex adaptive system takes larger steps (for mutations), however these steps also increase the chance of missing out on reaching peaks. Note that complex adaptive systems consist out of more entities and its thoughts need to be applied to groups of entities, for example, species, economics sectors and national systems rather than a single specimen or firm.

References

Boswijk, H. K. (1992). Complexiteit in evolutionair and organisatorisch perspectief, het zoeken naar balans tussen vermogens en uitdagingen. Rotterdam: Erasmus Universiteits Drukkerij.

Buchanan, M. (2004). It's the Economy, stupid. New Scientist, 182(2442), 34-37.

Colby, C. (1996, 7th January). Introduction to Evolutionary Biology. The Talks.Origin Archive. Retrieved from http://www.talkorigins.org/faqs/faq-intro-to-biology. html

Conway, J. (1970). The game of life. Scientific American, 223(4), 4.

Das, T. K., & Teng, B.-S. (2000). A Resource-Based Theory of Strategic Alliance. Journal of Management, 26(1), 31–61. doi: 10.1177/014920630002600105

Dawkins, R. (1989). The Selfish Gene. Oxford: Oxford University Press.

Dekkers, R. (2005). (R)Evolution, Organizations and the Dynamics of the Environment. New York: Springer.

Dekkers, R. (2009). Distributed Manufacturing as Co-Evolutionary System. International Journal of Production Research, 47(8), 2031–2054. doi: 10.1080/00207540802350740

Dekkers, R., & Kühnle, H. (2012). Appraising interdisciplinary contributions to theory for collaborative (manufacturing) networks: Still a long way to go? Journal of Manufacturing Technology Management, 23(8), 1090–1128. doi: 10.1108/17410381211276899

Fricker, A. R. (1996). Eine Methodik zur Modellierung, Analyse und Gestaltung komplexer Produktionsstrukturen. Aachen: RWTH Aachen.

Geritz, S. A. H., Metz, J. A. J., Kisdi, E., & Meszéna, G. (1997). Dynamics of Adaptation and Evolutionary Branching. Physical Review Letters, 78(10), 2024–2027. doi: 10.1103/PhysRevLett.78.2024

Gillespie, J. H. (1984). Molecular Evolution Over the Mutational Landscape. Evolution, 38(5), 1116–1129. doi: 10.2307/2408444

Gould, S. J. (1980). Is a new and general theory of evolution emerging? Paleobiology, 6(1), 119–130. doi: 10.1017/S0094837300012549

Guare, J. (1990). Six Degrees of Separation: A Play. New York: Vintage Books.

Heylighen, F. (2000, 6th August). Fitness Landscapes. Principia Cybernetica Web. Retrieved from http://pespmc1.vub.ac.be/REFERPCP.html

Jost, J. (2004). External and internal complexity of complex adaptive systems. Theory in Biosciences, 123(1), 69–88. doi: 10.1016/j.thbio.2003.10.001

Kauffman, S. A. (1993). The Origins of Order: Self-Organization and Selection in Evolution. New York: Oxford University Press.

Kauffman, S. A. (1995). At home in the universe: the search for laws of self-organization and complexity. New York: Oxford University Press.

Lorenz, E. N. (1963). Deterministic Nonperiodic Flow. Journal of the Atmospheric Sciences, 20(2), 130–141. doi: 10.1175/1520-0469(1963)020<0130:DNF>2.0.CO;2

Malik, F. (1992). Strategie des Managements komplexer Systeme. Bern: Haupt.

Max, E. E. (2001, 1st September). The Evolution of Improved Fitness. The Talks. Origin Archive. Retrieved from http://www.talkorigins.org/faqs/fitness/

McKelvey, B. (1999). Avoiding Complexity Catastrophe in Coevolutionary Pockets: Strategies for Rugged Landscapes. Organization Science, 10(3), 294–321. doi: 10.1287/orsc.10.3.294

Meszéna, G., Kisdi, É., Dieckmann, U., Geritz, S. A. H., & Metz, J. A. J. (2001). Evolutionary Optimisation Models and Matrix Games in the Unified Perspective of Adaptive Dynamics. Selection, 2(1–2), 193–210. doi: 10.1556/Select.2.2001.1-2.14

Mikulecky, D. C. (1996). Complexity, communication between cells and identifying the functional components of living systems: Some observations. Acta Biotheoretica, 44(3–4), 179–208. doi: 10.1007/BF00046527

Milgram, S. (1967). The Small World Problem. Psychology Today, 2, 60-67.

Morgan, G. (1997). Images of organization. Thousand Oaks: Sage Publications.

Nicolis, G., & Prigogine, I. (1977). Self-organization in nonequilibrium systems : from dissipative structures to order through fluctuations. New York: Wiley.

Noori, H., & Lee, W. B. (2002). Factory-on-demand and smart supply chains: the next challenge. International Journal of Manufacturing Technology and Management, 4(5), 372–383. doi: 10.1504/IJMTM.2002.001456

Prothero, D. R. (1992). Punctuated Equilibrium at Twenty: A Paleontological Perspective. Skeptic, 1(3), 38–47.

Reynolds, C. W. (1987). Flocks, Herds, and Schools: A Distributed Behavioral Model. Computer Graphics, 21(4), 25–34. doi: 10.1145/37402.37406

Stacey, R. (1996). Emerging Strategies for a Chaotic Environment. Long Range Planning, 29(2), 182–189. doi: 10.1016/0024-6301(96)00006-4

Surana, A., Kumara, S., Greaves, M., & Raghavan, U. N. (2005). Supply-chain networks: a complex adaptive systems perspective. International Journal of Production Research, 43(20), 4235–4265. doi: 10.1080/00207540500142274

The Economist. (2002, October 12th, 2002). A new workshop of the world. The Economist, 365, 59-60.

Watts, D. J., & Strogatz, S. H. (1998). Collective dynamics of 'small-world' networks. Nature, 393(6684), 440–442.

Wilson, E. O. (1998). Consilience: the unity of knowledge. New York: Alfred A. Knopf.

Worden, R. P. (1995). A Speed Limit For Evolution. Journal of Theoretical Biology, 176(1), 137–152. doi: 10.1006/jtbi.1995.0183

Wright, S. (1982). The shifting balance theory and macroevolution. Annual Review of Genetics, 16, 1–19. doi: 10.1146/annurev.ge.16.120182.000245

10 Organisations and Breakthrough

The previous chapters concentrated on reaching certain states of systems and process, through appropriate control processes and boundary control, embedded in the steady-state model and processes related to autopoiesis and complex adaptive systems. It seems unlikely that purely looking at control processes and steady-state processes will inform how to bring about change of structures and resources for (social) organisations. Control processes and the steady-state model allow hardly any response to changes in the environment due to the limitations in their capabilities to deal with variations, change and perturbations; even the conceptualisations of complex adaptive systems and allopoietic systems have limitations for the description of change in organisations. All these concepts originated from descriptions for systems of the lower levels in the systems hierarchy of Boulding (see Section 3.5), whereas organisations, as a human construct of the mind, are to be positioned at the ninth and tenth level of this hierarchy; this tells that some laws governing change of lower levels might apply and others not. To this purpose, this chapter expands the concepts of change to organisations, particularly building on the ones that are describing evolutionary (biological) processes; this includes using the system theories for autopoietic and complex adaptive systems, the topics of Chapter 8 and 9. Thus, this chapter will build and extend the theories from the previous chapters to organisations.

This means that a central tenet of this chapter is using for part principles of biological adaptation, characteristic for levels in the systems hierarchy of Boulding just below that of organisations, for describing adaptations by organisations (Dekkers [2005, pp. 145–149] makes the case for that). Note that the term adaptation is sometimes used as a synonym for natural selection, but most biologists discourage this usage. Adaptations are the the the results of processes of living organisms when coping with environmental stresses and pressures (see also Subsection 'Adaptive Processes in Section 5.2). It can be either structural or behavioural. Structural adaptations are special elements or subsystems of an organism that help it to survive in its natural habitat, for example, its skin colour, shape and body covering. Behavioural adaptations are changes in the way particular organisms behave to survive in its natural habitat. Organisms that are not suitably adapted to their environment will either have to move out of the habitat or become extinct. Thus, adaptations and developmental pathways are related to natural selection, but do not substitute natural selection.

Likewise for organisations, adaptations to changes in the environment, the latter called selectional pressure, play an important role. In the previous chapters, organisations as systems have been linked to autopoiesis and complex adaptive systems, indicating that adaptations have to follow developmental pathways, also called adaptive walks. Building on these concepts, those

© Springer International Publishing AG 2017
R. Dekkers, *Applied Systems Theory*,
DOI 10.1007/978-3-319-57526-1_10

responses that point to the introduction of processes and structures for change in organisations constitute the contents of this chapter (some of the concepts have been presented in Dekkers [2005]). Section 10.1 will briefly address the nature of growth and change for organisations; particularly, it will look at evolutionary processes for change. Section 10.2 will deal with processes of foresight, a capability that hardly exist at lower levels of the systems hierarchy of Boulding. Section 10.3 presents the breakthrough model as a model for internal processes to cope with change; in Section 10.4 this model is extended to the model for the dynamic adaptation capability. And finally, Section 10.5 will discuss the differences between the breakthrough and the steady-state model.

10.1 Adaptation by Organisations

That the nature of change for organisations differs from that of technical systems and biological systems has profound effects for viewing adaptation by organisations. Change, more precisely variations and perturbations, in technical systems mostly relates to steady-state processes for maintaining homeostasis (see Chapter 6). In technical systems, structural change is brought about by one-time external interventions; a designer (or engineer for that matter) determines the next contents and structure and implements these in the technical system. The outcome of a preliminary study into one-time interventions shows that these might have limited reach for organisations [Dekkers, 2005, p. 374]. In biological systems, the laws of natural selection and reproduction bind the evolution to latent change present in the contemporary structures (autopoiesis, i.e. memory decides on the evolutionary pathways). External events might have profound effects but adaptation depends on the capability to reproduce beneficial mutations. In contrast to biological and technical systems, organisations are formed as a mental construct of the human mind, though some of their manifestations are physical in nature; for example, assets, equipment and products. For these reasons, this section will look first at the how organisations can create mutations, how organisations compare to allopoietic systems and how organisations evolve over time.

Creation of Mutations

In addition to structural analogies, as propagated by Beer [1972] with his viable system model, which will briefly discussed in Section 10.4, the development of organisations might follow universal laws that arrive from the conversion of models from evolutionary biology. Hence, a reference model has been developed to describe the interaction between organisation and environment (see Figure 10.1), consisting of two intertwined cycles: the generation of variation and the selection by the environment [Dekkers, 2005, p. 150]. Some suggestions have been made to what to consider the equivalent of the genes and the genome on which the biological evolutionary mechanisms build. One possible view concentrates on the division of an organisation in departments, groups, individuals, etc. Morgan [1997, pp. 34, 43] did propose such when introducing the image of an organisation as an organism. Another view would be to look at organisations as a collection of resources with skills and knowledge present, expressing itself in the form of capabilities (and capabilities express themselves in function trajectories), like Nakane [1986]. The view on capabilities became later more popular with writings from Teece et al. [1997]. Dawkins [1989, p. 192] has proposed memes, as unit for imitation and recombination. Memes constitute elements of a culture or system of behaviour that may be considered to be passed on from one individual to another by non-genetic means, especially imitation. Dawkins extends this concept to a wide variety of ontological entities, such as ideas, artefacts, including people, products, books, behaviours, routines, knowledge, science, religion, art, rituals, institutions, and politics. In organisational studies memes enjoy a high degree of popularity. Differently, Kauffman et al. [2000] take technology as starting point for recombination. This way they connect the development of technology as genetic evolution to evolutionary biological models, especially fitness landscapes. Furthermore, The study of Nelson and Winter [1982] uses organisational routines as unit for selection. Knudsen [2002, p. 459] remarks that Nelson and Winter draw on the term routine as replicator (routines as the genes of the organisation) and as interactor (routines as recurrent patterns of behaviour among interacting social agents). The choice for which term should substitute the biological

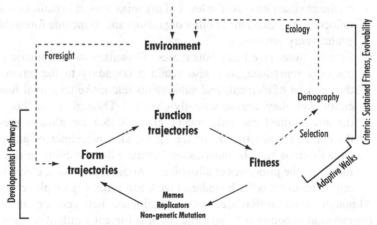

Figure 10.1 *Evolutionary mechanisms for organisations according to Dekkers [2005, p. 150]. Because organisations do not have genes; instead memes and replicators serves as input for genetic formation, which exists besides non-genetic formation. Pathways for development determine the form and function trajectories. These pathways also relate to organisations being a class of allopoietic systems. The selectional processes select beneficial phenotypes on fitness following adaptive walks based on the criteria of sustained fitness and evolvability. Organisations have the capability of foresight in contrast to organisms.*

genome in this reference model will depend on the objective of a particular study.

For the purpose of comprehending adaptations by organisations to the dynamics of the environment, the differences between organisms and organisations for evolutionary processes are summarised [Dekkers, 2005, pp. 145–155]:

- Organisations do not have the possibility for self-reproduction in contrast to living entities. Recombination may occur through the concepts of memes and replicators; such recombination as a genetic formation exists in addition to non-genetic mutation in organisations. Reproducing through recombination has very positive effects on finding fitness peaks in the adaptive landscapes as demonstrated through the NK-model developed by Kauffman (Section 9.3). Organisms have genes that allow recombination to occur by alleles. The direct deployment of the thoughts of genes to the domain of organisations carries the danger that any study will end up as a metaphor rather than an analogy. Therefore, the thoughts on organisations as allopoietic systems are more appropriate (see Section 7.6).

- Organisations have the capability for foresight; that capability is already latently present at the level of animals and present at the level of human beings. Through senses, organisms acquire information about the effects of actions and have the capability to learn by self-reference embedded in the structure of the entity (see Section 8.5). However, the evolution of organisms depends on the creation of mutations and selection of these by the environment. At the level of organisations, it becomes possible to influence the behaviour of other organisms and to include foresight in the evolutionary process.

- Organisations have fuzzy boundaries. Organisms as autopoietic systems not only reproduce, they also retain a boundary to the environment, they consist of elements and subsystems that make up a total functional entity and they are structurally-closed. Through these boundaries, the environment can only induce changes that are already present in the contents and structure of the entity, akin principles of autopoiesis. Organisations have boundaries too but have the capability to shift these, following the principles of allopoiesis. Additionally, some elements of an organisation cross the boundaries back and forth, e.g. employees.

Although these differences exist, analogies between organisms and organisation becomes only possible when sufficient similarities constitute a base for transferring the models of evolutionary biology to the domain of social organisations:

- Selection acts on mutations. Biological evolution generates a variety of phenotypes for organisms and the environment selects phenotypes for survival; phenotypes express the fitness an individual or population. Such a process exists also for organisations where the selection process finds itself in the competition for the customer base, the acquisition of resources, e.g. suppliers, and the acknowledgement of existence by society.

- Organisations and organisms are structurally closed with relations between subsystems and boundaries to the environment. The relationships between the components determine how the entity absorbs perturbations by the environment. Changes in the structure of organisations reside in the current structure and capabilities and depend less on principles of equifinality (see Subsection 'Equifinality, Homeostasis and Deductive Reasoning' in Section 3.3); this means that the design of organisations should account for the development of the current organisational structures and capabilities along its life cycle.

- Organisations have the possibility of self-reference and learning, also found at the fifth to eighth level of systems hierarchy of Boulding (see Section 8.5). The autopoietic principle of self-reference appears for both organisms and organisations; the latent changes are present in the structure of the entity. It is the environment that might induce these changes (or the internal processes in the case of organisations). Learning becomes possible because both organisations and organisms will deploy a set of sensors to acquire information about their behaviour and changes in the environment, although self-reference limits the possibilities to detect all changes and perturbations in the environment.

- Developmental pathways seem to exist for both organisations and organisms. In the case of organisms, they can increase their fitness by undertaking an adaptive walk in a fitness landscape where selection acts on the phenotypes. Organisations can also create mutations and then follow an adaptive walk to increase their fitness.

When putting the similarities and differences together, organisations can be best described as allopoietic systems that might follow similar laws of evolutionary development as organisms. That metaphor can be used to describe both the internal processes in organisations and the interaction with the environment, while accounting for the limitations of such a metaphor.

Organisations as Allopoietic Systems

In this perspective, organisations can be considered as a special class of allopoietic systems that have fuzzy boundaries and the capability for foresight [Dekkers, 2005, p. 148], see Figure 10.2. Allopoietic systems follow the evolutionary models for adaptation, derived from the theory of autopoiesis [Maturana and Varela, 1980], but these systems, and subsequently organisations, do not have the possibility for reproduction. Additionally, organisations have the possibility to shift their boundaries (mergers, outsourcing, alliances, etc.) due to interventions from actors within the system or design by external entities. Moreover, organisations have the possibility of foresight, which allows them to create purposeful mutations; the concepualisation of the learning organisation [e.g. Senge, 1992] is a recognition of this capability, preluded by the work of de Geus [1999].

Considering organisations as a special class of allopoietic systems concerns a structural comparison but does not necessarily imply isomorphism (see Subsection 'Isomorphism' in Section 3.4) in terms of the capabilities for foresight and shifting boundaries.

Another strand of research in the social-economical domain that draws on analogies with biology is called organisational ecology. This type of research investigates the factors that influence changes in the population, which leads to perspectives on populations of organisations and not on the individual firms themselves. Generally, these theories view organisations as relatively inert to environmental changes [Bruggeman, 1996, pp. 21–22], a perspective aligning with the principles of autopoiesis. This implies that most of the time organisations are not capable of substantially changing their structure in a way that results in successful and timely adaptation to new environmental conditions. This assumption is in line with the perspective of natural selection and, thus, the view of organisational ecology is that selectional processes create the diversity of forms and not the adaptive behaviour of individual organisations.

A crucial question in organisational ecology is which internal factors of a population and which environmental ones determine both the entry of new organisations and the survival, change, and failure of existing ones in product-market domains [Hjalager, 2000, p. 272]. Configurations of core features of the organisations are made to determine if certain forms and companies with the corresponding organisational form belong to the same

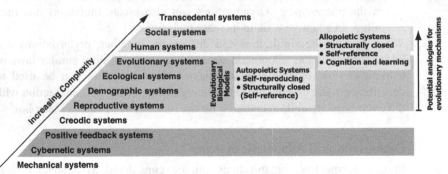

Figure 10.2 *Organisations as allopoietic systems depicted on the systems hierarchy of Boulding (see Figure 3.9). The domain of organisations moves at the ninth and tenth level indicating the importance of meaning, value systems and symbolisation. The domain of traditional systems theories and some other approaches in management science (e.g. information technology) find themselves at the second and third level. Models from evolutionary biology might bridge the gap between some of the approaches in management science and the actual organisational domain. However, differences exist between organisms and organisations mostly denoted by the difference between autopoietic systems and allopoietic systems. Additionally, the boundaries of an organisation are relatively open; companies might shift the boundaries and employees are part of other social organisations and contexts.*

population. Hannan and Freeman [1977, p. 935] state that forms are seen originally as the characterisation of the key elements within a decision-making framework. Forms have two purposes: to inform about the state of the external environment, and to activate responses to information. Hjalager [2000, p. 272] states that the organisational ecology has often been accused of ignoring firms' strategies and unique compositions of individual enterprises. However, this strand of research assumes that firms' choices on a population level determine the occurrence of crucial life events. Bruggeman [1996, p. 24] explains rational adaptation in organisational ecology as changes in the structure of individual companies in the case of substantial reorganisation and with that changes in the core features and form of the company.

Surely, the environment in which companies operate determines for a large part the prospects for the development of an organisation as an entity. For example, work forces might resist technological changes, reason to include employees in the (technological) development of a company when competition provides a base for technological progress; in turn, this depends also on the organisational culture whether such a style of leadership might hold (see Hofstede's [1994] assessment of socio-economic cultures for organisations). Competitiveness and innovation (as technological development) have a strong link, therefore both a prerequisite for development and growth. It means for companies that pluralistic approaches offer opportunities for development (pluralistic refers to the markets, products and technologies). Yet, Hannan and Freeman [1977, p. 933] state that the evolution of industries as aggregate of individual companies follows different dynamics than those of individual companies. According to them, events at the higher level cannot be reduced to events at the individual level. Following the major findings of organisational ecology, age and size of an organisation matter for increased chances of survival and additionally a strong link exists between competitiveness and innovation (or technological progress).

Evolution by Organisations

The evolution of the individual firm in terms of age and size has received attention in management literature. This subsection looks at three core concepts of life cycles for companies: the growth model of Greiner [1998], the life-cycle model of Lievegoed [1972], and the investigation in longevity by de Geus [1999]. All three have in common that they look at how organisations develop during ageing and growth.

The growth model of Greiner [1998], a reprint of an article in 1972, describes five phases of growth that a company goes through: creativity, direction, delegation, coordination and collaboration (see Figure 10.3). Each phase begins with a period of evolution, with steady growth and stability; it ends with a revolutionary period of substantial organisational turmoil and unrest in which organisations exhibit a change of management practices. Each evolutionary phase is characterised by a dominant management style used to achieve growth. Moreover, each revolutionary period is characterised by a dominant management problem that must be solved in order to enhance organisational performance and maintain continuity. Both axes in Figure 10.3 represent the two main dimensions for survival according to organisational ecology: age and size.

It is important to note that each phase in the development of a company emerges from the previous one and acts as a cause for the next phase. For each phase, managers are limited in what they can do for growth to occur. A company cannot return to previous practices; it must adopt new practices in order to move forward [Greiner, 1998, p. 56; Lievegoed, 1972]. Greiner describes the five phases as follows:

- Phase 1: Creativity. During the earliest stages of the organisational life cycle, the emphasis is on creating both a product and a market. The company is largely void of formal policies and structures, and often led by a (techno)entrepreneur. But as the organisation grows, production runs require more knowledge about the efficiency of manufacturing. Increased numbers of employees cannot be managed through informal communication alone. At this point, a crisis of leadership enfolds, because of the lack of managers that have the necessary knowledge and skills to introduce new business techniques. A new type of control structure is required.

- Phase 2: Direction. Those companies that survive the crisis of the first phase usually embark on a period of sustained growth, introducing a functional organisational structure. In most cases, departments arise, such as marketing and logistics, where teams of lower-level managers are treated more like specialists than as managers making decisions. Although

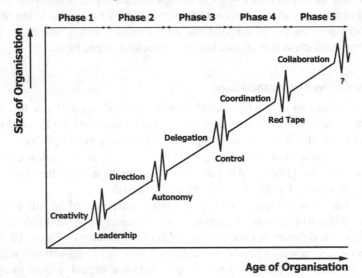

Figure 10.3 *The five phases of evolution according to Greiner [1998]. This classic approach distinguishes phases of growth interchanged by periods of turmoil moving into a next stage of development of a company.*

the directives and newly-deployed methods for managers channel the organisational resources more efficiently into growth, lack of autonomy on the lower levels becomes more and more problematic. Lower-level managers have to possess more direct knowledge, for example about markets and machinery, than their leaders at the top do. This introduces the second crisis, the crisis of autonomy.

- Phase 3: Delegation. The next period of growth evolves from the successful application of a decentralised organisational structure. The organisation will be divided into different units and the control paradigm becomes management by exception based on periodic reports from operations and order processing. This can only be done if operations and functions are narrowly described. A serious problem emerges eventually as top-level management feel that they are losing control over a highly diversified span of operations and market domains. The organisation falls into a crisis of control as top management seeks to regain control over the company as a whole. Those companies that move ahead find a solution in the use of newly-deployed techniques for coordination.
- Phase 4: Coordination. The evolutionary era of the coordination phase is characterised by the use of formal systems for achieving greater coordination and by top management taking responsibility for the initiation and administration of these systems. Those new coordination systems will allocate the organisation's limited resources more efficiently. The systems prompt field managers to look beyond the needs of their local units and will therefore make the company more externally oriented. Although these managers still have extensive responsibility for decision making, they learn to justify their actions more carefully to a watchdog audience at headquarters. Eventually, the company will become too large and complex to be managed by formal programmes for change and rigid control systems. Procedures take precedence over problem solving; a red-tape crisis is introduced.
- Phase 5: Collaboration. The last observable phase emphasises spontaneity in management action through teams and the skilful confrontation of interpersonal differences. Social control and self-discipline replace formal control. The collaboration phase builds around a more flexible and behavioural approach to management.

The question of what will be the next phase in response to the collaboration phase is difficult to answer. Greiner [1998, p. 63] imagines that the next phase will centre on the psychological saturation of employees who grow emotionally and physically exhausted from the intensity of teamwork and the heavy pressure for innovative solutions. Although companies may experience periods of evolution interchanged by revolutionary periods of substantial organisational turmoil and change, each phase builds on capabilities acquired in the past and on decisions taken rather than projections of the future on the present.

The view on life cycles of organisations has also been elaborated by Lievegoed [1972, pp. 54–85, 98–99]. He distinguishes only three phases in the life-cycle model for firms:

- Pioneering phase. The strength of a company in the pioneering phase is its potential and its powerful identity, concentrated in the founder or those who continue this style. Objectives and goals are visible at all levels within the company, everybody knows what to do and how to contribute to these objectives, even though the policy and strategy have not been formalised. The planning for the long term lacks but the organisation displays an enormous flexibility. The organisation is based on historical growth and tailored to the personal skills of the employees. Renewal and innovation happen through motivated employees that directly apply their own ideas. Managerial control processes are focused on direct contact with clients. The pioneer's model has its limitations in the health of the pioneer, the complexity of technology and the market in which the company operates.

- Differentiation phase. This phase finds its base in a hierarchical structure aiming at the expansion of the technical system, both for advancing technologies for operations, and improving the organisation. Specialists have entered the company and the expansion leads to increasing the layers of command. (Sub)optimisation of departments starts to take over and the attention of management shifts to control of the internal processes and structures, even into the direction of a mechanistic view on the labour force of the company. The rationalisation of the internal processes reflects also on the position of the customers. The market becomes more anonymous and the organisation moves away from the personal approach during the pioneering phase. Internal and external to the organisation the resistance builds up and companies find themselves more and more in conflicting situations.

- Integration phase. This phase calls for connecting all processes, departments and employees to a meaningful whole. Lievegoed states that this transformation should start at the top management level; eventually it should lead to a management style of coaching rather than directing. The internal organisation should allow participation of employees. It requires rethinking of all primary and control processes to suit the needs of customers and to appeal to the capabilities of employees. Decentralisation becomes a key-concept in this line of thought and the customers regain their position as focus of the internal processes.

The life cycle of companies calls for interventions to sustain the organisations [Whetten, 1980]; however, according to these models for the growth of firms, the interventions differ for each phase.

The same question, renewal of companies, is at the heart of the studies performed by Shell, tells de Geus [1999]. The shift from forecasting in the 1960s to scenario-analysis for strategic planning and the implementation of business concepts has driven this study by a practitioner rather than an academic. This concept of the living organisation has four main principles:

- The capability of foresight by an organisation to anticipate on the future through the development of scenarios strongly determines the possible reactions to the shifts taking place in the environment. De Geus [1999] makes it clear that such an activity should not reside within the financial or accounting Department; rather it requires participation by all actual decision makers to prepare for and to envision the future.

- Through learning, organisations develop an image of the effects of their actions and set the course for future actions. Thus, a continuous process of decision-making, studying effects of actions and evaluations enfolds which provokes learning cycles by which an organisation might increase its effectiveness.

- The organisation has an identity it wants to uphold and maintain, the so-called 'persona' (see also subsection 'Allopoietic Systems as Creation' in Section 8.6). Continuous managerial attention focuses at the behaviour and attitude of people in the organisation.

- A solid financial policy is not only governed by the circumstances of the day. A study by Laitinen [2000] confirms this thought. His comparative study shows that in the medium term, investment in product development and marketing and in the acquisition of additional resources, capabilities and market access proves the most successful strategy, whereas a strategy heavily based on negotiating finance contracts and restructuring was the most unsuccessful.

The concept of the living organisation has been strongly influenced by sociological themes of identity and learning. The idea – the organisation as a living entity – arrives from the works of Maturana and Varela [1980] on autopoiesis for organisms (see Chapter 8); organisations differ from organisms in the capability of foresight, the theme of the next section.

10.2 Processes of Foresight

Given the 'internal dynamics' of organisations and their interaction with the environment as allopoietic systems, the strategic process should provide a more flexible approach for finding optimal solutions; to this purpose this section will explore the themes of strategy and foresight in connection to the adaptation processes. After looking at what strategy constitutes, the concepts of dynamic strategies, forecasting, techniques for foresight and scenario planning will be looked at.

Strategy

Foresight, one of the principle differentiator between organisations and organisms (see Section 10.1), traditionally connects to the setting of strategy. However, there is much confusion about what strategy encompasses. It is already Mintzberg [1987, pp. 11–12] who remarks that it varies between plan and ploy; however, he characterised strategies as made in advance of actions and interventions to which they apply and as being developed consciously and purposefully. Others perspective on what strategy exactly is differ markedly. For example, Burgelman et al. [1995] divide strategy into a resource-based strategy and a product-market strategy. Furthermore, Porter [1996, p. 64] defines the essence of strategy as choosing a unique and valuable position rooted in systems of activities. And, Quinn [1981, p. 44] says: *a strategy is the pattern of plans that integrates an organisation's major goals, policies and actions sequences into a cohesive whole.* Putting all these definitions together, a strategy allocates the organisation's resources into a unique and viable posture based on its relative internal competencies and shortcomings, anticipated changes in the environment and contingent moves by opponents.

Whereas definitions for strategy vary, so do methods for setting the strategy. In the early days of strategic management, Ansoff [1965] introduced a matrix of four strategies, which became quite well known: market penetration, product development, market development and diversification. But this was hardly comprehensive. Fifteen years later, Porter [1980] introduced what became the best-known list of generic strategies: cost leadership, differentiation and focus. Even the Porter list was incomplete: while Ansoff focused on extensions of business strategy, Porter focused on identifying business strategy in the first place. At the same time, Abell [1980] used three dimensions to define strategy: customer groups, customer needs and technology or distinctive competencies. This array of approaches was extended by Mintzberg [1987] introducing differentiation strategies. Furthermore, adding to this variety, ten major schools characterise the strategy literature, according to Mintzberg and Lampel [1999]. All the different methods and schools of strategy were developed in a time of relative environmental stability.

Because of this relatively calm environment, there is a general lack of dynamics in most of these approaches to setting strategies. Most of the existing strategy schools are based on the assumption of competition in a stable and static environment, but technological advances and global changes have created a more dynamic, complex climate. Technology has lowered the market's entry boundary and geographical barriers are decreasing; think about companies such as Amazon and Google not being restricted to one country or market. To deal with these challenges, Porter [1980] has developed a model of forces affecting industry competition, subsequently threat of entry, powerful suppliers and buyers, substitute products and jockeying for position. Rivalry among existing competitors takes the familiar form of jockeying for position – using tactics such as price competition, product introduction and advertising. The forces described by Porter still exist, but they have changed in magnitude, creating a more dynamic environment.

Dynamic Strategies

These changes are forcing industries to react more quickly to changes in markets, in other words strategies need to be dynamic. In this context, there are two general interpretations for the dynamic strategy: constantly changing over time strategies or multiple strategies. Markides [1999, p. 63] finds that:

> *Designing a successful strategy is a never-ending, dynamic process of identifying and colonising a distinctive strategic position. Excelling in this position while concurrently searching for, finding, and cultivating another viable strategic position. Simultaneously managing both positions, slowly making a transition to the new position as the old one matures and declines and starting the cycle again.*

This phrase can be split into two parts. In the first part the assumption is that strategy presents a given position, which should be taken in by a firm. The second part is more concerned with the problems of competing today while preparing for tomorrow. The key is that strategy is a never-ending, dynamic process. This is a big difference from conventional conceptions of strategy. In the early days it was common that a strategy could be seen as a long-term process; meaning that once formulated strategies would serve a firm for many years. In an ever faster changing environment, dynamics are becoming an essential part of strategy. The more uncertain the future to deal with, the more sense it makes to create multiple scenarios and different strategies.

In this perspective, Beinhocker [1999, p. 106] recommends cultivating and managing populations of multiple strategies that evolve over time, because the forces of evolution acting on a population of strategies makes them more robust and adaptive. Note that his reasoning is based on the fitness landscapes,

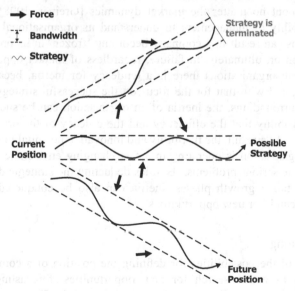

Figure 10.4 *Multiple strategies with bandwidths. Strategies are shaped by the forces of the environment and companies will have to adapt these during the course of time.*

presented in Section 9.3, and the similarity to evolution in populations where several alleles are present at the same time ensuring adaptation to changing environments. In high-velocity, intensely competitive markets, traditional approaches to strategy give way to competing on the edge creating a flow of temporary, shifting competitive advantages and strategies. Eisenhardt [1999], in her research on entrepreneurial and diversified businesses, demonstrates that successful firms in these markets have fast, and high-quality, strategic decision-making processes. This is also what Williamson [1999, p. 126] concludes: *the success rate of strategies can be greatly enhanced when they are not too specific*. A company must keep tactical opportunism within the bounds of its overall strategy, ruling out options that might cause it to deviate from its long-term strategies.

A dynamic strategy consists of multiple strategies; these strategies should grow within the set bandwidths; these bandwidths are imaginary boundaries to the application a singular strategy. Only in this way can we be more certain about the long-term prosperity of companies in a fast changing environment. This is visualised in Figure 10.4, in which one sees a current position from which multiple strategies are pursued. These strategies are not strictly formulated; they have some variance in the set bandwidth, i.e. to their application. When an organisation finds that one of the possible strategies is not viable that strategy is terminated, as shown. The pursuit of the remaining strategies could be done by implementing the variety of strategies in its overall business processes or create separate business units (the latter is a variant of Ashby's law of requisite variety, see Section 6.8).

However, as discussed in Section 10.1, many of the organisations' challenges are rooted more in past decisions than in present events or they come about no matter the market dynamics [Greiner, 1998]. Moreover, the inability of management to understand its organisational development problems can result in a company becoming 'frozen' in its present stage of evolution or, ultimately, in failure, regardless of market opportunities. In successful organisations there is a tendency for inertia, because of given success, and with that for the focus on the successful strategy. In relative stable surroundings, the inertia of an organisation can be successful; there is a possibility that the efficiency and the efficacy of the organisation will increase. As a result, the flexibility and innovative potential may reduce and, if an organisation is operating in a fast changing environment, this inertia can create serious problems. Besides deducting the strategic direction from organisational growth phases, inertia can also be conquered through the active search for new opportunities.

Forecasting

As one of the possibilities for defining the position of a company and the strategy as active search for new opportunities, forecasting dates back to the 1930s, according to de Geus [1999, p. 41]. The main objective of forecasting was to deal systematically with the future; in the centuries before, forecasting existed but was not yet incorporated into a process. A series of tools became available under the generic name of planning and managers used these tools during decision making. Mostly, forecasting took place in separate backrooms, planning departments and resided within the financial departments. This seemed logical because of the availability of figures, the data collection, and their objectives, resulting in balance sheets, profit-and-loss-accounts, budgets, etc. The planners handed their reports to management within the organisation supposing they would execute the plan. Essential to these planning processes, called forecasting, was the emphasis on the development of a plan based on historical figures, and only one route to achieve certain objectives.

This top-down approach continued until the early 1960s. After that more and more companies started with bottom-up planning. When asked for a forecast by management, for example, the planners went first to the district managers for predictions on sales figures for next year, two years or even five years. When all the data were collected, the planners added up the figures adjusting them for their own thoughts and generated budgets and forecasts. The predictions became part of the 'management by objectives' movement, led by the well-known management guru Drucker [1978, pp. 100–113]. At the end of the 1960s, forecasts became an internal contract, based on little external information and derived from the same introvert process. It created also a culture of handing down safe figures so that performance of individual managers could be ensured.

The forecasting process turned from a simple one-minute job for line management into a complex and time-consuming process, not only for the planning department but also for the whole company. Every choice had to be double-checked and agreed upon. In times of prosperity, this posed no problem to companies but in times of crises and turbulence, the process took long and led to totally wrong predictions also caused by the elapsed time. For example, that was the case during the oil-crisis in the 1970s [van der Heijden, 1996; de Geus, 1999].

Despite its ineffectiveness in dynamic, contemporary environments, Millett and Honton [1991] state that predicting the future by trend analysis is still the most popular technique for technology forecasting. The different techniques have all some common assumptions and features, namely:
- the future is a continuation of the recent past and can be expressed quantitatively, as human behaviour follows natural laws as in physics and chemistry;
- there is one future and it is predictable if you understand the underlying laws as shown in the trend data.

This does not represent a realistic view of the world. Most environments of companies are so complex that it will be impossible to understand all underlying laws, and, even so, all these laws governing human behaviour are so complex that there are too many exceptions to the rule; forecasting will only work in a perfect world.

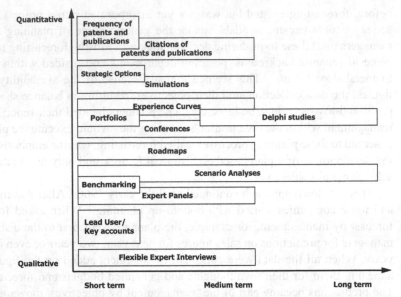

Figure 10.5 *Techniques for (technology) foresight mapped against time horizons [adapted from Lichtenthaler, 2005, p. 398]. The short-term generally indicates up to 5 years, medium term 5–10 years and long term more than 10 years, albeit that the exact time horizon might depend on a specific industry. The use of techniques is merely indicative.*

Techniques for Foresight

For the more complex and more dynamic environment, techniques have been developed. Albeit from a technological perspective, Lichtenthaler [2005] looks at the usefulness of the techniques for foresight from the perspective of the time horizon they apply to; Figure 9.5 shows a selection of these techniques with their application. In addition, Popper [2008] even reports 33 different techniques for foresight, although they are not classified to time horizon, but divided into qualitative, quantitative and semi-quantitative methods. Hence, there is a multiple of techniques available that support setting out strategies in dynamic and complex environments.

For the long run, Lichtenthaler [2005] identifies two techniques. The first technique is Delphi studies. The Delphi method is a systematic, interactive forecasting method that relies on a panel of experts. The experts answer questionnaires in two or more rounds. After each round, a facilitator provides an anonymous summary of the experts' forecasts from the previous round as well as the reasons they provided for their judgments. Thus, experts are encouraged to revise their earlier answers in light of the replies of other members of their panel. It is believed that during this process the range of the answers will decrease and the group will converge towards the 'correct' response. The second technique is scenario planning, a strategic planning method that some organisations use to make flexible long-term plans. It is scenario planning that is elaborated in the next subsection.

Scenario Planning

Scenarios and scenario planning might offer a way out of the complexity of factors acting on organisations and at the same time increase the capability of an organisation to deal with the effects of these factors. The use of scenarios might be traced back to the traditions of the Oracle of Delphi and the premonitions by Nostradamus. The first modern scenario developer was Kahn [Kahn and Wiener, 1967]. Working for the US Air Force, Kahn developed scenarios to imagine what the opponents might do and to prepare alternative strategies to react to the opponent. The close relationship between scenarios and strategies is one of the reasons that they are regarded as the same thing, which is not the case. During the 1960s, Kahn refined his scenario development tools to fit business prognostications. These scenarios are developed in three steps [Kahn and Wiener, 1967, pp. 5–10]:

- First, the Basic, Long Term Multifold Trends are described. These trends are derived from historical data, and do not change over the period that the scenario covers. They provide the basic structure on which the scenarios are built. Examples of these trends are birth rates, consumption of consumer goods, etc.
- After identifying the trends, the Surprise Free Projections and the Standard World are described. These projections are based on basic trends at the time of the development of the scenario. Experts assess these trends and draw plausible conclusions from them (for example, high and low trend developments of birth rates under different circumstances). These projections are intended to be used for further discussion, explanation of underlying assumptions and systematic consideration of major alternatives.
- The last step in developing the different scenarios is the introduction of the Canonical Variations. These are designed to raise certain issues. The introduction of these issues leads to scenarios that are out of the expected Surprise Free Projections. In these Canonical Variations issues, such as a sudden rise of costs for energy or crude oil due to a war, can be introduced.

Figure 10.6 *Different environments of an organisation. The transactional environment consists of players (e.g. customers) and competitors. The transactional environment is influenced by driving forces from the contextual environment.*

In their book, Kahn and Wiener [1967] use 13 basic trends and 8 canonical variations. Most of the time, these variations affect two or more basic trends. That way 9 different scenarios are developed. This approach to scenario planning was widely used until the work of Pierre Wack came to rise. He worked for the Royal Dutch/Shell in a newly formed department, called Group Planning [de Geus, 1999, pp. 58-60]. This group successfully predicted the oil crisis in 1973. After working in the Planning Department for several years, Wack came to the conclusion that the key point of scenario planning was not a clear picture of the future but what he describes as the gentle art of perceiving. He changed his efforts to enabling people to perceive different pictures of the future and act on changing circumstances.

In this perspective, scenarios should be seen as a hypothetical sequence of events with causal processes based on shared mental models. The sequence of events might help to generate a step-by-step description of multiple possible futures of the external world. Different scenarios may take place in the transactional and contextual environment as depicted in Figure 10.6; the transactional environment equals the definition of the environment of a system in Section 2.1, whereas the contextual environment is that part of the universe that connects to the environment of the system. According to van der Heijden [1996], scenarios are useful to create structure in events and patterns in the environment, to identify irreducible uncertainty, to confront different views with each other through dialectic conversation, to reveal individual knowledge of members of an organisation, to introduce external perspectives and to translate the above in a suitable form for strategic conversation.

For scenario development Ringland [1997] identifies three types of methods. Table 10.1 gives an overview of the methods and their steps:

- Trend-impact analysis (for example, used by the Futures Group). Trend-impact analysis is concerned with the effects of trends, for instance in markets or populations. The method focuses on isolating the important trends, similar to that used in what is more generally called scenario planning; however, the basic premise within trend analysis is to look for the unexpected, in other words what will upset the trends. In addition, the trend-impact analysis can provide multiple pictures of the future. This method of trend-impact analysis closely resembles the method developed by Kahn. Most of the time, the scenarios look alike and cover a high, a medium and a low trend of a certain event. The risk exists that a company will always choose the medium prediction just to be on the safe side Ringland, 1997, pp. 47, 92]. Such a deployment of predictions within scenario development resembles closely forecasting.
- Cross-impact scenario analysis (for instance used by Battelle). The analysis of scenarios based on cross-impact is used for complex systems. It concentrates on the ways in which external or internal forces may interact on an organisation to produce effects larger than the sum of the parts or to magnify the effect of one force because of feedback loops. The method is used to orient strategic thinking about new products, technologies and

Table 10.1 *Steps of the three scenario development methods. More or less the same steps are present in each of the methods.*

Scenario-writing (Global Business Network)	Cross-Impact Scenario Analysis (Battelle)	Intuitive Scenarios (Shell traditionally)
Identify focal issue or decision	Define and structure topic question	Analyse strategic concerns and decision needs
Identify key forces in corporate environment	Identify most important issues in response or topic questions	Identify key decision factors
List driving forces of the macro-environment	Select descriptors from most important issues	Identify key environmental forces
Rank driving forces by importance and uncertainty	Prepare descriptor white papers with projected alternative outcomes and a priori probabilities	
Select scenario logics structured by 2 axes and 4 quadrants	Cross-impact analysis	Analyse the key environmental forces
	Scenarios generated from cross-impact matrix (sorting of descriptor outcomes into alternative scenarios)	
Rashcut at least 4 scenarios and product narratives	Draw business-related implications from 5 scenarios and derive robust strategies	Define scenario logics (typically 2 critical issues)
	PC-based strategy simulations and scenario-sensitivity analysis, including disruptive events	Elaborate on two detailed descriptive scenarios
Draw implications and conclusions	Briefing discussions and implications focus groups	Draw implications for scenarios for strategic concerns and decision needs
		Make conclusions and recommendations
Select leading indicator sign posts for continued monitoring	Monitoring updates and revisions	

marketing towards most likely future market conditions, including the net effects of various customers, regulatory, competitive, economic and technological trends. Within the method, looking at the future anticipates long-term (beyond three years) customer behaviour when customers themselves cannot articulate their own future behaviour. The method generates alternative scenarios for long-term business environments. It also serves as a tool to simulate what-if questions to see how actions and events may change the baseline (most likely) scenarios. To this purpose, simulations test potential business investments and strategies. Most likely scenarios are compared with most desirable scenarios to identify critical success factors.

- Intuitive logistics (used by Royal Dutch/Shell and Global Business Network). The essence of this method is to find ways of changing mindsets so that managers can anticipate futures and prepare for them. The emphasis is on creating a coherent and credible set of stories of the future as the basis for testing business plans or projects, prompting public debate or increasing coherence. The term wind tunnel is used because intuitive scenarios can be seen as an analogy with wind tunnels in which strategy models can be tested under different circumstances.

Key factor in intuitive logistics is the recognition of events and how these events form causal relationships under different circumstances. This can be visualised by systems thinking in the way Senge [1992] propagates, with cause and effect diagrams. Other visualisations for this scenario planning method use hexagons that can, connected together like a jigsaw puzzle, explain underlying trends, patterns and structures.

The three methods support scenario development but not all do address multiple strategies. Each of them is a way of arriving at a scenario; only the latter method does not arrive at a one particular state and, therefore, is more dynamic in its application.

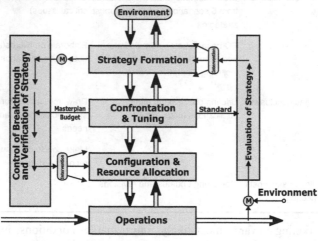

Figure 10.6 *Breakthrough Model. By scanning the environment new or adapted goals are set and the derived strategy acts as a reference for the review of tactical and operational decision making. The process of confrontation and tuning takes the possibilities into account leading to specific decisions on the utilisation of resources and structures for operations. Through the configuration and resource allocation process the actual implementation of the structural changes in operations takes place. The evaluation of strategies might create new input for the breakthrough processes. The verification enables companies to follow the progress of the breakthrough processes.*

10.3 Breakthrough Model

The breakthrough model comprises the overall processes necessary for implementing changes into the structure of organisations starting from strategies whether they are dynamic or informed by scenarios. This breakthrough model might apply to product development, process development, changes in organisational structures, etc., having a wide scope for any breakthrough or strategic renewal (Ravasi and Lojacono [2005, pp. 52–54] define strategic renewal as both structural transformation and as continuous innovation). Such adaptive processes continuously are at work, because selectional forces in the environment – either caused by competitive pressures or changing landscapes – force organisations to continuously generate beneficial mutations (by offering new or improved products or processes and by improved performance). However, the need to deploy this model will occur when the need for a certain output (or function) diminishes and an organisation needs to rethink its objectives, strategy and resource allocation.

The development of the original model for the breakthrough has been described in Dekkers [2005, p. 378], see Figure 10.6; it has also a strong resemblance to the viable system model (see Section 11.4). It is a developmental model for organisations when they grow, building from operations towards objectives and strategy (akin life-cycle models in Section 10.1). The model can be applied to any breakthrough for expanding entities and existing entities. Because of the link between internal structure and the mutations – these are embedded in the current components of a system, i.e. an organisation in this case – the setting of objectives and strategy formation cannot be detached from each other.

Strategy Formation

By its capability of foresight, an organisation strives to adaptation by the generation of potentially beneficial mutations. Selectional forces, external to the organisation, will determine whether these mutations prove deleterious or beneficial. Nevertheless, the creation of purposeful mutations might change the fitness landscapes in which companies operate, giving them the possibility to influence the selectional forces. Note that this contrasts to organisms, which are hardly capable to shape the directional selection of fitness landscapes. The creation of purposeful mutations bears a strong resemblance to the concept of destructive innovation by Sombart [1930] in 1902 [Reinert and Reinert, 2006, p. 72] and popularised by Schumpeter [1911, 1934, 1954], telling also that continuous innovations are a necessary source for sustained competitiveness of organisations.

These adaptations appear by adapting to changing performance requirements or by filling adaptive zones (new functions or new performance requirements). Adapting to changing performance requirements means that the organisations performs according to improved criteria, for example shorter delivery times of orders. However, the adaptations can also apply to existing products and services or features that were not present before. The changes can be incremental or radical. In adaptive zones, positive feedback operates allowing an accelerated growth before limitations or constraints are reached (balanced by negative feedback). The theory of fitness landscapes (Section 9.3) indicates that long jumps, i.e. radical innovation in either products and services or performance, might be deleterious. The mutations may occur at high speeds, even then each of the steps helps an organisation to improve its fitness and to explore the fitness landscape. These small steps and the exploration of the fitness landscape are brought about by the perception of the transactional and contextual environment in which an organisation operates.

By scanning the environment new or adapted goals are set and the derived policy acts as a reference for the review of tactical and operational decisions. In this sense, they are normative. However, the changes generated by organisations as allopoietic systems also reside in the components and internal structure of the organisation as a system. Allopoietic systems have self-cognition and self-perception as starting point for the processes of foresight, and, henceforth, for their strategy. Through active learning as present in the concept of the learning organisation [e.g. Senge, 1992], entities might adapt their perception and behaviour. Although internally normative, the strategy itself is relatively related to the self-perception of the organisation and, therewith, its perception of the environment.

Within the context of organisations as allopoietic systems, strategy formation consists of two major components. First, strategy formation comprises of the identification of objectives, the identification of means (incl. resources) and the development of principles. Objectives should be taken in the widest sense, e.g. market positioning for commercial organisations and addressing the needs of communities and citizens by governmental agencies. The second component concerns the selection of the objectives and the setting of priorities. Once the objectives and strategy relate to the deployment of resources, the process of tactical decision-making can take place (called 'confrontation & tuning' in the breakthrough model).

Confrontation and Tuning

When the strategy is set, decision-making takes place within the process of confrontation and tuning as optimisation of the deployment of the selected resources. This process takes the alternatives into account leading to specific decisions on the utilisation of resources and structures for operations. Characteristic is the iterative decision-making process. The strategy determines the direction and criteria for decision making, but it is also influenced by the assessment of the available resources. Likewise, iteration happen between 'confrontation and tuning' and 'configuration and resource allocation'. The decision-making includes the comparison of alternatives and considering the tactical level (deployment of resources). When there are no alternatives matching the requirement of the strategy, this process may result in the exploration of the environment again. This leads to a continuous stream of matching strategic requirements with the possibilities and capabilities for executing decisions.

Therefore, this calls for integrative decision making as an effort to reduce unnecessary iterations. Additionally, the decision making should cover the full scope of requirements set by the directives of the strategy. This may concern a broad range of aspects. Note that this type of decision making can also be found in Chapter 4 as non-programmed decision making; the analysis in the setting of the breakthrough model focuses on the validity and the effectiveness of the strategy. This means that root causes for reconsidering the validity and the effectiveness of the strategy result in new initiatives for product, services, processes and business models; these could include changes to the boundary of the organisation as a system of resources and collaboration with other organisations to achieve the organisational objectives and the implementation of a revised strategy. The integration of the decision into the structures of the system of resources and the processes will only succeed when an integral approach is chosen.

The output of this process in the breakthrough model is the feasibility of a course of action. First, the feasibility of the master plan addresses the strategic direction chosen during the strategy formation. And second, the feasibility of the master plan should fit with the capabilities of the systems of resources or projected amendments. The output of 'confrontation and tuning' serves also as a master plan for later verification of progress.

Configuration and Resource Allocation

Through the process of configuration and resource allocation the actual implementation of the structural changes in operational processes takes place. The configuration concerns the structure (some would call it the architecture) of operations. It may concern product or service development, the development of operations, the development of supply or the development of markets. Resource allocation directs towards linking the resources at disposal to the processes for the organisation; this covers both internal resources of an organisation and resources external to it.

During this phase of configuring the structure of operations and the allocation of resources to that structure, it might prove difficult to realise the chosen course of action. In such a case, iteration between the level of 'configuration and resource allocation' and the level of 'confrontation and tuning' leads to revaluation of earlier decision-making. The feasibility of earlier options might be questioned, which could result in a revaluation of objectives and strategy. This leads to iterations between the process of 'configuration and resource allocation' and the process of 'confrontation and tuning' until a feasible option is found; if this cannot be found the formation of strategy should be started again.

For the allocation of resources, an organisation has two possibilities, which depend on the inclusion or exclusion of external resources. The first option for the allocation of resources is by relying on the same system of resources but that implies a redesign of the processes for meeting existing or shifting performance requirements. Only through a redesign of the process a different performance will eventually become possible; most of time that will result in a reflection on which resources are needed to conform to performance criteria and integration requirements. Secondly, by adjusting the system boundaries, an organisation automatically affects the control in the boundary zones at least. More often, it will impact the total processes, including control mechanisms, present in the organisation. Whatever option chosen, only through further detailing it will become clear whether the solution is feasible and whether implementation of the solution is possible in the primary process.

Operations

The new solution, or for that matter the new structure, will lead to different set-up of the primary process, whether it concerns changes in input, resource allocation, process configuration or control processes. Sometimes, the primary process itself does not need adaptation but the control mechanisms do. For example, when there is a constant and predictable flow, it does not matter whether feedforward or feedback is used for managing the throughput. But when fluctuations of the input increase, feedforward may anticipate on demands for the capacity of the primary process; feedback will only lead to correcting after deviations have entered the process and, therefore, be rendered less effective. Also new input, such as new materials or components, might be the objective of changes made in the operational system. In any case, the processes in the breakthrough model aim at introducing a new recurrent process with related resources and configuration of these resources in the (operational) primary process.

After the implementation of changes in input, resource allocation, process configuration and control processes, the primary process will reach a steady state after a time lapse. This is caused by the new structure of the system of resources and the processes requiring fine-tuning. This might be partially encapsulated in the control mechanisms, such as present in the steady-state model, but it might also require modifications to make it work. Leonard-Barton [1987, p. 18] proposes that managing the integration of new technologies in the organisation yields better results than when companies adhere to an original strategy and implementation plan. She even states in another paper that implementation is innovation. Based on empirical research and a model for adaptation by Berman [1980], she [Leonard-Barton, 1988, p. 265] concludes that companies should allow adaptation cycles to actively link the actual implementation of technologies to the strategy. In the case of the introduction of an expert system at Digital, the success of the technology has depended on the interactive process to altering the technology to fit the organisation, and the simultaneously shaping of the user environment to exploit the full potential of the technology [Leonard-

Barton, 1987, p. 7]. Douthwaite et al. [2001] do also confirm the conclusion that the more complicated the technology, the more it requires interaction between the inventors, researchers, and the user environment. A simple top-down approach is not sufficient any more. This leads Leonard-Barton [1988] to the proposition of small and large adaptation cycles to exist within the organisations. Misalignment between the strategy and objectives are viewed as normal and the misalignment evokes an adaptation cycle where both the merits of the technology and the impact on the strategy are considered. The larger the adaptation cycle, the more factors it affects within companies. Within the breakthrough model these adaption cycles are found as iterations between strategy formation, confrontation and tuning, configuration and resource allocation, and operations. Each evaluation of performance of a change leads to considering the adaptation at aggregation strata given the impact of the misalignment, whether it concerns changes in input, resource allocation, process configuration and control processes; thus, introducing change may require active control and iterations as modifications to reach an optimal steady state.

In practice, it may have become more difficult to reach a steady state. The continuity of changes does not often allow an organisation to reach an optimal state of the primary process. Sometimes, and even more often nowadays, the next change is already conceived before the previous one has been implemented fully. Consequently, this requires organisations to view breakthrough as a continuous process rather than a process aimed at one-time interventions in conjunction with iterations as adaptive cycles to reach an optimal steady state for operations.

Verification of Master Plan

Despite the iterative character of breakthrough, control is necessary to adhere to the master plan set by the process of confrontation and tuning and to review actual progress against the objectives set by strategy formation. It should be noted that both require a different type of control processes. In this respect, two separate control mechanisms exist within organisations in the context of breakthrough and renewal: (i) verification of the master plan and (ii) evaluation of the strategy.

The first control mechanism, the process of verification, is based on the master plan, which describes the milestones to be achieved for the processes of configuration and resource allocation, and the actual configuration of operations to meet performance requirements. During configuration and resource allocation, milestones define whether progress and decisions align with the master plan. Deviations should result in preventive and corrective actions to prevent further deviations from the master plan (this plan might also contain milestones for other aspects, such as quality, which does not necessarily mean that milestones equal temporal deadlines); in this respect, there is some alikeness to feedforward (Section 6.4). During the implementation changes in input, resource allocation, process configuration and control processes in recurrent processes the master plan serves as an indicator for when the steady state will be reached. The verification process enables organisations to follow the progress of the breakthrough processes.

Evaluation of Strategy

In the case of the second control mechanism, the performance of an organisation is measured by its output and to what extent it leads to fulfilment of the function; this results in feedback towards strategy formation through the evaluation as the column on the right hand side in Figure 10.6. The evaluation of strategies may create new input for the breakthrough processes. This requires aggregation of any kind of the evaluation to assess the strategy, or certain aspects of it.

10.4   Model for the Dynamic Adapation Capability

Extending the breakthrough model with learning processes model shows the impact of improvements for the strategic renewal processes. In the breakthrough model of Figure 10.6 iterative processes from scanning the environment to the operational control at the lowest level of the model are already shown. The conceptualisation of these iterative process can be augmented with concepts from learning processes, akin the thoughts of the learning organisation (1992), and this results in the recognition of so-called innovation impact points ultimately leading to the model for the dynamic adaptation capability; these are the topics of the next subsections based on Dekkers [2005, pp. 248–250,].

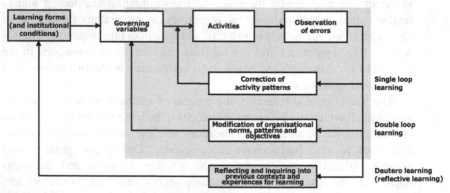

Figure 10.7 *Learning cycles from Argyris and Schön [1978](adapted from Moldaschl [1998, p. 18]). The first cycle, single-loop learning, directly corrects deviations of actions, mostly by compensating. Double-loop learning has a more preventive character, attempting to resolve the recurrence of aberrations. Deutero-learning directs itself to finding new pathways, including those for learning.*

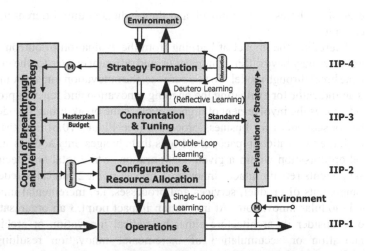

Figure 10.8 *Model for the innovation impact point. The breakthrough model shows the learning modes and the identified innovation impact points (IIPs). The higher the impact point, the more changes and innovations from lower levels affect organisational decision-making. Architectural, and often radical, changes and innovations come about through accumulation of minor changes and innovations.*

Learning Processes and Innovation Impact Points

Enhancing the breakthrough model with learning processes means distinguishing three types of learning; see Figure 10.7. Single-loop learning aims mostly at correcting mistakes, aberrations and failures directly. It does not exceed the current architecture of organisations, products, and processes. Double-loop learning reviews current solutions and possibilities to improve within an existing architecture (for any type of breakthrough). By reflective learning, also called deutero-learning, it becomes possible to search for integration of innovation, the so-called architectural innovations, and to adapt new models for governing business, including the anticipation of shifting performance requirements. The further the learning and the adaptation penetrates into the breakthrough model, the more the impact on the strategic decision processes, and the more the need for reflection to find optimal solutions.

The different types of learning are related to different levels of the breakthrough model. Single-loop learning restricts itself to the processes of 'operations' and 'configuration and resource allocation'. This because learning takes place without questioning the current architecture. Double-loop learning aims at changing the model of operations and output through architectural or radical innovation. It overturns concepts as they exist and is related to the processes of 'confrontation and tuning'. Deutero learning is found at the levels of strategy formation, by deriving goals and criteria from the environmental setting, and by using internal wisdom (or knowledge) to set priorities. Each of the types of learning processes hooks to a different level of the breakthrough model, and they indicate quite different modes of learning.

Therefore, the impact of learning from the evaluation of outcomes from processes may have different impacts on and different entries for the processes of the breakthrough model. To this purpose, the innovation impact point serves as an indicator for evaluating the ongoing innovation and renewal processes, and evokes the involvement of higher levels in the breakthrough model about the consequences for business processes (see Figure 10.8). At the lowest level, the innovation impact point 1 tells that changes only address standards and optimisation within a given configuration, while total architecture and components remain intact. Innovation impact point 2 indicates redesign of components of product, services and processes, i.e. incremental innovation and modular innovation. At innovation impact point 3 an organisation has to reconsider its architecture either by radical innovation or architectural innovation or accumulation of incremental innovation resulting in a breakthrough. Such decisions require fine-tuning of business requirements with the possibilities of innovation, e.g. dispersal in markets. These decisions may affect the strategy formation or even the objectives of an organisation (innovation impact point 3). Thus, the innovation impact points serves as indicator what potential impact adaptive cycles and learning process have.

Dynamic Adaptation Capability

In addition, the distinction of the innovation impact points shows us the importance of managing innovations and renewal, and the central role of confrontation and tuning. This should lead to the timely recognition of ongoing developments as drivers for business renewal. The central role of confrontation and tuning points to the capability of adapting to the dynamically changing environment. Fed by bottom-up innovations through the learning cycles and the technological improvements, improvements through collaboration and outsourcing, and driven by the dynamics of the market itself, continuous reflection on possibilities and opportunities leads to a continuous stream of innovations to the market. Product, services and process innovations match than with the changing customer demands. Through a stage-wise decision-making process the innovations will connect better to the actual market developments. The ability to maintain a scope of strategies with related innovations creates the opportunity for anticipation. Thus, companies avoid the static view of strategy which might result in missing opportunities and not recognising the value and impact of innovations.

This leads to the model for the dynamic adaptation capability as depicted in Figure 10.9. This model has two components: the dynamic capability and the internal innovation capability. Both these capabilities determine the dynamic adaptation capability of organisations. The dynamic capability has strong similarities to the concept as introduced by Teece et al. [1997, p. 515]: (a) the ability to renew competencies so as to achieve congruence with the changing business environment, and (b) the adaptation, the integration

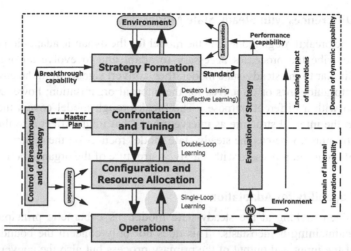

Figure 10.9 *Model for the dynamic adaptation capability. Expanding on the model of the innovation impact points, this particular model distinguishes the internal innovation capability and the dynamic capability with its external orientation.*

and reconfiguration of internal and external skills, resources and functional competencies to match the requirements of a changing environment. Both capabilities might be considered as major components of the complexity handling capability as defined by Boswijk (1992). We find the separation between these capabilities at one of the defined innovation impact points, at the level of *Confrontation and Tuning*. Above and inclusive this level the strategic adaptation takes place, whereas at lower levels in the breakthrough model and learning cycles a continuous flow of innovations is generated.

However, when the innovation impact point of the streams of innovations mainly moves at lower levels in this model, the innovations will not be assessed on its potential value for customers thus not leading to timely adaptations to the market. These adaptations might be necessary to guarantee a competitive position within the market. It also indicates in these cases that initiatives and product development might be obliterated by managerial levels; these innovations have to find their own course without management paying attention to the integration in the organisation and considering the value for (prospective) customers.

For example, during 2000, Daimler-Chrysler faced decision-making in this respect when they reviewed their policy on product development driven by cost reductions (announced publicly). Management officials announced that they wanted to maintain the continuous stream of new products in order to ensure a competitive position within the market. Within the model for the dynamic adaptation capability this last effort aligns with the characteristics of dynamic capability, although we have little information if stage-wise decision-making takes place during the confrontation and tuning process for this particular case.

10.5   Differences with Steady-State Model

The breakthrough model and the model for the dynamic adaptation capability describe the processes necessary to adapt and to evolve an organisation. Similar to the steady-state model, they convert signals from the environment into guidelines or directives for the internal organisation; however, they do so with a different purpose. The breakthrough model covers the changes in the internal structure as interventions (that might affect also the external structure), whereas the steady-state model focuses on the cybernetic control of recurrent processes within a given structure of the organisation.

Capability for Adaptation

When looking at the steady-state model, its cybernetic principles aim at maintaining homeostasis. This does not only concern the boundary zones at the input and output of the primary process but also the conversion from external standards into internal standards for the control mechanisms. When the environment indicates the reduced need, or even obsoleteness, for the output, the cybernetic mechanisms are unable to deal with this condition. The system of resources will collapse or disintegrate once the homeostasis cannot be maintained due to reduced need or obsoleteness of the output.

In this respect, the breakthrough model an the models for the dynamic adaptation capability describe the internal processes for finding new positions of homeostasis. At these points a new internal structure is needed, either for producing under different constraints, different output with similar resources or new output with new resources; this process of adaptation might also entail adapting resources to fit with new constraints or new output. This process of adaptation happens at the same time as the continuous improvement for meeting short-term performance requirements; in fact, these processes have a strong interrelation, certainly when looking at the theories of complex adaptive systems (see Section 9.3).

In addition to its focus, the breakthrough model and the model for the dynamic adaptation capability cover typically a multiple of aspects where the steady-state model describes one; note that for each investigation the aspects have to be redefined. For organisations decision making as embedded in the breakthrough depends on multiple criteria being assessed. These criteria constitute different aspects and, hence, the decision making should not only cover these aspects but also account for integral decision making (as synthesis). In any case, weighing of aspects and criteria involves the subjectivity of the actors involved and does not guarantee the best decision will be taken. The steady-state model singles out one aspect and optimises that aspect through its control mechanisms. In this respect, the scope of the steady-state model and the breakthrough model (and the model for the dynamic adaptation capability) differ widely.

The steady-state model and the breakthrough model (and the model for the dynamic adaptation capability) have in common that they both possess an evaluation process acting at the output of the primary process. Within the scope of the steady-state model that leads to corrections of the internal standards for the control process and externally to information about the control capability of the system of resources. The breakthrough model does evaluate too, but this might lead to revised objectives and strategies, even affecting the purpose of the organisation. Despite both models having evaluative processes, the purpose of these differ, though in practice they might be put together for use in the same reporting channels.

Linking Steady State to Breakthrough

In addition to he steady-state model and the breakthrough model (and the model for the dynamic adaptation capability) having evaluation processes in common, it might be possible to link the two models. The recurrent process of the steady-state model might occur at any of the phases of the breakthrough model. That means that we can describe the control of the internal processes for each of the levels by the primary process that takes place within that level and the control mechanisms to ensure its output. However, such an approach would need to take account of the iterative processes that are characteristic for the breakthrough model.

No matter, the steady-state model constitutes the process for the level of operations in the breakthrough model. The purpose of the breakthrough model is to adapt the steady state process of the recurrent processes, i.e. the direct execution of the primary process, to changing circumstances. Ultimately, the breakthrough process aims at establishing recurrent processes, whether through implementing changes in its process structure (primary and secondary) and control structure or through the reconfiguration of resources (abolishing some of the existing resources or by including supplementary or complementary resources)

10.6 Summary

Building on the concepts of previous chapters, organisations can be considered a special class of allopoietic systems. Consequently, they follow certain patterns for development that could be derived from the reference model presented in this chapter. Those patterns are also found in three approaches to the life cycle of organisations. The classification as allopoietic systems also implies that they are structurally-closed and uses self-reference as point of departure for interacting with the environment. That means that they are limited in the perception of changes, while subject to the dynamics of the environment.

The changing environment and the changes induced by organisations themselves require the implementation of strategies. Since the dynamics render the more static, canonical approaches defunct, it might be more appropriate for organisations to deploy dynamic strategies, foresight and scenario planning. Each of these strategies depends on how organisations perceive their environment, since they are allopoietic systems. The use of more dynamics forms of strategy should be used in conjunction with the breakthrough model.

That breakthrough model aims at implementing revised or new structures for operational processes; this might concern both internal structures (e.g. for operations and new product development) and external structures (for example, market segments). In the breakthrough model, by scanning the environment new or adapted goals are set and the derived policy acts as a reference for the review of tactical and operational decisions. The process of confrontation and tuning takes the possibilities into account leading to specific decisions on the utilisation of resources and structures for operations. Through the configuration and resource allocation process the actual implementation of the structural changes in operations takes place. The evaluation of strategies might create new input for the breakthrough processes. The verification enables companies to follow the progress of the breakthrough processes.

References

Abell, D. F. (1980). Defining the Business: The Starting Point of Business Strategy. Englewood Cliffs: Prentice-Hall.

Ansoff, H. I. (1965). Corporate Strategy: An Analytic Approach to Business Policy for Growth and Expansion. New York: McGraw-Hill.

Argyris, C., & Schön, D. A. (1978). Organizational learning: a theory of action perspective: Addison-Wesley.

Barábasi, A.-L. (2003). Linked: The New Science of Networks. New York: Persus Books.

Beer, S. (1972). Brain of the Firm - the Managerial Cybernetics of Organization. Chichester: John Wiley & Sons.

Beinhocker, E. D. (1999). Robust Adaptive Strategies. Sloan Management Review, 40(3), 95–106.

Berman, P. (1980). Thinking about Programmed and Adaptive Implementation: Matching Strategies to Situations. In H. Ingram & D. Mann (Eds.), Why Policies Succeed or Fail (pp. 205–227). Beverly Hills: Sage.

Boswijk, H. K. (1992). Complexiteit in evolutionair and organisatorisch perspectief, het zoeken naar balans tussen vermogens en uitdagingen. Rotterdam: Erasmus Universiteits Drukkerij.

Bruggeman, J. (1996). Formalizing Organizational Ecology. (Doctoral), University of Amsterdam, Amsterdam.

Burgelman, R. A., Maidique, M. A., & Wheelwright, S. C. (1996). Strategic Management of Technology and Innovation. Chicago: Irwin.

Das, T. K., & Teng, B.-S. (2000). A Resource-Based Theory of Strategic Alliance. Journal of Management, 26(1), 31–61. doi: 10.1177/014920630002600105

Dawkins, R. (1989). The Selfish Gene. Oxford: Oxford University Press.

de Geus, A. (1999). The Living Company. London: Nicolas Brealy.

Dekkers, R. (2005). (R)Evolution, Organizations and the Dynamics of the Environment. New York: Springer.

Dekkers, R., & Kühnle, H. (2012). Appraising interdisciplinary contributions to theory for collaborative (manufacturing) networks: Still a long way to go? Journal of Manufacturing Technology Management, 23(8), 1090–1128. doi: 10.1108/17410381211276899

Douthwaite, B., Keatinge, J. D. H., & Park, J. R. (2001). Why promising technologies fail: the neglected role of user innovation during adoption. Research Policy, 30(5), 819–836. doi: 10.1016/S0048-7333(00)00124-4

Drucker, P. F. (1978). Management in de Praktijk. Amsterdam: J. H. de Bussy.

Durrett, R. (2007). Random Graph Dynamics. Cambridge: Cambridge University Press.

Eisenhardt, K. M. (1999). Strategy as Strategic Decision Making. Sloan Management Review, 40(3), 65–72.

Fricker, A. R. (1996). Eine Methodik zur Modellierung, Analyse und Gestaltung komplexer Produktionsstrukturen. Aachen: RWTH Aachen.

Greiner, L. E. (1998). Revolutions as Organizations Grow. Harvard Business Review, 76(3), 55–67.

Guare, J. (1990). Six Degrees of Separation: A Play. New York: Vintage Books.

Hannan, M. T., & Freeman, J. (1977). The Population Ecology of Organizations. American Journal of Sociology, 83(4), 929–984. doi: 10.1086/226424

Hjalager, A.-M. (2000). Organisational ecology in the Danish restaurant sector. Tourism Management, 21(3), 271–280. doi: 10.1016/S0261-5177(99)00058-8

Hofstede, G. (1994). Cultures and Organizations. London: Harper Collins.

Kahn, H., & Wiener, A. J. (1967). The year 2000: a framework for speculation on the next thirty-three years. New York: McMillan.

Kauffman, S. A., Lobo, J., & Macready, W. G. (2000). Optimal search on a technology landscape. Journal of Economic Behaviour & Organization, 43(2), 141–166. doi: 10.1016/S0167-2681(00)00114-1

Knudsen, T. (2002). Economic selection theory. Journal of Evolutionary Economics, 12(4), 443–470. doi: 10.1007/s00191-002-0126-8

Laitinen, E. K. (2000). Long-Term Success of Adaptation Strategies: Evidence from Finnish Companies. Long Range Planning, 33(6), 805–830. doi: 10.1016/S0024-6301(00)00088-1

Leonard-Barton, D. (1987). The Case for Integrative Innovation: An Expert System at Digital. Sloan Management Review, 31(1), 7–19.

Leonard-Barton, D. (1988). Implementation as mutual adaptation of technology and organization. Research Policy, 17(5), 251–267. doi: 10.1016/0048-7333(88)90006-6

Lichtenthaler, E. (2005). The choice of technology intelligence methods in multinationals: towards a contingency approach. International Journal of Technology Management, 32(3–4), 388–407. doi: 10.1504/IJTM.2005.007341

Lievegoed, B. C. J. (1993). Organisaties in ontwikkeling: Zicht op de toekomst. Rotterdam: Lemniscaat.

Markides, C. C. (1999). A Dynamic View of Strategy. Sloan Management Review, 40(3), 55–63.

Maturana, H. R., & Varela, F. J. (1980). Autopoiesis and Cognition – The Realization of Living. Dordrecht: Reidl.

Milgram, S. (1967). The Small World Problem. Psychology Today, 2, 60–67.

Millett, S. M., & Honton, E. J. (1991). A Manager's Guide to Technology Forecasting and Strategy Analysis Methods. Columbus, OH: Batelle Press.

Mintzberg, H. (1987). The Strategy Concept I: Five Ps for Strategy. California Management Review, 30(1), 11–24. doi: 10.2307/41165263

Mintzberg, H. (1988). Generic Strategies: Towards a Comprehensive Framework. In R. B. Lamb & P. Shivastava (Eds.), Advances in strategic management (pp. 1–67). Englewood Cliffs: JAI Press.

Moldaschl, M. (1998). Kultur-Engineering and Kooperative Netzwerke. io Management, 67(6), 16–22.

Morgan, G. (1997). Images of organization. Thousand Oaks: Sage Publications.

Nakane, J. (1986). Manufacturing Futures Survey in Japan: A Comparative Survey 1983-1986. Tokyo: Waseda University.

Nelson, R. R., & Winter, S. G. (1982). An Evolutionary Theory of Change. Cambridge, MA: Belknap Press.

Popper, R. (2008). Foresight Methodology. In L. Georghiou, J. Cassingena, M. Keenan, I. Miles, & R. Popper (Eds.), The Handbook of Technology Foresight (pp. 44–88). Cheltenham: Edward Elger.

Porter, M. E. (1980). Competitive strategy: techniques for analyzing industries and competitors. New York: Free Press.

Porter, M. E. (1996). What Is Strategy? Harvard Business Review, 74(6), 61–78.

Quinn, J. B. (1981). Formulating strategy one step at a time. Journal of Business Strategy, 1(3), 42–63.

Ravasi, D., & Lojacono, G. (2005). Managing design and designers for strategic renewal. Long Range Planning, 38(1), 51–77. doi: 10.1016/j.lrp.2004.11.010

Reinert, H., & Reinert, E. S. (2006). Creative Destruction in Economics: Nietzsche, Sombart, Schumpeter. In J. G. Backhaus & W. Drechsler (Eds.), Friedrich Nietzsche (1844–1900): Economy and Society (pp. 55–85). New York, NY: Springer.

Ringland, G. (1997). Scenario Planning: Managing for the Future. Chichester: John Wiley & Sons.

Schumpeter, J. (1911). Theorie der wirtschaftlichen Entwicklung. Leipzig: von Duncker & Humblot.

Schumpeter, J. (1954). History of Economic Analysis. New York: Oxford University Press.

Schumpeter, J. A. (1934). The Theory of Economic Development: An Inquiry into Profits, Capital, Credit, Interest, and the Business Cycle. Cambridge, MA: Harvard University Press.

Senge, P. M. (1992). The Fifth Discipline. Kent: Century Business.

Sombart, W. (1930). Krieg und Kapitalismus. Leipzig: Duncker & Humblot.

Teece, D. J., Pisano, G., & Shuen, A. (1997). Dynamic Capabilities and Strategic Management. Strategic Management Journal, 18(7), 509–533.

van der Heijden, K. (1996). Scenarios, the art of strategic conversation. Chichester: Wiley.

Watts, D. J., & Strogatz, S. H. (1998). Collective dynamics of 'small-world' networks. Nature, 393(6684), 440–442. doi: 10.1038/30918

Whetten, D. A. (1980). Organizational Decline: A Neglected Topic in Organizational Science. Academy of Management Review, 5(4), 577–588. doi: 10.5465/AMR.1980.4288962

Williamson, P. J. (1999). Strategy as Options on the Future. Sloan Management Review, 40(3), 117–126.

11 Applications of System Theories

The previous chapters have elaborated the application of the concepts from systems theories to examples drawn from technical systems, biological systems and organisational systems; this chapter intends to have a further look at the applications. Beyond these three domains, there are also other domains that have benefited from systems approaches, such as psychology and communication. For example, applications of non-linear dynamic systems theory to psychology have led to advances in understanding neuro-motor development and advances in theories of cognitive development [Metzger, 1997]. More recent literature on systems thinking has a general (often philosophical) perspective, concerns computer systems or focuses on one highly specific problem. Heylighen [1991] sighs:

> *The fundamental concepts of cybernetics (ed.: incl. general systems theory) have proven to be enormously powerful in a variety of disciplines: computer science, management, biology, sociology, thermodynamics ... A lot of recently very fashionable approaches have their roots in ideas that were proposed by cyberneticians several decades ago: artificial intelligence, neural networks, complex systems, man-machine interfaces, self-organisation theories, systems therapy ... Most of the fundamental concepts and questions of these approaches have already been formulated by cyberneticians such as Ashby, von Foerster, McCulloch, Pask, ... in the forties and the fifties. Yet cybernetics itself is no longer fashionable, and the people working in those new disciplines seem to have forgotten their cybernetic predecessors.*

This fragment shows that systems theories have become part of science and practice; yet, progress is still made, especially in more advanced topics, such as autopoiesis [for instance, Steen, 2014] and complex adaptive systems [e.g. Mittal, 2013], and in specific domains, for example, supply networks [for example, Pathak et al., 2007] and social-ecological systems [for instance, Levin et al., 2013].

Each domain of application requires an extensive treatment to do justice to those that have been and are working on it; however, the sole purpose of this chapter is to indicate the applications in the three domains and possible avenues for the reader's interest. Section 11.1 will discuss systems engineering, a traditional field of application for system theories. Two topics on biological systems constitute Section 11.2; these have been selected from a wide field in the biological domain. Section 11.3 covers the application to organisations. Section 11.4 addresses some other (popular) system theories, particularly those used in the domain of organisations. Finally, Section 11.5 pays attention to how the concepts of systems theories can be used for research.

© Springer International Publishing AG 2017
R. Dekkers, *Applied Systems Theory*,
DOI 10.1007/978-3-319-57526-1_11

11.1   Systems Engineering

Systems engineering (or systems design engineering) as a field originated around the time of World War II, when the complexity of engineering projects increased. Large or highly complex engineering projects, such as the development of airplanes or warships, needed to be often decomposed into stages and managed throughout the life cycle of the product or system; later this approach became common for all kinds of complex systems, such as petrochemical plants and information systems. This approach to engineering systems is inherently complex, since the behaviour of and interaction between system components is not always clearly defined. Defining and characterising such complex systems is the primary aim of systems engineering.

For managing these inherently complex systems, there are several methods and tools frequently used by systems engineers (some of these appear in Figure 11.1):

- Elicitation of (functional) requirements.
- Functional analysis.
- Systems architecture and design.
- Interface specification and design.
- Communications protocol specification and design.
- Modelling and simulation.
- Acceptance testing and commissioning.
- Validation, verification and fault modelling.

These methods and tools are necessary because the design and engineering of systems, both large and small, can lead to unpredictable behaviour and the emergence of unforeseen system characteristics. Moreover, decisions

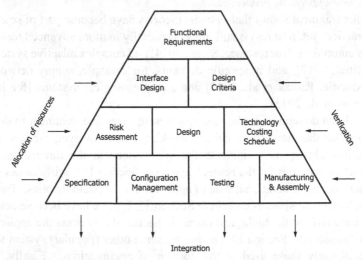

Figure 11.1 *Overview of methods and tools for systems engineering. Systems engineering provides processes that ensure the functional requirements are satisfied by the final product or service. It covers the range from functional requirements to production and deployment of complex systems, spanning the entire life cycle.*

made at the beginning of a project, which consequences are not clearly comprehended, can have enormous implications during the later phases of the life cycle of a system; systems engineering explores these issues and aims at making critical decisions to decrease these consequences. However, there is no single method that guarantees that decisions made today will still be valid when a system goes into service years or decades after it is first conception but there are techniques to support the process of systems engineering. Examples include the use of soft systems methodology (see Sections 5.8 and 11.4), system dynamics [Senge, 1992; Sterman, 2001] (see also Section 11.4) and the unified modelling language (see Section 5.8), each of which are being used to support the decision making process during product (and service) design and engineering.

Often, systems engineering involves the modelling or simulation of some aspects of the proposed system in order to validate assumptions or to explore theories. For example, highly complex systems such as aircraft are usually modelled and simulated before the maiden flight. In this way, the initial aerodynamic properties and control systems can be drafted initially and improved before the physical system itself is constructed. Since complex systems aircraft are often very expensive, this reduces the efforts and the difficulty of debugging the control system and the risk of crashing real aircraft with all potential complications. The use of advanced modelling and simulation software has created opportunities to reduce the engineering efforts during later stages of product (and service) design engineering and to predict behaviour of complex product (and service) systems more accurately.

However, despite all modelling and simulation, initial testing and commissioning are still required to reach acceptable levels of safety and performance in advanced product (and service) systems. Systems engineers perform validation and verification when a system has to have predictable behaviour. As case in point is medical support equipment, such as heart and lung machines, that usually consists of several parts, engineered by different companies. Validation and testing assures that normal operation and possible failures of each part will not harm patients. Other applications are communications systems and banking software, where failures can cause loss of property or liability. Test plans can often be adjusted to save significant amounts of efforts by testing partial systems or by including special features in a system to aid testing.

Because of its scope and because of the design of complex systems, many related domains use different techniques and methods useful for systems engineering. Some of those areas that contributed methods for systems engineering will follow now:

- Software engineering has more recently helped to shape modern systems engineering practice to a great degree. The techniques used in the handling of complexity of large software-intensive systems has dramatically reshaped the tools, methods and processes in systems engineering; examples of these tools, methods and processes are systems modelling language, capability maturity model integration, object-oriented analysis and design, requirements engineering, and formal methods and language theory).

- Control systems design. The design and implementation of control systems, used extensively in nearly every industry, is a large subfield of systems engineering. The cruise control of a car and the guidance system for spacecraft constitute two examples.
- Operations research. This is an interdisciplinary science that deploys methods such as mathematical modelling, statistics and algorithms to decision making in complex real-world problems, which are concerned with coordination and execution of the operations within an organisation. The eventual intention is to find the best possible solution to a problem, which either improves or optimises the performance of the organisation.
- Safety engineering. The techniques of safety engineering can be applied by non-specialists in designing complex systems to minimise the chance or the effect that the safety-critical failures can cause. Safety engineering helps to identify safety hazard areas of emerging designs and uses methods for mitigating the effects of safety-hazard failures that cannot be designed out of systems.
- Reliability engineering is the discipline of ensuring a system will meet the customer's expectations about a failure-free product life cycle. Reliability engineering applies to the entire system, including hardware and software. It is closely associated with maintainability engineering and logistics engineering. Two methods that are well known are the failure mode and effects analysis and fault tree analysis (see also subsection 'Analysing Problems' in Section 4.2). Reliability engineering relies heavily on statistics, probability theory and reliability theory for its tools and processes.
- Interface specification and design are concerned with making the subsystems desirably connect with and interoperate with other subsystems within the system and with external systems. Interface design also includes assuring that system interfaces should be able to accept new features, including mechanical, electrical, electronic and logical interfaces. The human-computer interaction is another aspect of interface design and is a vital part of modern systems engineering when considering the user of a system.

More recently, the methods of systems engineering have reached the field of biotechnology. Thus, systems engineering has a wide range of applications spanning domains where design and engineering activities play an important role.

11.2 Biological Systems

Most biological systems have even a higher degree of complexity than the technically complex systems outlined in the previous section. Complex systems research overlaps substantially with non-linear dynamics research, but complex systems specifically consist of a large number of mutually interacting agents, as is the case in biological applications. Especially, two areas of interest linked to systems theories have gained in ground in the past years: systems biology and ecosystems.

Systems Biology

Systems biology covers an emergent field that aims at understanding of biological systems as a whole. Since the days of Norbert Wiener, this holistic understanding has been a long-standing goal of biological sciences; this is a reversal of the early days of systems theories, when many concepts in systems theories had their foundation in concepts arriving from biology. For example, cybernetics lent some concepts, such as homeostasis and boundary control, from biology to complement its own control concepts. Molecular biology had just started at the same time and only phenomenological analysis was possible in that discipline of science. Only more recently, can the system level analysis be grounded on discoveries at molecular level. With the progress of the genome sequence project and a range of other molecular biology projects that accumulate in-depth knowledge of molecular nature of biological systems, scientists are now at the stage to seriously look into the possibilities of understanding biological systems as a whole.

What does it mean to understand at *system level* in systems biology? Unlike molecular biology, which focuses on molecules, such as the sequence of nucleotide acids and proteins, systems biology concentrates on systems that are composed of molecular components (either subsystems or elements as denoted in Applied Systems Theory). Although biological systems are composed of matters, the essence of a system lies in its dynamic behaviour and it cannot be described merely by enumerating elements of the system. Not only system structures, such as network topologies, are important but also the diversities and functionalities of elements. Both the structure of the system and the components plays an indispensable role forming the symbiotic state of the system as a whole. Within this context, (1) the understanding of a system's structure, such as gene regulatory and biochemical networks as well as physical structures, (2) the understanding of the dynamics of a system, both quantitatively and qualitatively, as well as construction of models with powerful prediction capabilities, (3) the understanding of control methods for the system and (4) the understanding of the design methods for the system, are key milestones to judge how much we understand the biological system [Kitano, 2002, p. 1662].

More recently, the prospect of designing biological systems has become feasible. Currently, this is mostly done by improving plants or animals through adding genes from other organisms, but the first simple from-scratch designs of biological functional modules are starting to appear. Examples are designed cells as thermometers and oscillators that are independent of the cell cycle. Even before all this became possible, the possibility of using methods from systems engineering to assist in reverse engineering for nature had attracted some biologists. One of the goals of systems biology is to understand a complex biological process in such sufficient detail to allow the building of a computational model. This would allow simulations of behaviour and lead to a quantitative understanding of function.

The implications of thinking in terms of systems are starting to take hold in research into systems biology. For example, the concept of modularity, which has served engineers and systems theorists well for some time, has been rediscovered for biology. Modularity is used as an equivalent to subsystems and aspectsystems. Many organisms consist of modules, both anatomically and in their metabolism. Anatomical modules are usually segments or organs. Classical biology already had this concept on a rather macroscopic scale, without explicitly calling it by this name. Now researchers see a modular framework for biology, treating subsystems of complex molecular networks as functional units that perform identifiable tasks perhaps even able to be characterised in familiar engineering terms [Lauffenburger, 2000]. This coincides with the concept of systems in systems theory (system, modularity), where scientists think in terms of classes of systems, defined by a certain set of common characteristics, which can be handled by a common set of methods. It would also be the base for future developments to more complex models, once the cellular and sub-cellular levels can be described in sufficient detail. This could be seen as a macro-scale extension to the modular concepts and as an application of systems engineering practice to biological engineering.

One major goal of these efforts is a better understanding of how cells work through modelling (see Section 3.4). This is different from the way biologist defined models in the past, using pure descriptions of concepts and ideas as models. The most feasible application of systems biology research is to create a detailed model of cell regulation, focused particularly on transduction cascades and molecules to provide system-level insights into mechanism-based drug discovery. Such models may help to identify feedback mechanisms that offset the effects of drugs and predict systemic side-effects. Some of the possibilities for application are: drug design, personalised drugs, i.e. built for purpose, medicines free of side effects, developed for (or at least adapted to) individual patients, directed, reliable manipulation of gene information (e.g. treatment of tumours or hereditary diseases) and more. Such a systemic response cannot be rationally predicted without a model of intracellular biochemical and genetic interactions. With such models another transfer from engineering practice would become possible: newly designed drugs could be tested in simulations before going into clinical testing.

One of the more recent advances in systems biology is that the complexity, which is unarguably present in biological systems, is often not a complexity of function. It is rather a complexity of regulation that is necessary to ensure that a relatively simple function can be maintained robustly in spite of severe perturbations from the environment (robustness); compare this with Ashby's law of requisite variety (Section 6.8). In other words, the objective of this complexity is to guarantee that the core function will generate reliable output; the system complexity is built in to provide for simple behaviour (please note the parallel with the concepts of autopoiesis in Section 8.1). This is in sharp contrast to the popular chaos and complexity theories, which associate complexity with fractals and edge-of-chaos, originating in simple systems (see Chapter 9). This distribution of complexity can also be observed on a level of aggregation even lower than that of cell functions. As the various genome projects are showing, there are more regulatory sections to a genome than there are for metabolic functions and a lot of sections have no essential function at all (or not yet discovered). If this inference proves to be generally true, it could be speculated that the compositional complexity of cells is designed chiefly to enable cells to maintain simple functions reliably in uncertain and variable environments (robustness and sensitivity). Another aspect of complexity at the genetic level is contained in the realisation that there is no strict demarcation between information storage and functional units. Gene regulation is embedded in basic processes within cells, though complex in their interactions for maintaining a steady state.

Biological Ecosystems

Whereas systems biology focuses on micro-level, may be building up to organisms, ecosystems consist of the biological communities that occur in some locales and the physical and chemical factors that make up their non-living or abiotic environments. There are many examples of ecosystems – ponds, forests, estuaries and grasslands. A principle of ecology is that each living organism has an on-going and continual relationship with every other element that makes up its environment. An ecosystem can be defined as any situation of interaction between a range of organisms (species) and their environment. Such boundaries are not fixed in any objective way, although sometimes they might be obvious, as with the shoreline of a small pond, but even there some species might cross this boundary back and forth. Usually the boundaries of an ecosystem are chosen for practical reasons having to do with the goals of the particular study (commensurate with the definition of systems in Section 2.1).

The study of ecosystems mainly consists of the study of certain processes that link the living, or biotic, components to the non-living, or abiotic, components. The ecosystem is composed of the entirety of life (called the biocoenosis as closely integrated community of different organisms) and the medium that life exists in (the biotope – the region or habitat). Within the ecosystem, species are connected and dependent upon one another in the food chain; and they exchange energy and matter between themselves and with their environment. Energy transformations and biogeochemical cycling are the main processes that comprise the field of ecosystem ecology.

Within the domain of ecosystem ecology, there are different kinds of studies. The studies of ecology happen at the level of the individual, the population, the community and the ecosystem itself. The studies of individuals are concerned mostly about physiology, reproduction, development or behaviour, while studies of populations usually focus on the habitat and resource needs of individual species, their group behaviours, population growth and what limits their abundance or causes extinction. The studies of communities examine how populations of many species interact with one another, such as predators and their preys or competing species that share common needs or resources. Ecosystem ecology puts all of this together, which means trying to understand how the system operates as a whole. This means that, rather than worrying mainly about particular species, the study of ecosystems tries to focus on major aspects. These aspects include the amount of energy that is produced by photosynthesis, how energy or materials flow along the many steps in a food chain and what controls the rate of decomposition of materials or the rate at which nutrients are recycled in the system. Ecosystems have energy flows and ecosystems cycle materials. These two processes are linked, but they are not quite the same:

- Energy enters the biological system in the form of light, or photons, and is transformed into chemical energy in organic molecules by cellular processes including photosynthesis and respiration, and ultimately is converted to energy in the form of heat. This energy is dissipated, meaning it is lost to the system as heat; once it is lost, it cannot be recycled. Without the continued input of solar energy, biological systems would quickly shut down. The earth is an open system with respect to energy.

- Elements such as carbon, nitrogen, or phosphorus enter living organisms in a variety of ways. Plants obtain these elements from the surrounding atmosphere, water or soils. Animals may also get elements directly from the physical environment, but usually they obtain these mainly as a consequence of consuming other organisms. These materials are transformed biochemically within the bodies of organisms, but sooner or later, due to excretion or decomposition, they are returned to an inorganic state. Often bacteria complete this process, through the process called decomposition or mineralisation. During decomposition these materials are not destroyed or lost, so the earth is a closed system with respect to elements (with the exception of a meteorite entering the system now and then).

Hence, the earth as a system is open with respect to energy but closed with regard to its elements. The elements are cycled endlessly between their biotic and abiotic states within the ecosystem earth. Those elements whose supply tends to limit biological activity are called nutrients. So that means that a continuous chain of (re)cycling elements drives ecosystems, such as the earth, driven by the openness with regard to the aspect energy.

In reality, the organisation of biological systems is more complicated than can be represented by a simple 'chain'. There are many food links and chains in an ecosystem and all of these linkages are called a food web. Such food webs can be very complicated, where it appears that 'everything is connected to everything else' and it is important to understand what are the most important linkages in any particular food web. Biosphere II demonstrated how fragile the balance can become (see Figure 11.2). This grand experiment attempted to replicate natural ecosystems inside a self-contained world. However, the system started to fail several months into the experiment. All parts of the ecosystem were in jeopardy because the experiment's designers had overlooked the importance of every part in the ecosystem, including the microbes. This only demonstrates how complex adaptive systems, such as ecosystems, are actually complex and difficult to grasp.

11.3 Organisations

A third domain to which systems theories have been applied is the processes, structures and adaptation of organisations. Some advocate that the nature of management may be conceptualised from a perspective of systems theories as the process by which an organisation generates a global representation of its own processes; this modelling is also found the soft systems methodology, see Section 5.8, and is latently present in the viable system model from Beer [1972], see Section 11.4. In other words, organisations depend upon modelling their own structures from their own perspective; this is an allopoietic perspective of self-cognition and self-reflection (Section 8.6). Thus, modelling allows organisations to perform distinctive activities, such as foresight, monitoring, evaluation and control to ensure continuity for the

Figure 11.2 *The dome of the project Biosphere II (picture edited from: http://upload.wikimedia. org/wikipedia/commons/d/d3/Biosphere2_1.jpg, accessed: 9ᵗʰ July, 2014). The Biosphere II was an experiment conducted in the early 1990s. It was supposed to be a self-contained ecosystem with a team of scientists locked into it for 2 years. The 3.15 acre facility, made of glass and space-frame, was the largest total enclosed ecosystem ever built. All of the living things inside were taken directly from Biosphere I (i.e. the Earth). All seven ecosystems of Earth existed within the confines of Biosphere II. They were a rainforest, a desert, a savannah, a marsh, a farmland (in an area called the Intensive Agriculture Biome) and a 'human habitat'. Thus, it contained soil, air, water, animals and plants. About 4,000 plants and animals were introduced to Biosphere II and its ocean contained 3,400 cubic metres of water. It was hoped that these provisions would give the ecosystems enough material to be self-sustaining.*

system of resources from the perspective of the stakeholders and to fulfil functions as adopted by the environment. This means that the purposes to which these activities are directed are a product of the interaction between an organisation and its environment, particularly stakeholders. This is a consequence of the way that organisation will adapt with the purpose of sustaining themselves and growing in the specific context in which they are operating; note that this can lead to very different management processes and structures in different environments. Within the application of systems theory for organisations three main streams can be distinguished that describe these processes and structures for adaptation to specific contexts: management cybernetics, analysis and design of organisational systems, organisations as allopoietic systems and evolution of organisations.

Management Cybernetics

Management cybernetics, or also called organisational cybernetics and cybernetic management, is the application of cybernetic laws to all types of organisations and institutions created by human beings and to interactions with and within them. For example, Beer's [1972] cybernetic management theory (based on the viable system model, see Section 11.4) is not limited only to industrial and commercial enterprises. It also relates to the management of all types of organisations and institutions in the profit and non-profit sectors: from individual enterprises to large multinationals in the private and public

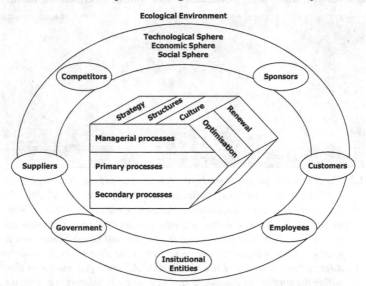

Figure 11.3 *Adapted St. Gallen Management Model. The original model [Schwaninger, 2001, p. 1212] has been complemented by including managerial processes and using terminology that is also found in Applied Systems Theory (primary and secondary processes). There is also a strong parallel with concepts about stakeholders in Section 7.6.*

sector, and from associations to political bodies. In addition to the viable system model, this approach to management has become most known through the St. Gallen management model [Schwaninger, 2001]; see Figure 11.3. Pruckner [2002] describes how the the viable systems model influenced the development of this approach. Most characteristically, cybernetic management takes as premise that it should not only cover general management issues and actions by top managers, but also that every individual encounters analysis and decision making; thus, actions are not restricted to those of top managers. In this sense, management cybernetics is a way of considering and thinking about issues that can be used to analyse the thinking, communication, acting and functioning of human beings themselves and to give them an effective meaning within the context of organisations.

Analysis and Design of Organisations

Another strand of systems thinking using cybernetic principles applied to organisations has focused on the design of organisational structures. This stream builds on the steady-state model (Chapter 7) and the breakthrough model (Section 10.3) as notional concepts for analysis and design. The approach to organisational design is depicted in Figure 11.4. In the perspective of this methodology, the design of the organisation should combine processes and resources within the system from a strategic point of view (i.e. the re-design of an organisation might cause a breakthrough); this means grouping the tasks and activities into a so-called organelle structure to match criteria. Examples of organelle structures are production lines, production cells, group technology and job-shops. Each of these performs differently across a range of criteria, such as control of quality, flexibility, reliability of delivery, utilisation of resources. Also, characteristics of product and services play a role in the design of the organelle structure; take the production of fossil fuels as example compared the production custom-made furniture. Therefore, the design of organelle structures strongly depends on the set or imposed

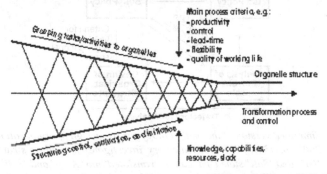

Figure 11.4 *Design process for the organelle structure [Dekkers, 2005, p. 433]. The organelle structure affects both the grouping of tasks in the primary process as well as the control processes. By subsequent integration and iteration, the design of the organelle structure meets performance requirements.*

performance criteria. More specifically, strategic choices relate the organelle structure to external performance criteria dictated by product-market combinations and internal performance criteria. Hence, there are organelle structures that range from the functional structure (job-shop) to the product flow organisation with their impact on design requirements for organisational structures. Factually, the organelle structure as core concept represents the trade-off between the requirements for exerting control, the capabilities of an organisation and the utilisation of resources.

In addition to the organelle structure, the structure of the hierarchy represents the management of the resources. To that purpose, leadership issues, span of control and communication structures play a paramount role in the choice for the most adequate structure. Also, the hierarchical structure accounts for communication, coordination and control related to the organelle structure and the control structure (i.e. the steady-state model and the breakthrough model). However, the choices for the hierarchical structure may be subject to biased views within the organisation.

The design of an adequate organisational structure should incorporate the opportunities provided by product and process characteristics in addition to meeting all performance requirements. In this context, it should be noted that the management of resources incorporates the breakthrough processes, the primary process and the control processes. Each of these processes deploys resources, with specific skills and knowledge, to achieve outcomes whether it concerns the manufacturing of products and the provision of services or the transformation from signals into interventions (the domain of control processes). Optimisation by management, the hierarchy, concentrates on all available resources for the primary process and control processes to reach organisational objectives.

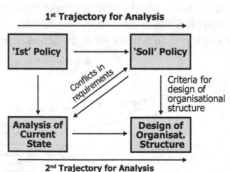

Figure 11.5 *Simplified version of the methodology for (re)design of organisations [Dekkers, 2005, p. 434]. The first trajectory investigates prevailing strategies for the 'Ist'- and 'Soll'-state ('Ist' can be translated into 'As-Is' and 'Soll' into 'Ought-To-Be'). The second trajectory analyses the current organisational structure (primary process, control process, organelle structure, hierarchy) and arrives at a redesign of the integral organisational structure. The two trajectories are intertwined through the criteria for analysis and redesign.*

The design methodology follows two principles. First, an organisation is analysed, and after the analysis of signals of weakness and symptoms (see Section 4.2), the design follows the requirements derived from the strategy of the organisation; see Figure 11.5. Changes in the (corporate) strategy, external developments, internal performance information or any combination of these factors set new requirements for an organisation. External developments may concern market research, technological changes or other external information that influence the business processes and organisational structures. The internal information refers to data about the performance, the structure and the activities of the organisational unit. The changes in general policy, external developments, and internal information should lead to either a radical or an incremental upgrade of the organisation. Second, the performance requirements reflect on the decision making during design and detailing of solutions; see Sections 4.4 and 4.5 for the generic approaches to decision making and detailing of solutions. Moreover, the design approach relies on a step-wise approach: first, the setup of the primary process is considered, then the design of the control processes, followed in iterations by consecutively the organelle structure and the hierarchical structure (see Figure 10.4); this approach is akin the controlled convergence method presented in Section 4.3. During each stage, potential performance of possible solutions is compared with design requirements.

Organisations as Allopoietic Systems

No matter how they are structured, organisations can be considered as special class of allopoietic systems that have fuzzy boundaries and the capability for foresight [Dekkers, 2005, p. 397]; see also Sections 8.6–8.7. The usual organisation science perspective that an organisation adapts to its environment, or at least influenced by it, is fundamentally turned around. An 'autopoietic' organisation, on the contrary, is self-referentially closed. It only perceives its environment as a projection of its self-identity. It only functions in order to survive and to maintain its identity. An example of such thinking is found in the book about the living organisation by de Geus [1999], as mentioned in Section 8.6. He refers to the existence of an organisational identity even if it is present in the actions of the individual that constitute that organisation. That it might have far-reaching consequences is brought to our attention by Bakan [2004] who characterises firms as psychopathic, mainly to indicate the lengths to which an organisation might go to preserve itself. Thus, organisations as allopoietic systems have also a tendency to maintain self-identity, even if this has adverse effects for stakeholders (incl. society).

In the conceptualisation of autopoiesis systems are both open and closed and this applies to organisations, too. Autopoietic organisational systems interact with their environment, which consists of other systems (i.e. are open interactively), see Section 8.2. But they are also closed by the boundaries of meaning as the meaning creation takes place through the system's auto-referencing [Hernes and Bakken, 2003, p. 1516]. The system can only make sense of the outside world through the observation of its own experiences. As mentioned in Section 8.6, the concept of the learning organisation, coined by de Geus [1999] and expanded by Senge [1992], has become a popular way to describe the interaction between organisation and its environment and the learning experiences of organisations. In some way, this is analogous to the steady-state model (Chapter 7) and the breakthrough model (Section 10.3). In the steady-state model learning is present through the direct evaluation of the output, the information from the environment and the assessment for revaluing the standards and determining the control capability of the system of resources. The breakthrough model takes this evaluation even further by determining new structures to fulfil functions or reprogram functions; the distinction of the innovation impact points for the model for the dynamic adaptation capability underline this point (Section 10.4). The interaction through the operational processes as throughput characterises the organisation as an open system while the structural changes and the perception of the environment denote the organisation as a closed system.

Evolutionary Approaches for Organisations

In addition to the more structural approaches of cybernetic management and analysis and design of organisations, economists and management scientists have embraced the core thoughts of evolutionary approaches, either explicitly or implicitly; this way they also included core thoughts of systems theory. Such was already true for the early contributions by Veblen [1898] and Schumpeter [1911], even though Schumpeter was highly critical of attempts to apply theories from the natural sciences to economics [Fagerberg, 2003, p. 127, 144]. Later on, evolutionary approaches experienced a revival with the writings of Nelson and Winter [1982] and Hannan and Freeman [1977]; especially, Nelson and Winter denounced pursuing biological analogies, for their own sake or for the purpose of developing a general evolutionary theory applicable to both natural and social sciences [Nelson and Winter, 1982, p. 11]. Since then, an increasing stream of publications has employed evolutionary approaches, following the founders in avoiding to use analogies from evolutionary biology.

Within this context, it is useful to examine some of the crucial differences between economic and biological evolution (see for example Dekkers [2005], Eldredge [1997], Hodgson [2005] and van den Bergh and Gowdy [2000]):
- Whereas in biology the genotype-phenotype distinction is very clear, in economics and management science no such distinction exists. For there is no singular equivalent in economics to the most basic unit of selection, i.e. the gene. Related to this is the fact that the distinction between 'ontogeny' – development of an organism – and 'phylogeny' – 'family tree' or evolutionary history of a group of organisms – has no counterparts in economics. Both these differences relate to the fact that biological evolution is genetic evolution, whereas social-economic evolution is a combination of genetic and non-genetic evolution, in which the latter

is dominant in the short run. Nevertheless, some authors have tried to strictly impose the genotype-phenotype analogy to economics (Boulding [1981], Faber and Proops [1990]). It should be noted, though, that in biology the notion of sociobiology has been brought to the fore (e.g. [Wilson, 1998]; this indicates that non-genetic evolution plays a role in biological systems, too.

- Ideas and artefacts, including people, products, books, behaviour, routines, knowledge, science, religion, art, rituals, institutions and politics are all concrete and durable information carriers and can act as 'genes' if this is relevant for the study concerned. Some authors prefer to refer to cultural and economic genes as 'memes', a term originally proposed by Dawkins [1989] and by others examined from various disciplinary perspectives [e.g. Aunger, 2000]. According to Norgaard [1994, p. 87], 'one type of gene is no more real than the other'. This suggests that there is no objection against choosing the 'gene' in economics ad hoc, i.e. depending on the context or type of analysis. Note that if macro-evolution and higher level sorting exist in biological systems, the gene is not the exclusive unit of selection, and selection is not the only mechanism of durable change, in biological evolution anyway. Therefore, the lack of an equivalent to the gene in economic evolution would not be such a serious criticism after all (e.g. Hodgson [1993]).

- Lamarckian or goal-oriented evolution occurs at various levels in economic systems: individuals, groups and sectors. This is due to social, organisational, group and individual learning and search, notably through education and research. In biological systems and most animal species such learning is largely absent, and mutations are mainly random and certainly not the result of purposeful search. The distinction between selection as social learning (selection and diffusion) and individual learning (which is very limited for most species) is clear-cut in biology. In contrast to biology, in economics such a distinction is blurred, as technologies developed and lessons learned as knowledge in one sector can be easily transferred to other sectors and contexts. In summary, characteristic for economic-cultural evolution is that information can be purposefully accumulated (learning), that changes (mutations) can be purposefully stimulated and that innovations can be diffused very easily across sectors.

- The biological sexual recombination mechanism to generate new genetic structure has no direct economic analogue. Even though some may associate acquisitions and mergers with recombination in evolutionary biology, the combined systems of resources should be approached from a supplementary and complementary view (see also Section 9.6 for the application of this thinking to networks of agents); this combines economies of scale with synergy, something that will be poorly described by biological concepts. Nevertheless, in economic systems inheritance, as a concept derived from evolutionary biology, can occur in different ways

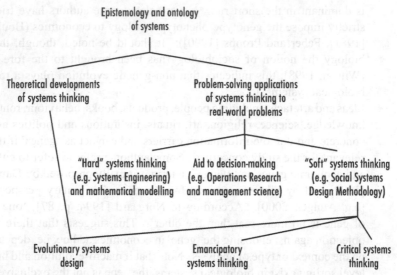

Figure 11.6 *Overview of the systems movement (adapted from Laszlo and Krippner [1998]). Systems theories have evolved from the movement of defining them (epistemology and ontology, characteristic for the search into the early General Systems Theory) into two main strands. The first one of theoretical development resulted in interests in evolutionary systems design, which can be linked to the science of complexity. The second stream focused on the application and divided quickly into three directions: hard systems thinking, support to decision-making and soft systems thinking. Ultimately, soft systems thinking triggered critical systems thinking (using the theories for social problems), emancipatory systems thinking (similar) and evolutionary systems design.*

and on different levels of aggregation. For example, ideas in economics and about technology often suddenly increase in value when combined. In fact, major innovations often result from combining existing insights, concepts, technologies or institutions. This suggests that recombination as an abstract concept, may be valuable to evolutionary economic reasoning for a limited range of applications.

Apparently, the structure of organisations, embedded in the economic environment, differs from those of organisms. This might indicate that comparisons and analogies apply more to the governing principles for the evolution of organisations and economies than to the resulting structures; O'Shea [2002] provides a similar argument evolutionary approaches for new product innovation, based on Bergson [1911]. For example, comparing an organisation with the human body would be Hodgson's [2002, p. 263] literary ornament and add little to the understanding of the evolution of organisations. The stance that limitations apply to analogies is supported by the systems hierarchy of Boulding (Section 3.5); in view of the levels of Boulding, evolutionary models aimed at describing the evolution of species (levels of genetic-societal systems, animals, humans) do not directly apply to the evolution of organisations (level of social organisations). Hence, it seems plausible to direct the use of analogies towards those of governing principles for the domain of organisations.

11.4 Other Systems Theories in Brief

During and after the development of the general systems theory, in the 1950s and 1960s particularly, scientists have developed applications of system theories. Initially, the focus of system theories was on epistemological and ontological issues; for example, what are the definitions of systems and how are they described. That resulted in books, such as those by von Bertalanffy [1973] and West Churchman [1979]. However, in parallel a stream existed that was more directed at solving problems; a case in point is the development of control theory. These differing foci lead to the distinction of different strands in system theories, see Figure 11.6. Some of these strands are aimed developing theoretical concepts and others are more oriented towards applications and solving problems. The latter happened especially in the domain of organisations and social organisations. Moreover, some management scientists consider system theories a basic tool for studying organisational entities. Note that applied systems theory, as presented in this work and being a methodology for analysis and design of organisations (see Section 11.3), combines different strands; these are 'hard systems' thinking, systems engineering methodology, 'soft systems' thinking and social systems design methodologies. In addition to applied systems theory, the main conceptualisations that evolved as a collection of thoughts about systems are: system dynamics, soft systems methodology by Peter Checkland [1981], viable system model by Stafford Beer [1972], metasystem transition theory and critical systems thinking.

System Dynamics

System dynamics is an approach to understanding the behaviour of complex systems over time. It is mainly based on internal feedback loops and time delays that affect the behaviour of the entire system. Generally, it is applied to analyse any dynamic system that is characterised by interdependence among elements, mutual interaction between actors and elements, feedback loops and circular causality. Circular causality means that the effect of an event or variable returns indirectly to influence the original event itself by way of one or more intermediate events or variables; for example, event A causes event B, consequently event B causes event C and eventually event C influences the original event A. The application of system dynamics is often supported by simulation software.

The approach of system dynamics for simulation begins with defining problems dynamically and proceeds through mapping and modelling stages to steps for building confidence in the model and its policy implications.

BOX 11.1: LEVERAGE POINTS FOR INTERVENTION IN A SYSTEM

Based on the core concepts of system dynamics, Meadows [2008] has identified twelve leverage points for interventions in a system based on feedback. They are presented in order of increasing effectiveness; some have been adapted to fit with the terminology in this book. The interventions are:

- CONSTANTS, PARAMETERS AND VALUES (AS PROPERTIES OF SYSTEMS). Such interventions include subsidies, taxes and standards.
- SIZE OF BUFFERS AND OTHER STABILISING STOCKS. This might concern the buffers and overflow valves (see steady-state model in Section7.5) and should be considered in relation to the flowing elements.
- STRUCTURE OF MATERIAL STOCKS AND FLOWS. This is the structure of a system and how processes are interrelated; examples are transport networks and population age structures.
- LENGTH OF DELAYS. This refers to how fast control mechanisms respond relative to the rate of changes in the system of resources and the processes.
- STRENGTH OF NEGATIVE FEEDBACK LOOPS. This concerns the magnitude of the response by control mechanisms relative to the deviations they are trying to correct against.
- GAIN AROUND DRIVING POSITIVE FEEDBACK LOOPS. This covers the strength of positive feedback mechanisms; the stronger the positive feedback relative to the negative feedback, the higher the chances of oscillation occuring.
- STRUCTURE OF INFORMATION FLOW. This intervention considers which actors do and do not have access to which type of information.
- RULES OF THE SYSTEM. This refers to responses that are triggered, such as incentives, punishment and constraints.
- POWER TO ADD, CHANGE, EVOLVE OR SELF-ORGANISE SYSTEM STRUCTURE. These interventions reflect the capability of stakeholders to change parts of the system.
- GOAL OF THE SYSTEM. The key question is here whether stakeholders cannot only change the system but also influence its purpose.
- MINDSET OR PARADIGM. Going beyond the purpose of the system, this reflects the transcendental system level in the hierarchy of Boulding.
- POWER TO TRANSCEND PARADIGMS. Again, this reflects the transcendental system level in the systems hierarchy of Boulding.

The latter four interventions are strongly related to the concept of boundary critique discussed in Sections 8.6 and 11.4.

Mathematically, the basic structure of a formal system dynamics computer simulation model is a system of coupled, non-linear, first-order differential (or integral) equations. Simulation of such systems is easily accomplished by partitioning simulated time into discrete intervals and stepping the system through time one interval at a time. Conceptually, feedback is at the heart of the system dynamics approach. Diagrams of loops of information feedback and circular causality are tools for conceptualising the structure of a complex system and for communicating model-based insights. The concept of endogenous change is fundamental to the system dynamics approach. It dictates aspects of model formulation: exogenous disturbances are seen at most as triggers of system behaviour; the causes are contained within the structure of the system itself. These ideas are captured in Forrester's [1961, 1969, 1971] organising framework for system structure as well as the work of de Rosnay [1975]. The system dynamics approach emphasises a continuous view. The continuous view strives to look beyond events to see the dynamic patterns underlying them. Moreover, the continuous view focuses not on discrete decisions but on the policy structure underlying decisions; this is exemplified by the leverage points for intervention in a system; see Box 11.1. Events and decisions are seen as surface phenomena that ride on an underlying tide of system structure and behaviour.

Soft Systems Methodology

The soft systems methodology [Checkland, 1981; Checkland and Scholes, 1990] is an approach for tackling soft, ill-defined real-world problems by formulating the concept of a purposeful human activity system (see Figure 11.7). A human activity system is a notional purposive system, which could in principle be found in the real world; for describing the root definition of

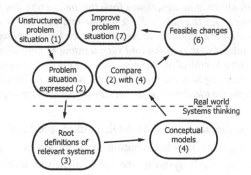

Figure 11.7 *Soft systems methodology. The methodology is based on a seven-stage process that moves from clarifying an unstructured or messy problem situation through designing ideal or conceptual human activity systems that would help improve the situation. These conceptual models are then compared with the problem situation in order to identify desirable and feasible change. The methodology integrates thinking about the logic of how to improve a situation with what is socially and politically feasible.*

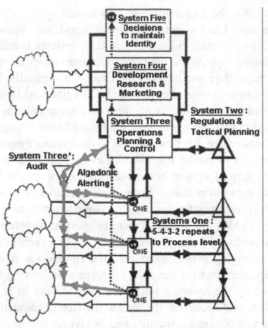

Figure 11.8 *Viable system model (source: http://en.wikipedia.org/wiki/Viable_system_ model#mediaviewer/File:Vsm.gif). System One comprises all activities that are undertaken in the organisation, i.e. its operations or primary process; it also might include control processes for the transformation processes. System Two symbolises all the activities and resources involved in the coordination between the operative units. System Three stands for all the activities and resources that focus on the optimising of the operations of the individual systems. Part of System Three are the individual operative units of the organisation that interact with the environment; it also includes all activities and resources that observe the environment, gain experience from this interaction and support strategies to be developed for the future. System Four indicates all the normative rules and regulations that apply in the organisation, such for example as the entrepreneurial ones relating to the creation and safeguarding of both identity and quality, the ones relating to ethical attitudes and to statutory and contractual provisions, and the ones relating to mandatory instructions. System Five coordinates all efforts to maintain the identity of an organisations (akin living organisations [de Geus, 1999])*

such systems, the acronym CATWOE is used, see Section 5.8; noteworthy is that the language in which human activity systems are modelled is in terms of verbs. Such systems are notional in the sense that they are not descriptions of actual real-world activities but are constructs of the mind for solving problems. Those descriptions facilitate discussions about possible changes, which might be introduced into a real-world problem situation. The soft systems methodology provides a way of getting from 'finding out' about a problem situation to 'taking action' to alleviate it; this methodology also stresses the interaction with stakeholders in an implicit manner.

One of the major strands of application of soft systems methodology is business modelling and support for developing information systems. The literature shows a number of methodologies based on soft systems methodology relevant to business modelling in information system development. Examples of them are the information system analysis methodology [Wilson, 1990], the methodology for functional analysis of office requirements [Schäfer, 1988], the compact methodology [CCTA, 1989] and the multiview methodology [Avison and Wood-Harper, 1990]. However, the flexibility in the way of operationalising and the low level of formality of the soft systems methodology modelling language limit the application of it in practice.

The Viable System Model

Arriving at a very different approach to systems thinking, management cybernetician Stafford Beer [1959, 1966, 1979] spent many years researching the necessary and sufficient conditions for an organisation as a complex system to be viable. As one of the key figures in the systems theories movement he determined that viability was maintained by engaging in different activities, keeping them from interfering with each other, managing them together, focusing on the future and doing so in the context of an identity within which the interests of the whole over time could be considered. In his perspective, this is how the human nervous system works and how successful collective enterprises work, too; see notes on comparison of biological concepts with organisational entities in Section 11.3. Most of all, his viable system model uses the resemblances in both governing laws and structure for organisations and organisms as point of departure.

The viable system model, see Figure 11.8, represents this thinking and consists of five essential functions or systems. These management functions Systems One through Five are repeated at different levels: the individual, the work group and each successive category, as long as it remains relevant (this has some similarity with aggregation strata, see Section 3.1). The five crucial systems of the viable systems model act in a similar, holistic, way in each 'cell'. They are connected together in the same way as the various organ systems in the human being and are responsible for performing the following tasks: (1) executing processes, (2) coordinating, (3) optimising, (4) observing and drawing conclusions and (5) deciding on and keeping track of values and ensuring identity. This leads to the following brief descriptions. System 1 stands for what is done in the organisation as transformation process and System 2 for how it is coordinated. System 3 stands for operative corporate management, System 4 for strategic corporate management and System 5 for normative corporate management. The only criterion for using the model is that the System One units, which these management functions support, must produce something of value for the environment such that it could be, in its own right, a viable system (this has similarity with the steady-state model, see Chapter 6, whereas the processes of maintaining identity and development of strategies resemble that of the breakthrough model). When using the viable system model, it is often helpful to consider one level of recursion (see Section 9.5) as the 'system in focus' and to explore the levels of recursion immediately above and below it, again similar to the distinction of aggregation strata in Section 3.1.

The viable system model has been used to both diagnose existing organisational structures and to design new ones. Many applications of this model have been undertaken, by Beer and others, in business, government, non-profit organisations and non-organisational systems [Espejo and Harnden, 1989]. It also provides a useful template against which to consider alternative structures and new challenges the system is facing, such as integrating its internal and its external knowledge or monitoring the evolution of its identity in a changing market.

MetaSystem Transition Theory

Very differently, the metasystem transition theory is the name for a particular cybernetic philosophy about the evolutionary process by which higher levels of complexity and control are generated, propagated by Joslyn and Heylighen [1995]. According to Joslyn and Heylighen, it also includes views on philosophical problems and makes predictions about the possible future of mankind and life. Their goal is to create, on the basis of cybernetic concepts, an integrated philosophical system, or 'world view' (also called 'Weltanschauung'), proposing answers to the most fundamental questions about the world, ourselves and our ultimate values.

Three concepts dominate metasystem transition theory. The first one of the central concepts is that of evolution in the most general sense, which is produced by the mechanism of variation and selection (i.e. following mostly a Darwinian perspective on selection); for the application of such thinking, see for example Dekkers [2005, pp. 67–75], who uses the analogy between evolution of organisms and evolution of organisations. The second is control, defined in a cybernetic sense, and asserted as the basic mode of organisation in complex systems (see Chapters 6 and 7 for the basic concepts of control mechanisms). This brings us to the third concept for metasystem transition theory, that of the metasystem transition, or the process by which control emerges in evolutionary systems. This third concept implies that the creation of variants calls for requisite mechanisms of dealing with variety (akin Ashby's law of requisite variety, see Section 6.8). For those that are interested, this also corresponds with the increasing number of dimensions for fitness in evolutionary systems; see Dekkers [2005, pp. 126–128] for a more detailed discussion. These three concepts help to explain the emergence of control during the development of systems, according to metasystem transition theory.

As an illustration of metasystem theory Turchin [1977] shows that the major steps in evolution, both biological and cultural, are metasystem transitions of a large scale. The concept of metasystem transition allows introducing a kind of objective quantitative measure of evolution and distinguishes between evolution in the positive direction, progress, and evolution in the negative direction, regress (cf. the direction of evolution). For example, here is the sequence of metasystem transitions which led, starting from the appearance of organs for motion, to the appearance of human thought and human society: movement as the control of position, irritability (simple reflex) as the control of movement, (complex) reflex as the control of irritability, associating (conditional reflex) as the control of reflex, human thinking as the control of associating and culture as the control of human thinking. It is possible to explain all those transitions in evolution as logic sequences from a metasystem transitions perspective.

Critical Systems Thinking

As the final strand of system theories discussed in this section critical systems thinking and the methodologies associated with it have been developed for the analysis of complex societal problems and interventions to resolve such problems (note that some points related to critical systems thinking appeared in Sections 8.6 and 8.7). Early approaches employing system thinking, such as operations research, system analysis and systems engineering (Section 11.1), are suitable for tackling certain well-defined problems, but have limitations for complex problems involving people with a variety of viewpoints and frequently at odds with one another. Systems thinkers responded with approaches such as system dynamics and organisational cybernetics to deal with complexity, soft systems methodology and interactive planning to handle subjectivity and critical systems heuristics to help the disadvantaged in situations involving conflict. Because of the corresponding enlargement of the context of problems when applying systems theories, it is critical systems thinking that aims at providing a more holistic picture from a stakeholders' perspective.

Critical systems thinking draws on the combination of social theory and systems thinking [Jackson, 2001]. Social theory provides material for the enhancement of existing and the development of new systems approaches. Not all the fine theoretical distinctions drawn by social scientists make a difference when applied in the real-world, but some are of considerable importance and must be regarded as crucial for systems practice. Social theory also provides the means whereby systems practitioners can reflect on and learn from their interventions. Within this perspective, systems thinking can assist in the task of translating the findings of social theory into a practical form and encapsulating those in well-worked out approaches to intervention. The success of systems thinking in linking theory and practice provides a model, which can be used and applied to disciplines generally.

One of the core concepts of critical systems thinking is an approach called critical systems heuristics that refers to the concept of the critical employment of boundary judgments [Ulrich, 1983, pp. 225–314], also called boundary critique. It says that the practical implications of a proposition (the 'difference' it makes in practice) and thus its meaning as well as its validity depend on how we bound the system of concern, i.e., that section of the real world which we take to represent the relevant context. The judgment of the merits of a proposition (it being preferred above some alternative proposition or its 'rationality') will depend heavily on this context; this is because the context determines what 'facts' (for example, consequences) and 'values' stakeholders and individuals will identify and how they will assess them.

Box 11.2: EXAMPLE OF BOUNDARY CRITIQUE – HOUSING SERVICES FOR THE ELDERLY

This example about applying the boundary critique is described in Midgley et al. [1998].

CASE DESCRIPTION

In that paper they describe how they were called in for the multi-agency development of housing services for older people. In the remit of the project it was not only about providing 'brick and mortar' but also a wider scope that also included adaptations of existing housing so that the older people could stay in their home. In that perspective the project covered public housing provision, housing associations, the voluntary sector, privately rented accommodation, owner occupied housing and related support services. Such a wide ranging coverage also implies a broad range of stakeholders; some of these stakeholders might be willing to seek influence of the solution at the expense of others, particularly the group for which it is all meant, the elderly people themselves.

APPLICATION OF BOUNDARY CRITIQUE

The authors show how through stages they achieved involvement of stakeholders that would be marginalised otherwise. Those stakeholders that are marginalised are found at the distinction between what they call the primary boundary and the secondary boundary; in addition, these stakeholders are found in the beyond the primary boundary. By interviewing all stakeholders they could clarify the planning provision itself would define a too narrow focus for the project. That allowed during the second phase of the project, the actual defining and organisation of services to achieve a wider focus of these services than a provision that would otherwise have resulted from only involving stakeholders within the primary boundary. However, it is the provision of feedback by parties that do not directly participate in the project and that cannot be associated with its outcomes that make it possible to set different boundaries; however, the condition is that this 'mediating' function is accepted by all stakeholders.

With respect to this crucial issue of boundary judgments, experts are no less lay people than citizens or other stakeholders in processes of societal change. Surfacing and questioning boundary judgments thus provides people with a means to counter unqualified rationality claims on the part of experts or decision makers – as well as other people – by demonstrating they way they may depend on debatable boundary judgments. The boundary critique demonstrates how systems thinking immediately translates into methodologically cogent forms of argumentation, i.e. they can make a difference between valid and invalid claims. The concept allows identifying invalid claims by uncovering underpinning boundary judgments other than those intended (or pretended) by the proponent. It explains why and how people and stakeholders in change processes are capable of contesting propositions and of advancing counter-propositions, without risking of being immediately convicted of lacking competence. Box 11.2 gives a concise description of how this was used for the development of housing services for older people in the United Kingdom. Hence, the concept of boundary critique indicates that critical systems thinking aims largely at resolving social problems through elevating conflicting arguments (for the purpose of discussion between various stakeholders).

11.5   Research Methods

Thus, critical systems thinking and the other strands of research show how closely system theories are related to research. First, the distinction of subsystems and aspects are ontological considerations for the objectives of a study; ontology refers to how things really are, what has to be considered in science. Thus, the distinction of systems, elements and environment determines what is to be studied. In addition, the three types of abstraction processes in Section 3.1 support further analysis of systems and aspects. Second, from an epistemological point of view systems theories provide a positivist view, though, for example, soft systems methodology and critical systems thinking also embrace interpretivism; epistemology in the philosophy of science is about what is true and what is not true. Therefore, systems thinking is closely related to ontology and its epistemological stance is dependent on how systems theories are used.

Furthermore, the concepts of systems theories can be used for analysis of findings for qualitative and to a lesser extent quantitative research. For example, the types of abstraction in Section 3.1 can be used for deriving inferences and synthesised findings. This type of analysis is related to coding that is often associated with qualitative research, particularly grounded theory [Glaser and Strauss, 1967]; for example Ryan and Bernard [2003] present twelve different techniques for coding of qualitative research. The codes need to be transferred to findings and possibly to the formation of tentative theories. Figure 11.9 shows this process of arriving at findings; each of the steps represents an abstraction for which either classification, aggregation or

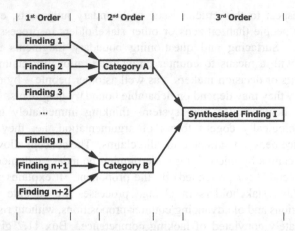

Figure 11.9   Aggregation process for findings.

generalisation can be used. The same holds for true for the analysis of results from quantitative research. By applying the concepts of abstraction the construct validity and the internal validity of the research can be increased; internal validity refers to the inferences about causal relationships and construct validity indicates the degree to which inferences can legitimately be made from the operationalisation of theoretical constructs (i.e. distinction of systems, environment, elements, subsystems and aspectsystems). Thus, the basic concepts of systems theories and the abstraction mechanisms can increase the construct and internal validity of research.

Particularly, this can be applied to the case study methodology for solving problems. An example is the use of Applied Systems Theory for modelling business processes; cases in point are the processes for innovation in a Chinese company [Dekkers, 2009] and the study of outsourcing processes in five companies [Dekkers, 2011]. Note that in some cases the generic models are tuned to the objectives of a specific study. Also in Dekkers and van Luttervelt [2006, p. 13] the steady-state model is used for how networks can be designed. Because the steady-state model, the breakthrough model and the model for the dynamic adaptation capability provide notional representations of business processes they can be adapted to the specific study and also serve as thematic classification of results.

Finally, it should be noted that some of the concepts are studies on their own. A case in point is general systems theory; even though this goes back to the writings of von Bertalanffy [1973], still some are looking to what and it how it can be applied (e.g. Pickel [2007] and Rousseau [2015]). Also, the interdisciplinary character of systems theories is subject of investigation (for instance, Rousseau et al. [2016]). Thus, also the conceptualisation of systems theories is still not settled and in flux; this also indicates that new applications of systems theories may be found.

11.6 Concluding Remarks

This chapter has shown that system theories have a wide variety of applications, spanning from technological systems to societal systems to research. In some fields the system theories have integrated into comprehensive approaches whereas in others they constitute an upcoming paradigm, e.g. systems biology. The range of applications extends beyond biological, technological and organisational systems from which most of the examples have been drawn in the text of this book. A case in point for stretching beyond those examples is that some of the theories and applications address societal challenges; an early example is system dynamics for limitations to societal growth and later methods are soft systems methodology and critical systems thinking. Although the basic concepts and some methodologies have been existent for a while, the concepts undergo further development and extension to new applications.

Because the system theories have been applied to so many different domains, it is also those domains that inform the further development of system theories. While not addressed in this chapter explicitly, the extension to complex (adaptive) systems (Chapter 9) and autopoietic systems (Chapter 8) are instances of the further development. These developments make it possible to understand better complex and non-linear behaviour. Also, mechanisms for networked structures that display that same complex and non-linear behaviour (see Section 9.6) can be better understood by these new developments. However, those developments sometimes do not deliver on promises made, sigh Richardson et al. [2001, p. 7] for the domain of organisations. Therefore, the concepts and theories for systems are subject to further development, though sometimes they might be considered to be in stages of infancy.

In this respect, systems theories also underpin inter-disciplinary approaches. The theories draw concepts from different and broad-ranging disciplines and they find their application in other domains. This is to be considered the foremost characteristic of inter- and trans-disciplinary approaches (following the terminology of Aboelela et al. [2006]). And it makes system theories a true domain for consilience by analysis and synthesis, the latter advocated by Wilson [1998, p. 68]. However, we have not reached a stage where inter-disciplinary systems thinking serves a bridge between disciplines. Even within disciplines strands have emerged that hardly refer to each other. While inter-disciplinarity poses its challenges, it is also makes system theories an exciting domain, for both theoretical and practical developments. However, there is still a long way to go in developing both its applications as well as its theoretical foundations.

References

Aboelela, S. W., Larson, E., Bakken, S., Carasquillo, O., Formicola, A., Giled, S. A., . . . Gebbie, K. M. (2006). Defining Interdisciplinary Research: Conclusions from a Critical Review of the Literature. Health Services Research, 42(1, Part 1), 329–346. doi: 10.1111/j.1475-6773.2006.00621.x

Aunger, R. (2000). Darwinizing culture: the status of memetics as a science. Oxford: Oxford University Press.

Avison, D. E., & Wood-Harper, A. T. (1990). Multiview: an exploration in information systems development. Oxford: Blackwell.

Bakan, J. (2004). The Corporation: The Pathological Pursuit of Profit and Power. London: Constable.

Beer, S. (1959). Cybernetics and Management. New York: Wiley.

Beer, S. (1966). Decision and control: the meaning of operational research and management cybernetics. London: Wiley.

Beer, S. (1972). Brain of the Firm - the Managerial Cybernetics of Organization. Chichester: John Wiley & Sons.

Beer, S. (1979). The Heart of Enterprise. Chichester: Wiley & Sons.

Bergh, J. C. J. M. v. d., & Gowdy, J. M. (2000). Evolutionary Theories in Environmental and Resource Economics: Approaches and Applications. Environmental and Resource Economics, 17(1), 37–57. doi: 10.1023/A:1008317920901

Bergson, H. (1911). Creative Evolution. New York: Dover.

Bertalanffy, L. v. (1973). General System Theory. New York: George Braziller.

Boulding, K. E. (1981). The Economy of Love and Fear: A Preface to Grants Economics. Belmont, CA: Wadsworth.

Carasquillo, O. (2001). System Dynamics Modeling: Tools for Learning in a Complex World. California Management Review, 43(4), 8–25. doi: 10.2307/41166098

CCTA. (1989). 'Compact' Manual (Version 1.1). Norwich.

Checkland, P. (1981). Systems Thinking, Systems Practice. Chichester: John Wiley & Sons.

Checkland, P., & Scholes, J. (1990). Soft Systems Methodology in Action. Chichester: John Wiley & Sons.

Dawkins, R. (1989). The Selfish Gene. Oxford: Oxford University Press.

de Geus, A. (1999). The Living Company. London: Nicolas Brealy.

de Rosnay, J. (1975). Le Macroscope: Vers une vision globale. Paris: Éditions du Seuil.

Dekkers, R. (2005). (R)Evolution, Organizations and the Dynamics of the Environment. New York: Springer.

Dekkers, R. (2009). Endogenous innovation in China: the case of the printer industry. Asia Pacific Business Review, 15(2), 243–264. doi: 10.1080/13602380802396466

Dekkers, R. (2011). Impact of strategic decision making for outsourcing on managing manufacturing. International Journal of Operations & Production Management, 31(9), 935–965. doi: 10.1108/01443571111165839

Dekkers, R., & van Luttervelt, C. A. (2006). Industrial networks: capturing changeability? International Journal of Networking and Virtual Organisations, 3(1), 1–24. doi: 10.1504/IJNVO.2006.008782

Eldredge, N. (1997). Evolution in the marketplace. Structural Change and Economic Dynamics, 8(4), 385-398. doi: 10.1016/S0954-349X(97)00020-9

Espejo, R., & Harnden, R. (1989). The viable system model : interpretations and applications of Stafford Beer's VSM. Chichester: Wiley & Sons.

Faber, M., & Proops, J. L. R. (1990). Evolution, Time, Production and the Environment. Berlin: Springer.

Fagerberg, J. (2003). Schumpeter and the revival of evolutionary economics: an appraisal of literature. Journal of Evolutionary Economics, 13(2), 125–129. doi: 10.1007/s00191-003-0144-1

Forrester, J. W. (1961). Industrial Dynamics. Waltham, MA: Pegasus Communications.

Forrester, J. W. (1969). Urban Dynamics. Waltham, MA: Pegasus Communications.

Forrester, J. W. (1971). World Dynamics. Cambridge, MA: Wright-Allen Press.

Glaser, B. J., & Strauss, A. L. (1967). The Discovery of Grounded Theory. Chicago: Aldine.

Hannan, M. T., & Freeman, J. (1977). The Population Ecology of Organizations. American Journal of Sociology, 83(4), 929–984. doi: 10.1086/226424

Hernes, T., & Bakken, T. (2003). Implications of Self-Reference: Niklas Luhmann's Autopoiesis and Organization Theory. Organization Studies, 24(9), 1511–1535. doi: 10.1177/0170840603249007

Heylighen, F. (2013). Building a Cybernetic Philosophy with Cybernetic Tools: the Principia Cybernetica project. Principia Cybernetica Web. Retrieved from ftp:// ftp.vub.ac.be/pub/papers/Principia_Cybernetica/Papers_Heylighen/Unifying_ Cybernetics_PCP.txt

Hodgson, G. M. (1993). Economics and Evolution - Bringing Life Back into Economics. Cambridge: Polity Press.

Hodgson, G. M. (2002). Darwinism in economics: from analogy to ontology. Journal of Evolutionary Economics, 12(3), 259–281. doi: 10.1007/s00191-002-0118-8

Hodgson, G. M., & Knudsen, T. (2005). The Nature and Units of Social Selection (0424). Jena: Papers on Economics & Evolution.

Jackson, M. C. (2001). Critical systems thinking and practice. European Journal of Operational Research, 128(2), 233–244. doi: 10.1016/S0377-2217(00)00067-9

Joslyn, C., & Heylighen, F. (1995, 29th June). Metasystem Transition Theory. Principia Cybernetica Web. Retrieved from ftp://ftp.vub.ac.be/pub/projects/ Principia_Cybernetica/PCP-Web/MSTT.html

Kitano, H. (2002). Systems Biology: A Brief Overview. Science, 295(5560), 1662–1664. doi: 10.1126/science.1069492

Laszlo, A., & Krippner, S. (1998). Systems Theories: Their Origins, Foundations, and Development. In J. S. Jordan (Ed.), Systems Theories and A Priori Aspects of Perception (pp. 47–74). Amsterdam: Elsevier Science.

Lauffenburger, D. A. (2000). Cell signaling pathways as control modules: Complexity for simplicity? Proceedings of the National Academy of Science, 97(10), 5031–5033.

Levin, S., Xepapadeas, T., Crépin, A.-S., Norberg, J., de Zeeuw, A., Folke, C., . . . Walker, B. (2013). Social-ecological systems as complex adaptive systems: modeling and policy implications. Environment and Development Economics, 18(2), 111–132. doi: 10.1017/S1355770X12000460

Meadows. (2008). Thinking in Systems. White River Junction, VT: Chelsea Green Publishing.

Metzger, M. A. (1997). Applications of nonlinear dynamic systems theory in developmental psychology: Motor and cognitive development. Nonlinear Dynamics, Psychology, And Life Sciences, 1(1), 55–68. doi: 10.1023/A:1022323926870

Midgley, G., Munlo, I., & Brown, M. (1998). The Theory and Practice of Boundary Critique: Developing Housing Services for Older People. Journal of the Operational Research Society, 49(5), 467–478. doi: 10.1057/palgrave.jors.2600531

Mittal, S. (2013). Emergence in stigmergic and complex adaptive systems: A formal discrete event systems perspective. Cognitive Systems Research, 21, 22–39. doi: 10.1016/j.cogsys.2012.06.003

Nelson, R. R., & Winter, S. G. (1982). An Evolutionary Theory of Change. Cambridge, MA: Belknap Press.

Norgaard, R. (1994). Development Betrayed: The End of Progress and a Coevolutionary Revisioning of the Future. London: Routledge.

O'Shea, A. (2002). The (R)evolution of New Product Innovation. Organization, 9(1), 113–125. doi: 10.1177/1350508402009001351

Pathak, S. D., Day, J. M., Nair, A., Sawaya, W. J., & Kristal, M. M. (2007). Complexity and Adaptivity in Supply Networks: Building Supply Network Theory Using a Complex Adaptive Systems Perspective*. Decision Sciences, 38(4), 547–580. doi: 10.1111/j.1540-5915.2007.00170.x

Pickel, A. (2007). Rethinking Systems Theory. Philosophy of the Social Sciences, 37(4), 391–407. doi: 10.1177/0048393107307809

Pruckner, M. (2002). Management Cybernetics and St. Gallen. Lampeter.

Richardson, K. A., Cilliers, P., & Lissack, M. (2001). Complexity Science: A "Gray" Science for the "Stuff in Between". Emergence, 3(2), 6–18. doi: 10.1207/S15327000EM0302_02

Rousseau, D. (2015). General Systems Theory: Its Present and Potential. Systems Research and Behavioral Science, 32(5), 522–533. doi: 10.1002/sres.2354

Rousseau, D., Blachfellner, S., Billingham, J., & Wilby, J. (2016). A Research Agenda for General Systems Transdisciplinarity. Systema, 4(1), 100–110.

Ryan, G. R., & Bernard, H. R. (2003). Techniques to Identify Themes. Field Methods, 15(1), 85–109. doi: 10.1177/1525822x02239569

Schäfer, G. (1988). Functional Analysis of Office Requirements: a multiperspective approach. Chichester: Wiley & Sons.

Schumpeter, J. (1911). Theorie der wirtschaftlichen Entwicklung. Leipzig: von Duncker & Humblot.

Schwaninger, M. (2001). System theory and cybernetics: A solid basis for transdisciplinarity in management education and research. Kybernetes, 30(9/10), 1209–1222. doi: 10.1108/EUM0000000006551

Senge, P. M. (1992). The Fifth Discipline. Kent: Century Business.

Steen, L. (2014). The Meaning of System: Towards a Complexity Orientation in Systems Thinking. International Journal of Systems and Society (IJSS), 1(1), 21–34. doi: 10.4018/ijss.2014010103

Turchin, V. (1977). The Phenomenon of Science: A Cybernetic Approach to Human evolution. New York: Columbia University Press.

Ulrich, W. (1983). Critical heuristics of social planning: A new approach to practical philosophy. Bern: P. Haupt.

Veblen, T. (1998). Why is economics not an evolutionary science? Quarterly Journal of Economics, 12, 373–397.

West Churchman, C. (1979). The Systems Approach: Revised and Updated. New York: Dell.

Wilson, B. (1990). Systems: Concepts, Methodologies and Applications. Chichester: Wiley & Sons.

Wilson, E. O. (1998). Consilience: the unity of knowledge. New York: Alfred A. Knopf.

Index

Symbols

5 whys technique 90, 112
80-20 rule. *See also* Pareto analysis
787 Dreamliner 104

A

A330 104
A330-200Lite 104
A350 104
abduction. *See* abductive reasoning
abductive inference. *See* abductive
    reasoning
abductive reasoning 42, 54, 61, 87
Abell, Derek F. 246
Abilene paradox 100, 106, 113
Aboelela, Sally W. 295
abstraction 41, 42, 43, 63, 79, 127, 293
accuracy 69, 70
Ackoff, Russell Lincoln 16, 70
action learning 141
adaptation 197, 235
adaptation cycle 259
adaptation syndrome 179
adaptive behaviour 211, 240
adaptive dynamics 74
adaptive process 117, 119, 120, 122–
    123, 131
adaptive radiation 30
adaptive walk 220, 223, 235, 237, 239
aerospace engineering 2
agent-based model 227–229, 228
aggregation 44, 47, 51, 65, 79, 260, 275,
    284, 293
aggregation stratum xii–xiii, 49, 69, 89,
    94, 107, 109, 113, 124, 127, 133,
    134, 143, 259, 290
Airbus 104
algorithm of inventive problem solving
    96
allele 217–224, 219–221, 221, 238

allopoiesis 193, 204–206, 238
allopoietic system xxiv, 208, 212, 215,
    237, 238, 239, 239–241, 256, 260,
    265, 278, 281, 281–282
Altshuller, Genrich 95
American government 105
analogue models 72
analogy 63, 65, 67, 207, 238, 240, 282,
    290
analytic hierarchy process 102
Ansoff, H. I. 246
Antarctic 36
anthropology 5, 197
architectural innovation 261, 262
Argyris, C. 260
Aristotle 1, 5
ARIZ. *See* algorithm of inventive prob-
    lem solving
Arthur, W. Brian 59
artificial intelligence 6, 212
artificial life 6
artificial neural networks 212
artificial system 8, 214
Ashby, William Ross 4, 146, 169, 248,
    269, 274, 290
as-is 55
ASME Mapping Standard 139–140
aspectsystem viii–ix, x, xi, xii, 15, 26, 27,
    29, 38, 43, 63, 88, 107, 294
Atlantic Ocean 26
attractor 199, 215, 225–226
Aunger, Robert 283
Australia 77
Austrian School of Economics 68
autocatalytic set 195
autopilot 155
autopoiesis xxiii–xxiv, 11, 12, 173,
    193–210, 219, 230, 231, 235, 236,
    238, 239, 245, 269, 275, 281
autopoietic system xxiii–xxiv, 57, 215,
    238, 295

© Springer International Publishing AG 2017
R. Dekkers, *Applied Systems Theory*,
DOI 10.1007/978-3-319-57526-1

Printed in the United States
By Bookmasters